AF616643

ADVANCES IN QUANTUM CHEMISTRY

VOLUME 10

CONTRIBUTORS TO THIS VOLUME

B. H. BRANDOW
PHILIP COPPENS
PER KAIJSER
JACOB KATRIEL
HANS KLEINPOPPEN
RUBEN PAUNCZ
BERNARD PULLMAN
VEDENE H. SMITH, JR.
EDWIN D. STEVENS

ADVANCES IN

QUANTUM CHEMISTRY

EDITED BY

PER-OLOV LÖWDIN

DEPARTMENT OF QUANTUM CHEMISTRY
UPPSALA UNIVERSITY
UPPSALA, SWEDEN
AND
QUANTUM THEORY PROJECT
UNIVERSITY OF FLORIDA
GAINESVILLE, FLORIDA

VOLUME 10—1977

ACADEMIC PRESS New York · San Francisco · London
A Subsidiary of Harcourt Brace Jovanovich, Publishers

ACADEMIC PRESS, INC.
111 Fifth Avenue, New York, New York 10003

United Kingdom Edition published by
ACADEMIC PRESS, INC. (LONDON) LTD.
24/28 Oval Road, London NW1

LIBRARY OF CONGRESS CATALOG CARD NUMBER: 64–8029

ISBN 0–12–034810–1

PRINTED IN THE UNITED STATES OF AMERICA

CONTENTS

List of Contributors ix
Preface xi

Accurate X-Ray Diffraction and Quantum Chemistry: The Study of Charge Density Distributions
Philip Coppens and Edwin D. Stevens

I. Introduction 1
II. Definition of Density Functions 2
III. Theory of X-Ray Scattering and the Derivation of Electron Densities from X-Ray Scattering Amplitudes 4
IV. Experimental Errors 7
V. Some Experimental Densities 8
VI. Theoretical Densities 20
VII. Comparison of Theoretical and Experimental Densities. Thermal Smearing 22
VIII. Functions Based on Experimental Densities 25
IX. Concluding Remarks 32
References 33

Evaluation of Momentum Distributions and Compton Profiles for Atomic and Molecular Systems
Per Kaijser and Vedene H. Smith, Jr.

I. Introduction 37
II. Dirac–Fourier Transformation of Position Space Wavefunctions and One-Particle Density Matrices 40
III. Dirac–Fourier Transformation of Atomic Orbitals 44
IV. Evaluation of the Momentum Density $\rho(\mathbf{p})$ 58
V. Evaluation of the Directional Compton Profile $J(\mathbf{q})$ 62
VI. The Spherically Averaged Momentum Density $\overline{\rho(p)}$ and the Isotropic Compton Profile $\overline{J(q)}$ 71
VII. Concluding Remarks 73
References 74

Analysis of Electron–Atom Collisions
HANS KLEINPOPPEN

I. Introduction 77
II. Polarization of Impact Radiation 78
III. Spin Effects in Electron–Atom Collisions 92
IV. Electron–Photon Coincidences and Angular Correlations 114
V. Coherent Electron-Impact Excitation of Atomic Hydrogen 131
VI. Electron Scattering from Laser-Excited Atoms 132
References 139

Theoretical Interpretation of Hund's Rule
JACOB KATRIEL AND RUBEN PAUNCZ

I. Introduction 143
II. Scaled Atomic Orbitals 147
III. Optimized Atomic Orbitals 153
IV. Inclusion of the Correlation Correction 156
V. The $1/Z$ Perturbation Theory 157
VI. Role of Inner Shells 160
VII. The Pair Distribution Function 162
VIII. An Interpretation of Hund's Rule 176
Appendix A: Screening in $(n, l)^k$-Type Configurations 180
Appendix B: The Pair Distribution Function 181
References 184

Linked-Cluster Perturbation Theory for Closed- and Open-Shell Systems
B. H. BRANDOW

I. Introduction 188
II. Degenerate Brillouin–Wigner Perturbation Theory 190
III. The Linked-Cluster Expansion for Closed-Shell Systems 192
IV. Physical Interpretation and General Philosophy of Applications 201
V. Further Closed-Shell Results 204
VI. Open-Shell Systems—A General Formalism for Model Hamiltonians and Model Operators 220
VII. Formal Derivation of Effective Spin Hamiltonians 237
References 246

Quantum-Mechanical Approach to the Conformational Basis of Molecular Pharmacology
BERNARD PULLMAN

I. Introduction 251
II. Histamine 259
III. Antihistamines 268

IV. Serotonin and Related Indolealkylamines 275
V. Phenethylamines and Phenethanolamines 281
VI. Acetylcholine 289
VII. Conformation of Acetylcholine in Solution 297
VIII. Acetylcholine Derivatives and Analogs 302
IX. Conclusions 314
Addendum 323
References 323

Subject Index 329
Contents of Previous Volumes 335

LIST OF CONTRIBUTORS

Numbers in parentheses indicate the pages on which the authors' contributions begin.

B. H. BRANDOW, Theoretical Division, Los Alamos Scientific Laboratory, University of California, Los Alamos, New Mexico (187)

PHILIP COPPENS, Chemistry Department, State University of New York at Buffalo, Buffalo, New York (1)

PER KAIJSER,* Department of Chemistry, Queen's University, Kingston, Ontario, Canada (37)

JACOB KATRIEL, Department of Chemistry, Technion Israel Institute of Technology, Haifa, Israel (143)

HANS KLEINPOPPEN, Institute of Atomic Physics, University of Stirling, Stirling, Scotland (77)

RUBEN PAUNCZ, Department of Chemistry, Technion Israel Institute of Technology, Haifa, Israel (143)

BERNARD PULLMAN, Institut de Biologie Physico-Chimique, Fondation Edmond de Rothschild, Paris, France (251)

VEDENE H. SMITH, JR., Department of Chemistry, Queen's University, Kingston, Ontario, Canada (37)

EDWIN D. STEVENS, Chemistry Department, State University of New York at Buffalo, Buffalo, New York (1)

* *Present Address:* Quantum Theory Project, University of Florida, Gainesville, Florida.

PREFACE

In investigating the highly different phenomena in nature, scientists have always tried to find some fundamental principles that can explain the variety from a basic unity. Today they have not only shown that all the various kinds of matter are built up from a rather limited number of atoms, but also that these atoms are constituted of a few basic elements of building blocks. It seems possible to understand the innermost structure of matter and its behavior in terms of a few elementary particles: electrons, protons, neutrons, photons, etc., and their interactions. Since these particles obey not the laws of classical physics but the rules of modern quantum theory of wave mechanics established in 1925, there has developed a new field of "quantum science" which deals with the explanation of nature on this ground.

Quantum chemistry deals particularly with the electronic structure of atoms, molecules, and crystalline matter and describes it in terms of electronic wave patterns. It uses physical and chemical insight, sophisticated mathematics, and high-speed computers to solve the wave equations and achieve its results. Its goals are great, but perhaps the new field can better boast of its conceptual framework than of its numerical accomplishments. It provides a unification of the natural sciences that was previously inconceivable, and the modern development of cellular biology shows that the life sciences are now, in turn, using the same basis. "Quantum biology" is a new field which describes the life processes and the functioning of the cell on a molecular and submolecular level.

Quantum chemistry is hence a rapidly developing field which falls between the historically established areas of mathematics, physics, chemistry, and biology. As a result there is a wide diversity of backgrounds among those interested in quantum chemistry. Since the results of the research are reported in periodicals of many different types, it has become increasingly difficult for both the expert and the nonexpert to follow the rapid development in this new borderline area.

The purpose of this serial publication is to try to present a survey of the current development of quantum chemistry as it is seen by a number of the internationally leading research workers in various countries. The authors have been invited to give their personal points of view of the subject freely and without severe space limitations. No attempts have been made to avoid overlap—on the contrary, it has seemed desirable to have certain important research areas reviewed from different points of view. The response from the authors has been so encouraging that an eleventh volume is now being prepared.

The editor would like to thank the authors for their contributions which give an interesting picture of the current status of selected parts of quantum chemistry. The topics covered in this volume range from the interpretation of Hunds rule, over studies of Compton profiles, charge-density distributions, electron-atom collisions, and linked-cluster expansions in perturbation theory to the analysis of the conformational basis of molecular pharmacology by using quantum mechanical methods. Some of the papers emphasize studies in fundamental quantum theory, and others applications to comparatively complicated systems.

It is our hope that the collection of surveys of various parts of quantum chemistry and its advances presented here will prove to be valuable and stimulating, not only to the active research workers but also to the scientists in neighboring fields of physics, chemistry, and biology, who are turning to the elementary particles and their behavior to explain the details and innermost structure of their experimental phenomena.

PER-OLOV LÖWDIN

ADVANCES IN QUANTUM CHEMISTRY

VOLUME 10

Accurate X-Ray Diffraction and Quantum Chemistry: The Study of Charge Density Distributions

PHILIP COPPENS and EDWIN D. STEVENS

Chemistry Department
State University of New York at Buffalo
Buffalo, New York

I. Introduction . . . 1
II. Definition of Density Functions . . . 2
III. Theory of X-Ray Scattering and the Derivation of Electron Densities from X-Ray Scattering Amplitudes . . . 4
IV. Experimental Errors . . . 7
V. Some Experimental Densities . . . 8
A. Density in the Bonding Regions . . . 9
B. Single, Double, and Triple Bonds . . . 14
C. Density in the Lone-Pair Regions . . . 15
D. Density in Solids Containing 3d Transition Metals . . . 17
E. Experimental Densities in Very Small Molecules . . . 18
VI. Theoretical Densities . . . 20
A. Deformation Density and Quality of the Wavefunction . . . 20
VII. Comparison of Theoretical and Experimental Densities. Thermal Smearing . 22
VIII. Functions Based on Experimental Densities . . . 25
A. Discussion of Methods . . . 25
B. Net Atomic and Molecular Charges . . . 26
C. Molecular Dipole Moments . . . 29
D. Orbital Exponents . . . 29
E. Angular Deformation of Pseudoatoms and Two-Center Population Parameters . . . 32
IX. Concluding Remarks . . . 32
References . . . 33

I. Introduction

The measurement of charge densities in solids by means of X-ray diffraction is a developing field, the progress of which depends on systematic analysis and elimination of experimental errors, on the application of new experimental techniques, and on the development of interpretive methods of analysis. It is this last aspect that requires close cooperation between the experimentalists and theoreticians. Results obtained in recent years indicate that experimental densities are a sensitive test of theoretical methods. They cannot, for example, be reproduced with limited basis set calculations. On the other hand, experimental densities are by necessity thermally averaged, so that the thermal smearing of electron densities has to be analyzed before any quantitative comparison can be accomplished.

One may question whether comparison between theory and experiment should be at the level of electron densities or whether other numerical quantities derivable from the one-electron density function offer better perspectives. Examples of such functions are net atomic charges, atomic deformation functions, electrostatic potentials, molecular dipole and higher moments. Each has specific problems: in some cases there is an ambiguity of definition (such as net atomic charge) and, in others, the experimental accessibility is questionable (molecular dipole and higher moments). Furthermore, even in molecular crystals, matrix effects due to intermolecular forces may play a role and may account for some discrepancies between calculations on an isolated molecule and measurements on a molecular crystal. It is in these respects that interaction between theory and experiment may be extremely fruitful.

No attempt will be made here to present a comprehensive review, rather we will describe some recent experimental and theoretical results in the hope of establishing a firm link between experimental and theoretical quantum chemists in this field. The complementary techniques of Compton-profile analysis (momentum density), polarized-beam neutron diffraction (magnetization density), and gas-phase electron diffraction (radial density distribution in small molecules) should be mentioned here.

II. Definition of Density Functions

The total electron density contains a large contribution from the core electrons, which are relatively unperturbed by the chemical environment. More sensitive functions can be obtained by subtraction of all or parts of the unperturbed spherical atom density, as first proposed by Roux and Daudel (1955) in their study of the bonding in the Li_2 molecule. A sum of spherical ground-state atoms is subtracted from the total density:

$$\rho_{\text{deformation}}(\mathbf{r}) = \rho(\mathbf{r}) - \sum_{\substack{\text{all} \\ \text{atoms}}} \rho_{i,\ \text{spherical atom}}(\mathbf{r} - \mathbf{r}_i). \tag{1}$$

Since Eq. (1) represents the deformation of a superposition of spherical atoms (the "promolecule"; Hirshfeld and Rzotkiewicz, 1974) upon molecule formation, it is referred to as the *deformation density*. Alternatively, valence state or prepared atoms may be subtracted out in Eq. (1), as has been done for diatomic molecules by Bader *et al.* (1967a,b). However, we will take the position with Ransil and Sinai (1967) and with Rosenfeld (1964) that the

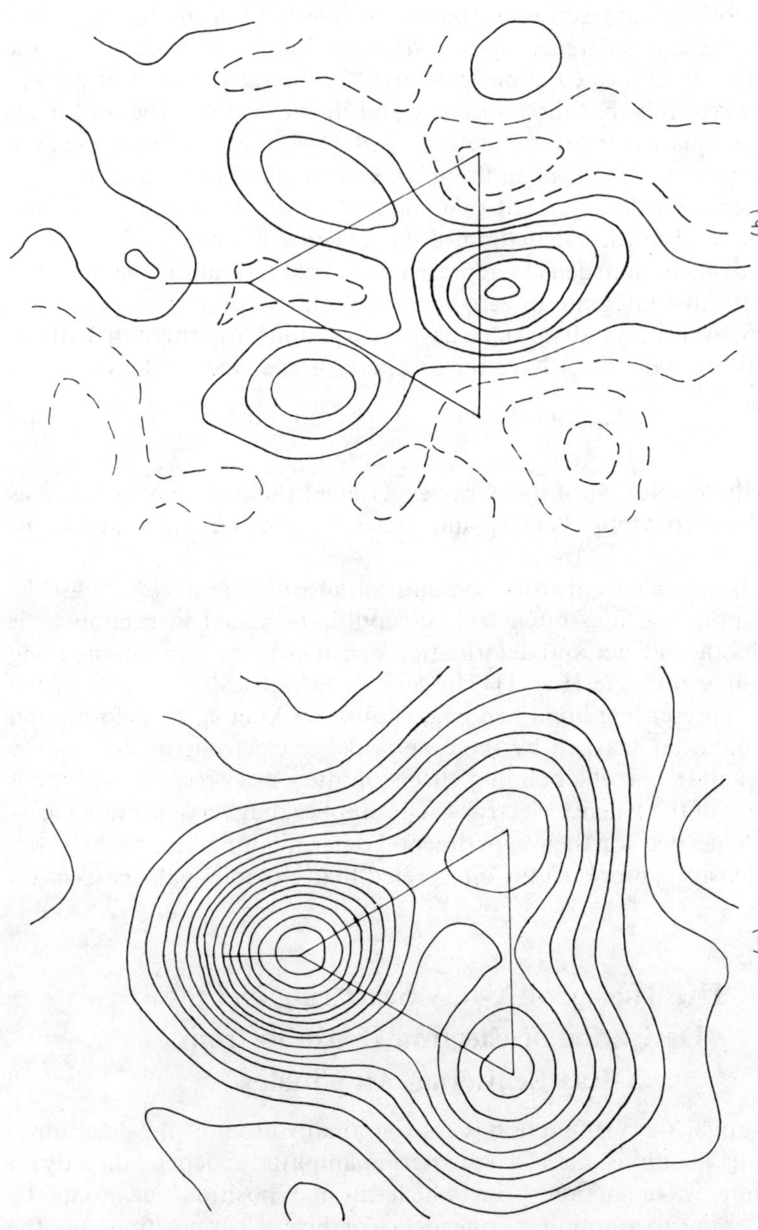

Fig. 1. Ethylenimine ring of ethylenimine quinone. (a) The $\Delta\rho_{\text{valence}}$; contours at $0.25e\ \text{Å}^{-3}$. (b) The $\Delta\rho_{\text{deformation}}$; contours at $0.05e\ \text{Å}^{-3}$. (X-Ray data from Ito and Sakurai, 1974.)

deviation from spherical symmetry to prepare an atom for bond formation is an intermediate step in bonding and that the total deformation is properly represented by the subtraction of spherically averaged ground-state atoms.

In a theoretical difference map, a decision has to be taken as to the quality of the atomic calculation to be used in the second term of Eq. (1). Consistency requires that the molecular and the atomic densities in Eq. (1) are of similar quality. Work by Stevens *et al.* (1971) on cyclopropane indicates the importance of this condition. In experimental deformation maps, however, there is little argument that the best available atomic wavefunction is to be used in the calculation of the deformation maps.

As the deformation density represents the redistribution on molecule formation, it must integrate to zero when the integration is performed over all space (or over a crystallographic asymmetric unit). An alternative difference function that does not have this property is the valence density:

$$\Delta\rho_{\text{valence}}(\mathbf{r}) = \rho(\mathbf{r}) - \sum_{\substack{\text{all} \\ \text{atoms}}} \rho_{i,\,\text{core}}(\mathbf{r} - \mathbf{r}_i). \tag{2}$$

Within the resolution of the X-ray experiment the core may be taken as in the nonbonded atom (Bentley and Stewart, 1974; Groenewegen *et al.*, 1971).

Although the valence density contains all information necessary for the analysis of bonding, it is often less susceptible to visual inspection as is illustrated by the valence and deformation densities in the ethylenimine ring of ethylenimine quinone (Fig. 1). The lack of cylindrical symmetry of the ring bonds, representing bond bending, is quite striking in the deformation map but somewhat masked by the other valence electrons in the valence density. It is clear that the bonding effects of interest correspond only to a small fraction of the valence electrons. This implies that very accurate measurements are needed for their experimental determination and that the deformation density is more readily interpreted than the total valence electron distribution.

III. Theory of X-Ray Scattering and the Derivation of Electron Densities from X-Ray Scattering Amplitudes

Even though X-ray diffraction is conventionally used for the determination of atomic positions, the X-ray scattering amplitudes depend directly on the electronic wavefunction from which atomic positions can only be derived under the assumption of coincidence of the nuclear positions and the centroids of electronic charge. According to the quantum-mechanical

Waller–Hartree theory (Waller and Hartree, 1929), the total scattering including incoherent processes expressed in units of the scattering of a classical electron is given by

$$I_{\text{total}}(\mathbf{S}) = \int \psi^* \left| \sum_i \exp(2\pi i \mathbf{S} \cdot \mathbf{r}) \right|^2 \psi \, d\tau$$
$$= \sum_i \sum_j \psi^* |\exp(2\pi i \mathbf{S} \cdot \mathbf{r}_{ij})| \psi \, d\tau, \tag{3}$$

where $\mathbf{r}$ is a vector defining position in real space, $\mathbf{S}$ is a reciprocal lattice vector of magnitude $2 \sin \theta/\lambda$, ψ is the electronic wavefunction, the sum is over all electrons and $\mathbf{r}_{ij} = \mathbf{r}_i - \mathbf{r}_j$. The total scattering defined by Eq. (3) is a two-electron property. Its coherent component is given by

$$I_{\text{coherent}}(\mathbf{S}) = \left| \int \psi^* \left| \sum_i \exp(2\pi i \mathbf{S} \cdot \mathbf{r}_i) \right| \psi \, d\tau \right|^2. \tag{4}$$

When integration is performed over all coordinates but those of the ith electron, one obtains

$$I_{\text{coherent}}(\mathbf{S}) = \left| \int \rho(\mathbf{r}) \exp(2\pi i \mathbf{S} \cdot \mathbf{r}) \, d\tau \right|^2 \tag{5}$$

in which $\rho(\mathbf{r})$ is the one-electron density function. For a system undergoing thermal vibrations, the coherent scattering intensity contains both an elastic and an inelastic component. The former is given by replacing $\rho(\mathbf{r})$ in Eq. (5) by the thermally averaged density $\langle \rho(\mathbf{r}) \rangle$:

$$I_{\text{coherent, elastic}}(\mathbf{S}) = \left| \int \langle \rho(\mathbf{r}) \rangle \exp(2\pi i \mathbf{S} \cdot \mathbf{r}) \, d\tau \right|^2. \tag{5a}$$

Thus, the elastic scattering amplitude is given by

$$F(\mathbf{S}) = \int \langle \rho(\mathbf{r}) \rangle \exp(2\pi i \mathbf{S} \cdot \mathbf{r}) \, d\tau = \mathscr{F}\{\rho(\mathbf{r})\}, \tag{6}$$

where $\mathscr{F}$ is the Fourier transform operator. Since the electron density in a crystal is the convolution of the density in the unit cell and a periodic lattice function, one obtains

$$F(\mathbf{S}) = \mathscr{F}\{\langle \rho_{\text{unit cell}} \rangle * \delta(\mathbf{r} - u\mathbf{a}_1 - v\mathbf{a}_2 - w\mathbf{a}_3)\}$$
$$= \mathscr{F}(\langle \rho_{\text{unit cell}} \rangle) \mathscr{F} \delta(\mathbf{r} - u\mathbf{a}_1 - v\mathbf{a}_2 - w\mathbf{a}_3), \tag{7}$$

where u, v, w are integers and $\mathbf{a}_1$, $\mathbf{a}_2$, and $\mathbf{a}_3$ represent the lattice translations. The Fourier transform of a Dirac δ function is again a delta function with periodicity along the reciprocal lattice axes $\mathbf{a}_1^*$, $\mathbf{a}_2^*$ and $\mathbf{a}_3^*$ defined by

$\mathbf{a}_i \cdot \mathbf{a}_j^* = \delta_{ij}$. Thus $F(\mathbf{S})$ is only observable at distinct points $\mathbf{H}$ in reciprocal space representing the elastic Bragg reflections at which points

$$F(\mathbf{H}) = \int_{\substack{\text{unit}\\\text{cell}}} \langle \rho(\mathbf{r}) \rangle \exp(2\pi i \mathbf{H} \cdot \mathbf{r})\, d\tau. \tag{8}$$

The electron density $\langle \rho(\mathbf{r}) \rangle$ is given by the inverse Fourier transform

$$\langle \rho(\mathbf{r}) \rangle = \frac{1}{V} \sum_H F(\mathbf{H}) \exp(-2\pi i \mathbf{H} \cdot \mathbf{r}), \tag{9}$$

where the integral has been replaced by a sum over all reciprocal lattice points for which $\mathbf{S} = \mathbf{H}$, and V is the volume of the crystallographic unit cell. Expression (8) implies that the structure factor amplitude is generally complex, i.e., for the calculation of $\rho(\mathbf{r})$ both magnitude and phase are required. Only the former can be measured experimentally. However, when the crystal contains a center of symmetry, one may choose the origin, such that $\rho(\mathbf{r}) = \rho(-\mathbf{r})$, which gives

$$F(\mathbf{H}) = \int_{\substack{\text{unit}\\\text{cell}}} \langle \rho(\mathbf{r}) \rangle \cos 2\pi \mathbf{H} \cdot \mathbf{r}\, d\tau. \tag{8a}$$

For centrosymmetric structures, therefore, the phase problem reduces to a question of the sign of the structure factor. As experience has shown that this sign can be calculated reliably from a spherical atom model for all but the very weakest reflections, centrosymmetric crystals are much more suitable for charge density studies than acentric crystals.

Expression (6) gives the total time-averaged electron density provided that the amplitudes $F(\mathbf{H})$ are on an absolute scale. This may be done experimentally (Stevens and Coppens, 1975), but is more commonly achieved by adjustment of the scale factor k, defined by $F_{\text{obs}}(\mathbf{H}) = kF_{\text{calc}}(\mathbf{H})$, using a least-squares procedure and a spherical atom model.

To obtain the difference densities, Eqs. (1) and (2), the density for either spherical atom or core is to be subtracted:

$$\Delta\rho = \frac{\rho_{\text{obs}}}{k} - \frac{1}{V} \sum_H F_{\text{calc}}(\mathbf{H}) \exp - (2\pi i \mathbf{H} \cdot \mathbf{r}) \tag{10}$$

which requires positional and thermal parameters in the calculation of F_{calc}. These cannot be obtained from the conventional least-squares fit of a spherical model to the X-ray data because the asphericity of the molecular electron density introduces bias in the refined parameters. At best, unbiased structural parameters are obtained from an independent elastic neutron diffraction experiment, in which case the function $\Delta\rho$ is often referred to as an X-N map. Useful results may be obtained also with parameters based on high-order X-ray data (e.g., see Stevens and Hope, 1975; Stewart and Jensen, 1969), making the reasonable approximation that valence scattering is concentrated in the low-order region. Furthermore, data refinements based on aspherical atoms show promise in eliminating bias in all but the hydrogen atom structural parameters. Such difference densities based completely on X-ray data will be referred to here as X-X densities. These X-X densities are often adequate in describing the charge distribution in the bond regions, but may underestimate the density associated with the lone-pair electrons (Wang *et al.*, 1976; Coppens and Lehmann, 1976).

IV. Experimental Errors

It is obviously essential that measurement errors be estimated, so that only statistically significant observations will be interpreted. The subject has been treated in a number of publications (Cruickshank, 1949; Cruickshank and Rollet, 1953; Coppens and Hamilton, 1968; Ruysink and Vos, 1974; Coppens, 1974; Rees, 1976), and only the main conclusions necessary for a proper assessment of the results will be treated here. As described in Section III, the deformation density is defined as

$$\Delta\rho = \frac{\rho_{\text{obs}}}{k} - \rho_{\text{calc}},$$

where all densities are again time-averaged.

Thus,

$$\begin{aligned}\sigma^2(\Delta\rho) &= \sigma^2\left(\frac{\rho_{\text{obs}}}{k}\right) + \sigma^2(\rho_{\text{calc}}) + \text{cov}\left(\frac{\rho_{\text{obs}}}{k}, \rho_{\text{calc}}\right) \\ &= \frac{\sigma^2(\rho_{\text{obs}})}{k^2} + \left(\frac{\rho_{\text{obs}}}{k}\right)^2\left[\frac{\sigma(k)}{k}\right]^2 + \sigma^2(\rho_{\text{calc}}) \\ &\quad + \text{cov}\left(\frac{\rho_{\text{obs}}}{k}, \rho_{\text{calc}}\right).\end{aligned} \tag{11}$$

The fourth term representing the covariance arises when the scale factor k and the parameters on which ρ_{calc} is based are obtained in the least-squares refinement. If we use the symbol ρ'_{obs} for the scaled experimental density, we get

$$\sigma^2(\Delta\rho) = \sigma^2(\rho'_{\text{obs}}) + \rho'^2_{\text{obs}}\left[\frac{\sigma(k)}{k}\right]^2 + \sigma^2(\rho_{\text{calc}}) + \text{cov}\left(\frac{\rho_{\text{obs}}}{k}, \rho_{\text{calc}}\right). \tag{11a}$$

Thus, there are four contributions, the first of which is fairly constant over the unit cell (except at or near crystallographic symmetry elements). A typical value of σ (ρ'_{obs}) with good quality data presently obtainable is $0.0075 e a_0^{-3}$ ($0.05 e$ Å^{-3}). The last three terms show much larger fluctuations as a function of position. The second term representing the scale factor error will have its main effect where ρ'_{obs} is large, i.e., near the atomic positions, but will be small elsewhere. The same is the case for the error contribution represented by the third term. An error in the thermal vibration parameters used to obtain ρ_{calc}, for example, will have a severe influence within about 0.3 Å of the nuclear position (Stewart, 1968). Thus, in current work *errors near the atomic positions are exceptionally large*, and no significance should be attached to the experimental deformation density being positive or negative at the nuclear positions. Maps of the distribution of the errors in the unit cell have been calculated in several cases [e.g., chromium hexacarbonyl, Rees and Mitschler (1976); sodium azide, Stevens *et al.*, (1977, Fig. 9d); orthorhombic sulfur, Coppens *et al.*, 1977)].

In the absence of a detailed error distribution map, errors away from the atomic positions may be approximated by

$$\sigma^2(\Delta\rho) = \sigma^2(\rho'_{obs}) + \sigma^2(\rho_{cal}) \sim 2\sigma^2(\rho_{obs}) \qquad (12)$$

if (a) the structure contains mainly first-row atoms and (b) the determinations of the observed density and of the structure parameters (e.g., the neutron experiment) are of comparable accuracy (Coppens, 1974). Such considerations typically lead to $\sigma(\Delta\rho) \sim 0.01 e a_0^{-3}$ (i.e., $0.07 e$ Å^{-3}) in recent work.

V. Some Experimental Densities

The experimental density $\langle\rho(\mathbf{r})\rangle$ is time averaged over the vibrational modes in the crystal. Since zero-point motion will persist even at liquid helium temperatures, no quantitative comparison between theory and experiment is possible unless either the thermal motion is deconvoluted from the experimental maps or the theoretical functions are thermally smeared. The former approach requires the use of a model describing the atomic deformations, because it involves introduction of details not represented in the experimental measurements. The latter method is therefore preferable and is further discussed in Section VII.

In this section we will give a qualitative discussion of some recent experimental results and consider deformation densities in the bond and lone-pair regions. We will specifically avoid interpreting the regions around the atomic nuclei, which, as already discussed are notoriously unreliable.

A. Density in the Bonding Regions

With few exceptions, deformation densities show extra density near the midpoints of covalent bonds. Such charge concentration on the nuclear axis contributes to the balance of the electrostatic forces acting on the nuclei as required by the Hellmann–Feynman principle (Berlin, 1951; Ransil and Sinai, 1967; Bader *et al.*, 1967a,b).

A detailed analysis of the Hellman–Feynman field by Hirshfeld and Rzotkiewicz (1974), based on the diatomic molecule wavefunctions of Bader and co-workers, indicates the importance of a polarization of inner valence and core electrons, which, together with the π-density in molecules such as N_2 and O_2, form the binding part of the deformation density, opposing the destabilizing influence of the σ-density beyond the radial node of the 2s electrons and thereby reducing the electrostatic importance of the density on the bond axis near its midpoint.

In the experimental results, density accumulation has generally been observed in C—C and C—N bonds, with a tendency for a quantitative decrease when more electronegative atoms are involved. Experimental densities in C—O bonds are generally lower than in C—C bonds (e.g., see Coppens *et al.*, 1969; Griffin and Coppens, 1975), whereas little or no density is accumulated in the NO bonds in *p*-nitropyridine N-oxide (Wang *et al.*, 1976; Coppens and Lehmann, 1976) and in the chlorine molecule (Stevens, 1977).

The deformation density in a typical C—O bond (in glycylglycine) is shown in Fig. 2a. Unlike in C—C bonds (Fig. 2b), the density here appears slightly polarized toward the more electronegative atom. Diffraction studies on a number of compounds have convincingly shown the bending on the bonds in strained small-ring compounds. Examples are tricyanocyclopropane (Hartmann and Hirshfeld, 1966), tetracyanoethylene oxide (Matthews and Stucky, 1971), and ethylenimine quinone (Ito and Sakurai, 1974). An X—X deformation map on the latter compound is shown in Fig. 1. These studies provide unequivocal confirmation of theoretical considerations proposed first by Coulson and Moffit (1949) and expanded more rigorously in later work (Stevens *et al.*, 1971; Martenson and Sperber, 1970; for a detailed review, see Newton, 1976).

In bonds between first-row atoms the deformation density peaks near the bond midpoint, whereas the valence density peaks at nitrogen or oxygen atoms or forms a continuous ridge between atoms as in diamond (Figs. 3a and 3b) (Y. W. Yang and P. Coppens, unpublished results; Yang, 1976). This is not the case in the second-row analog of diamond, silicon, for which unusually accurate measurements made on perfect crystals with the Pendellösung technique are available (Tanemura and Kato, 1972; Aldred and Hart, 1973). Here both the deformation and the valence density show their

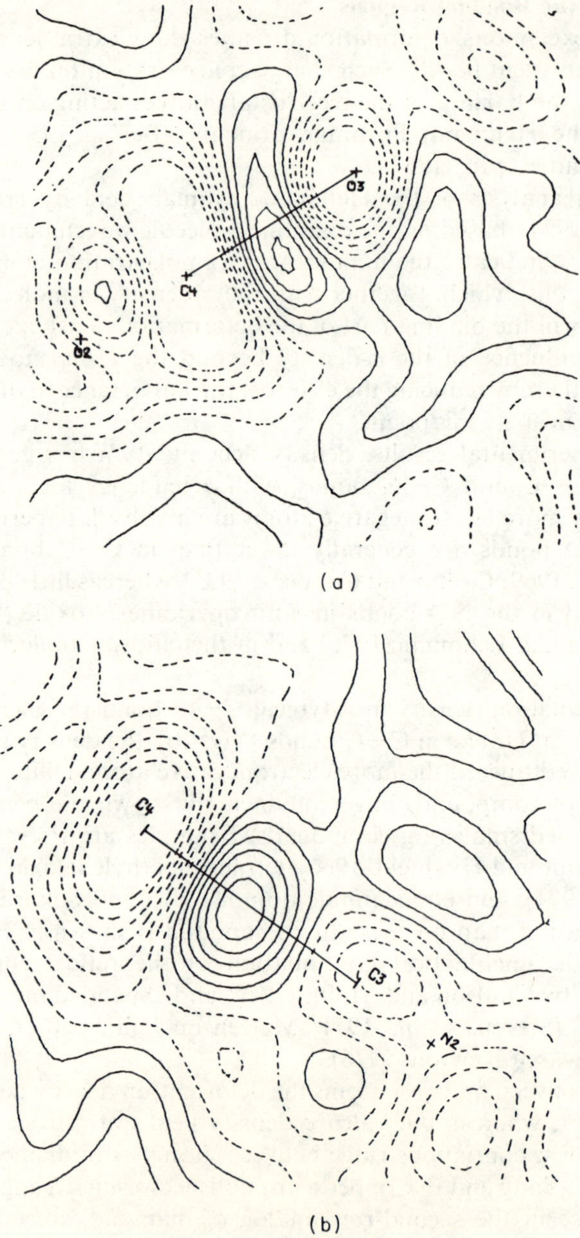

Fig. 2. The $\Delta\rho_{\mathrm{deformation}}$ of glycylglycine. Sections are perpendicular to the plane of the carboxyl group and contain (a) the C(4)—O(3) bond and (b) the C(3)—C(4) bond. Contours at $0.05e\ \text{Å}^{-3}$. (From Griffin and Coppens, 1975.)

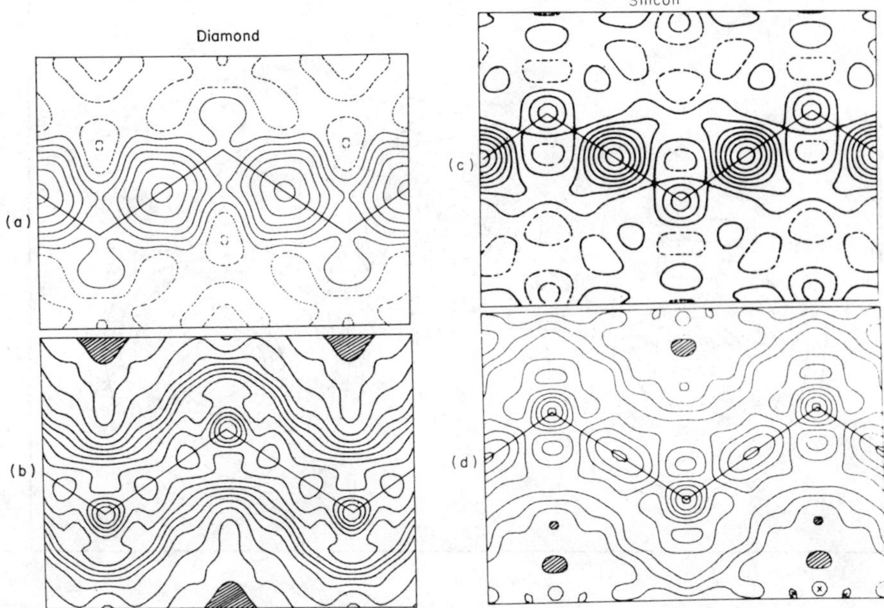

Fig. 3. (a) The $\Delta\rho_{\text{deformation}}$ in the plane of the C—C bonds in diamond, contours at $0.05e$ Å^{-3}; (b) $\Delta\rho_{\text{valence}}$ in the same plane, contours at $0.10e$ Å^{-3}. (c) The $\Delta\rho_{\text{deformation}}$ in the plane of the Si—Si bonds in silicon, contours at $0.05e$ Å^{-3}; (d) $\Delta\rho_{\text{valence}}$ in the same plane, contours at $0.10e$ Å^{-3}.

maxima between the connected atoms (Yang and Coppens, 1974) (Figs. 3c and 3d). The difference with diamond must be a result of the larger radius of the silicon M valence shell as compared with the diamond L shell. The spherical low-density region at about 0.4 Å from the Si nucleus (Fig. 3d) corresponds to a radial node of the M shell (the second node of 3s and the node of 3p almost coincide) and is also observable in a map synthesized from free-atom valence densities at the same thermal amplitudes. This indicates that in the Si crystal the electron distribution near the nuclear region is similar to that in the free atom.

The density maps in silicon have been reproduced in pseudopotential calculations adjusted to fit the experimental results (Walter and Cohen, 1971; Chelikowsky and Cohen, 1974). A still different picture is obtained in the S—S bond in orthorhombic sulfur which shows not one but two maxima along the line connecting the sulfur atoms (Fig. 4a). Examination of the plane bisecting the two S—S bonds reveals that the two bond maxima are part of a sphere of contracted density surrounding the sulfur atoms at about 0.8 Å radial distance (Fig. 4b). Thus, the density map can be best interpreted

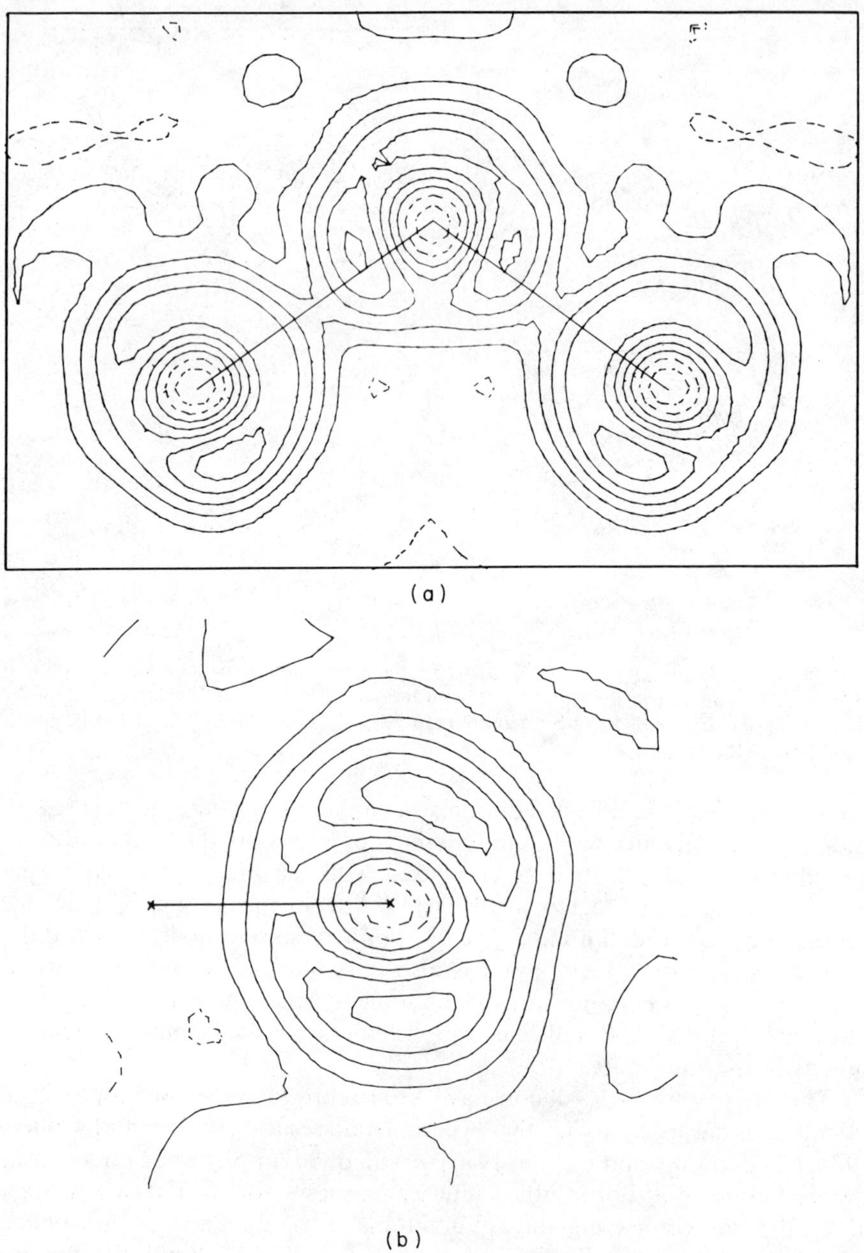

Fig. 4. (a) The $\Delta\rho_{\text{deformation}}$ of orthorhombic sulfur in a plane defined by 3 adjacent sulfur atoms. $T = 100°K$. $\Delta\rho$ has been averaged over the four crystallographically independent planes of the S_8 ring. (b) Average $\Delta\rho_{\text{deformation}}$ of a section through sulfur and bisecting the two S—S bonds; Contours at $0.10e$ Å^{-3}.

as consisting of a "normal" bond between two radially modified sulfur atoms (Coppens *et al.*, 1977). Such an observation suggests deriving radial distribution parameters (such as orbital exponents) from the experimental densities, a possibility which is discussed further in Section VIII,D. No theoretical work on the S_8 molecule is available at the time of writing, but a recently published pseudopotential valence density on Cl_2 (Dixon and Hugo, 1975) shows a similar "camel-back" feature.

On the other hand, a deformation density, based on Dunning and Winter's calculation (1971) of H_2O_2 in which an optimally contracted Gaussian basis O(9s 5p 1d/4s 3p 1d) and H(4s 1p/2s 1p) is used, shows practically no density in the bonding region (Fig. 5), nor does a slightly improved calculation, O(14s 5p 1d/7s 3d 1d), H(5s, 1p/3s 1p), at the same geometry which leads to a practically identical deformation density (Rys *et al.*, 1975).

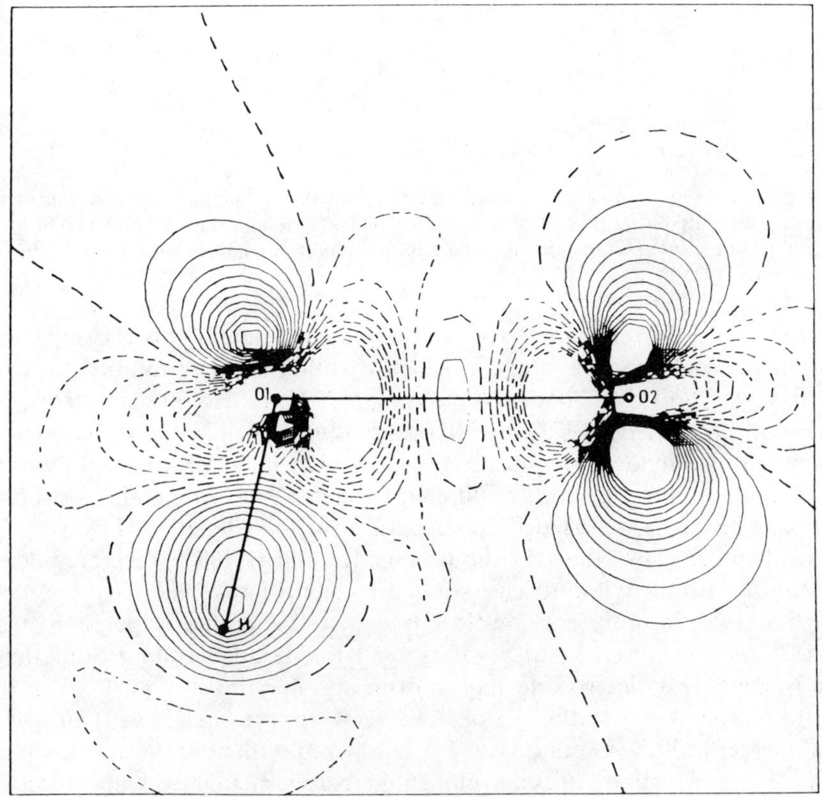

Fig. 5. Theoretical at rest deformation density of hydrogen peroxide from the wavefunctions of Dunning and Winter (1971). Contours at $0.10e\ Å^{-3}$. Very negative contours near atoms are omitted.

The lack of charge build up in the bonding region raises questions about the Hellmann–Feynman equilibrium requirements. It may also be noted that the theoretical difference density on Cl_2 disagrees considerably with the experimental distribution in the solid (Stevens, 1977).

B. Single, Double, and Triple Bonds

Differences among single, double, and triple bonds are readily apparent in sections perpendicular to the bonds through their midpoint. The lack of cylindrical symmetry of the double bond in tetracyanoethylene (Fig. 6)

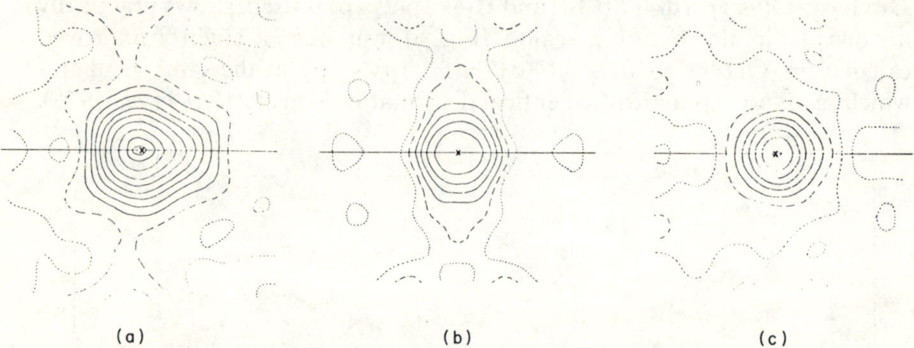

Fig. 6. Sections of $\Delta\rho_{deformation}$ in cubic tetracyanoethylene perpendicular to the bonds through their midpoints. (a) The C≡N bond; (b) the C=C bond; (c) the C—C bond. Straight lines indicate the intersection with the plane of the molecule. Contours at 0.10*e* $Å^{-3}$. (From Becker *et al.*, 1973.)

(Becker *et al.*, 1973) as opposed to the single and triple bonds is compatible with theoretical concepts and confirmed by other studies, notably those on C—C, C=C, and C≡C bonds by Ruysink (1973) and Helmholdt (1975); see also Helmholdt *et al.* (1972). Furthermore, the density in the triple bond is clearly more extended than in the single bond as may be expected. By itself the distinction among bonds of different strength is not of great importance as it can be easily deduced by comparison of bond lengths. However, the detailed information that can be obtained is illustrated by the cumulenic system in tetraphenylbutatriene, studied by Berkovitch-Yellin *et al.* (1976), which shows the adjacent multiple bonds to be mutually perpendicular (Fig. 7) and the central bond to have some triple-bond character. Sometimes bond length and electron density information may be combined, as in the (room temperature) study of α-oxalic acid (Coppens *et al.*, 1969), in which the long central C—C bond [1.537 Å compared with 1.50 Å expected for a $C(sp^2)$—$C(sp^2)$ bond] appears elongated perpendicular to the molecular plan, thus suggesting a π-contribution and a relatively weak σ-contribution which may be partially antibonding.

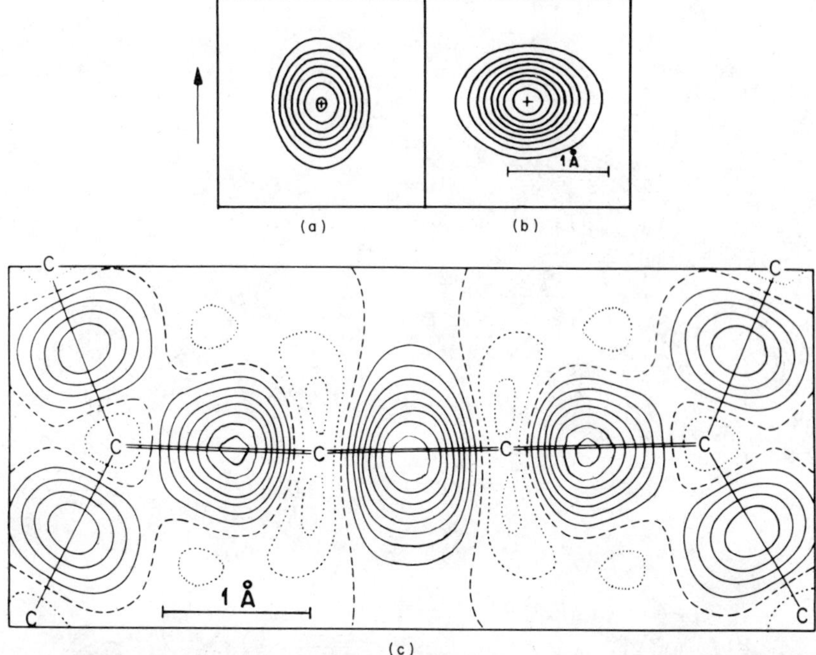

Fig. 7. Deformation density of tetraphenylbutatriene. Sections perpendicular to the outer (a) and inner (b) bonds of the butatriene chain through their midpoints. The arrow is perpendicular to the butatriene plane. (c) Deformation density in the butatriene plane. Contours at $0.10e$ Å^{-3}. (From Berkovitch-Yellin *et al.*, 1976.)

C. Density in the Lone-Pair Regions

A second well-known cause of atomic asymmetry is the double occupancy of nonbonding atomic orbitals. Density will be observed in the deformation maps if such atomic orbitals are not spherically symmetric, i.e., do not have pure s character. Such peaks do not represent the total lone-pair density, but rather its part in excess of spherical atomic distribution. The location of the lone-pair peaks will depend on the nature of the corresponding nonbonding molecular orbital. Some experimentally observed peaks are illustrated in Fig. 8. In the CO group, lone-pair density is found at roughly 120° from the CO bond axis or in a broad unresolved peak, whereas in the C≡O molecule (as a ligand in organometallic complexes) and in the cyano group, single peaks at 180° are observed. In the nitroxide group in *p*-nitropyridine N-oxide, however, the lone pair is clearly at a 90° angle to the bond axis (Fig. 8). The corresponding hybridizations, given in Table I, indicate decreasing s character in the σ-bonding orbital in the direction

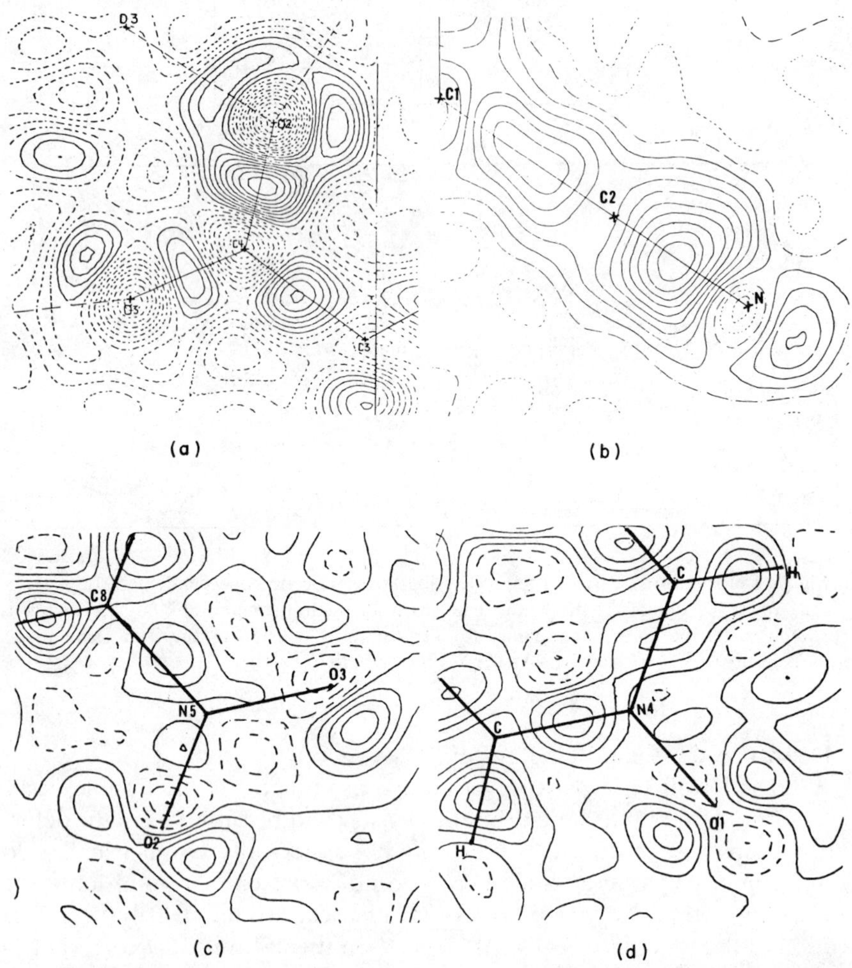

Fig. 8. Lone-pair deformation densities. (a) Carboxyl group of glycylglycine. A second lone-pair peak on O(3) is located out of the plane; contours at 0.05e Å^{-3} (Griffin and Coppens, 1975). (b) The C≡N group of tetracyanoethylene; contours at 0.10e Å^{-3} (Becker *et al.*, 1973). (c) Nitro group of *p*-nitropyridine N-oxide; contours at 0.10e Å^{-3} (Coppens and Lehmann, 1976). (d) Nitroxide group of *p*-nitropyridine N-oxide; contours at 0.10e Å^{-3}. (Coppens and Lehmann, 1976).

(carbonyl) → NO (nitroxo), i.e., when the atom to which oxygen is bound becomes more electronegative. This is in agreement with concepts developed by Bent (1961) who has argued that more p character will be found in bonding hybrids to more electronegative atoms. To achieve the hybridization of the CO oxygen, one-third of an s electron has been promoted to a 2p orbital, which according to estimates by Slater would require about 65 kcal/mole. This energy must be retrieved by more efficient bonding of the σ sp^2 bonding hybrid, which is more readily accomplished with the more diffuse, carbon atom bonding hybrids.

TABLE I

APPROXIMATE HYBRIDIZATIONS FROM DEFORMATION DENSITY MAPS

Bonds	σ-Bonding hybrid	Lone-pair hybrid
C≡N, C≡O	sp	sp
C=O	sp^2	sp^2, sp^2
N—O	p	s, p

The difference density in Cl_2 similarly indicates pure p character of the asymmetric lone-pair electrons and, therefore, p character of the σ-bonding hybrids in agreement with the general trends observed.

D. Density in Solids Containing 3d Transition Metals

Experimental measurements must be increasingly accurate when heavier atoms are involved, i.e., when a larger fraction of the electrons is in unperturbed inner-shell orbitals. Nevertheless, a few studies are now available indicating that meaningful information can be obtained. This is of importance as the bonding situation in many organometallic compounds and in alloys is controversial and less easily studied by theoretical methods.

The experimental density in benzenechromium tricarbonyl (Rees and Coppens, 1973) shows the lone-pair density at both ends of the carbonyl ligand, also observed in a very recent study on chromium hexacarbonyl (Rees and Mitschler, 1976), and a large residual ring of density around the Cr atom. Although its experimental significance has been questioned by the authors, a somewhat similar feature appears in the theoretical map (Guest *et al.*, 1975).

More regularly arranged asymmetric features are observed in $[Co(NH_3)_6][Co(CN)_6]$ (Iwata and Saito, 1973) and in γ-Ni_2SiO_4 (Marumo *et al.*, 1974). In both cases 8 peaks are located at 0.45 Å from the metal atom at the corners of a cube, suggesting excess electrons in the t_{2g} orbitals of the octahedral crystal field. Such results are reminiscent of more accurate unpaired electron densities which can be obtained on ferro- and antiferromagnetic materials with the polarized neutron technique (e.g., see Moon, 1972; Tofield, 1975).

A potentially most interesting group are compounds containing polynuclear metal clusters, in which metal–metal distances are often shorter than in metals and alloys. Experimental work here is often hampered by crystal quality and sample instability; the only study presently available is on μ-acetylene dicyclopentadienyldinickel, $(h^5C_5H_5Ni)_2CH{\equiv}CH$ (Wang and Coppens, 1976), which contains a short Ni—Ni distance of 2.345(2) Å. The density in the bond region indicates a straight rather than a bent bond, with a double maximum much more pronounced than in the sulfur–sulfur bond. The density in the acetylene molecule, which is in a cis-bent pseudoexcited state is significantly perturbed; its maximum is displaced away from the Ni—Ni region and no longer coincides with the center of the C—C bond.

A vanadium–vanadium bond with considerable covalent character is found in the alloy V_3Si, which becomes superconducting at about 17 K (Staudenmann *et al.*, 1976). The vanadium atoms form one-dimensional chains along which density is concentrated, whereas the bonding between Si and V seems to a large extent ionic. With the exception of benzenechromium tricarbonyl, no theoretical densities are available for any of these compounds, and additional theoretical work would be most useful.

E. Experimental Densities in Very Small Molecules

Before continuing with a discussion of theoretical deformation densities and their implications, we will summarize studies on very small molecules, which are also amenable to advanced theoretical calculations. The number of studies is limited because a small molecular weight generally implies that the compound is a liquid or gas at room temperature, causing additional experimental complications. The exceptions are ionic crystals containing molecular ions, of which the azide ion in KN_3 (Stevens, 1977) and NaN_3 (Fig. 9) (Stevens, 1973; Stevens and Hope, 1977), the cyanide ion in $NaCN \cdot 2H_2O$ (Bats, 1977), and the thiocyanate ion in NaCNS and NH_4CNS (Bats and Coppens, 1977) have been studied. Except for the chlorine molecule discussed above and a study of formamide (E. D. Stevens, unpublished results), no compounds that are not solid at room temperature have been analyzed accurately.

It is in this area that experiment and advanced theoretical methods can most fruitfully interact, and where effects of thermal smearing, basis set composition, and configuration interaction may be studied.

We will proceed with a discussion of theoretical deformation densities before reviewing the comparisons of experiment and theory that have been made up to the time of writing.

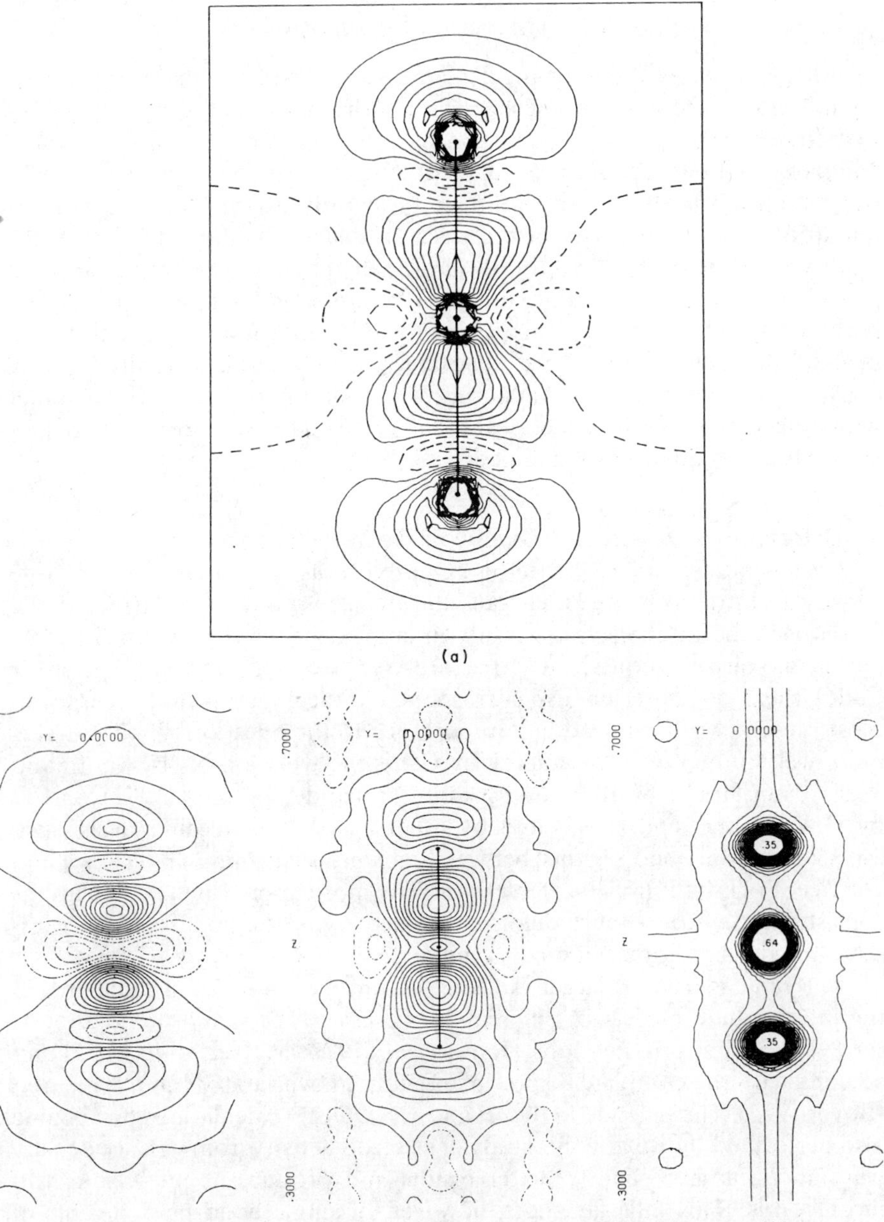

Fig. 9. Electron deformation density in the azide ion. (a) Theoretical at rest; contours at $0.10e$ Å^{-3}. (b) Theoretical at experimental thermal motion and experimental resolution; contours at $0.05e$ Å^{-3}. (c) As measured in NaN_3; contours at $0.05e$ Å^{-3}. (d) Error distribution in the experimental density; contours at $0.015e$ Å^{-3}, lowest contour is $0.015e$ Å^{-3}. Highest contours omitted and labeled with maximum values.

VI. Theoretical Densities

Although many theoretical calculations of increasing quality have been completed in recent years, few electron density maps, and especially deformation densities have been published, even though their calculation is rapid compared with the effort needed to obtain the self-consistent field (SCF) wavefunction. As observed by Smith and Richardson (1967) in their pioneering study of the sensitivity of the electron density to improvement of the quality of the wavefunction, (total) charge densities from the various approximations are nearly the same, but their resemblance is only superficial as becomes apparent upon calculation of the difference function. As discussed in the following, the difference (deformation) function is a sensitive test of convergence of molecular calculations, which should be considered together with other criteria such as energy and one-electron density properties including dipole and quadrupole moments.

A. Deformation Density and Quality of the Wavefunction

"What quality wave function is required to assure an electronic charge density accurate to 2–5%? Generally the answer is not known" (Cade, 1972, p. 4). Since the total electron density in a covalent bond between first-row atoms is typically about $2e$ Å^{-3} (i.e., $0.3ea_0^{-3}$) (see some examples given by Cade), the 2–5% criterion also corresponds to the experimental accuracy in most present work, thus adding importance to the question posed. What is quite well known from the survey by Cade, the work on N_2 by Smith and Richardson (1967), on B_2H_6 and N_2 by Laws and Lipscomb (1972), on CO by DeWith and Feil (1975) and by others, is that large differences exist, especially in the bond regions, between semiempirical, minimal basis set and Hartree–Fock limit (SCF) calculations. The more approximate calculations consistently fail to produce deformation density in the bonds and overestimate the density in the lone-pair regions.

The same is true for larger molecules such as cyanuric acid calculated in the intermediate neglect of differential overlap (INDO) and minimal basis set (STO-3G) approximations (Jones *et al.*, 1972; McIver *et al.*, 1971) and adenine with the complete neglect of differential overlap (CNDO/2) method (Boyd, 1972). The extended Hückel and CNDO/2D calculations on adenine also performed by Boyd in his study of electron density from valence density wavefunctions give slightly better results and produce about $0.1e$ Å^{-3} in most bonds. This falls far short, however, of single bond peak heights of 0.5–$0.6e$ Å^{-3} obtained in low-temperature diffraction experiments, so that even these densities cannot be considered reliable. Some degree of convergence toward the Hartree–Fock density is only obtained at the double-zeta plus polarization level. Cade, in the absence of a truly advanced calculation or experimental information, has used the restricted Hartree–Fock (RHF)

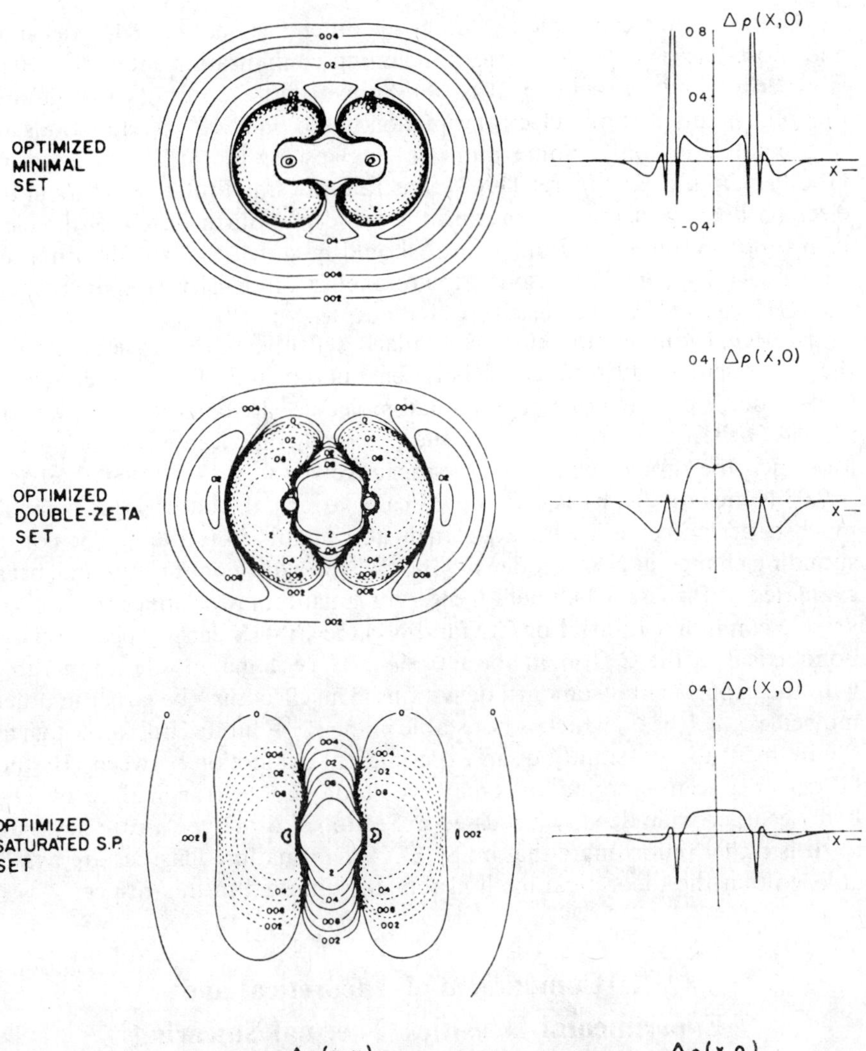

Fig. 10. Variation of the molecular charge density with quality of the wavefunction for $O_2(X^3\Sigma_1^-$, $R = 2.282a_0)$; $\Delta\rho = \rho_{\text{Hartree-Fock}} - \rho_{\text{approx. HF}}$. (From Cade, 1972.)

density as a reference state and plotted the difference between the density from more approximate calculations and the RHF densities as a theoretical error function (Cade, 1972). His results clearly indicate the inadequacy of the approximate calculations in representing the deformation density in the bonding and lone-pair regions (Fig. 10). As bond density is consistently observed in experimental maps, this also implies that experimental densities clearly discriminate against less advanced calculations.

This leads to the question whether the difference between RHF densities and those based on wavefunctions including configuration interaction (CI) is experimentally detectable. It is well-known that according to Brillouin's theorem, errors in a one-electron function based on RHF wavefunctions are of second-order only. Some numerical values for the He atom given by Goodisman and Klemperer (1963) illustrate the small relative errors in the electron density at the nucleus and the r and p^4 (fourth power of the electron's momentum) operators. But, it should be noted that the deformation density is only a small fraction of the total electron density, so that errors in the RHF density can be relatively significant.

Molecular dipole moments are similarly sensitive to the wavefunction as they represent small differences between nuclear and electronic contributions. The CI dipole moments for small molecules are consistently lower by 0.1–0.4 D than those calculated in the HF limit, an extreme case being CO for which the dipole moments are respectively +0.28 D (O negative) and −0.08 (C negative) for the HF and CI calculations, the latter value being in much better agreement with experiment (Grimaldi *et al.*, 1967). The corresponding change in electron density has, for the wavefunctions quoted, been evaluated at Professor Daudel's Center for Quantum Mechanics (P. Becker, private communication). For CO (and N_2) the CI–HF density peaks off the bond axis near the C atom at about $0.06e$ Å^{-3}, i.e., a magnitude comparable with the experimental standard deviations. The effect may be larger in other molecules, such as F_2 which is not stable in the RHF limit. Thus, although at the moment the possibility of an experimental distinction between HF and CI densities seem marginal, this may even today not be true in all cases. The hydrogen peroxide density discussed in Section V,A may be a case in point.

It is rather unfortunate that no SCF-Xα deformation densities are available, so that the theoretical method remains untested in this respect.

VII. Comparison of Theoretical and Experimental Densities. Thermal Smearing

In a number of cases reasonable *qualitative* agreement has been observed between experiment and medium quality calculations. Examples include the following: (*a*) the double-zeta deformation density on cyclobutane and a low-temperature X-ray study on tetracyanocyclobutane, which both show bending of bonds (Harel and Hirshfeld, 1975); (*b*) a double-zeta calculation on glycine, including the electrostatic effect of the crystalline environment in a point charge approximation and an X—N room temperature density (Almlöf *et al.*, 1973); (*c*) a Hartree–Fock–Slater calculation on the nitrate ion and a room temperature X—N density on uronium nitrate (DeWith *et al.*, 1975), which both lack density in the NO bonds and show lone pairs much as in *p*-nitropyridine N-oxide; (*d*) a minimal basis set calculation on benzenechromium tricarbonyl (Guest *et al*., 1975) and a liquid nitrogen temperature X—N study (Rees and Coppens, 1973), which both show lone-pair electrons at the carbonyl oxygen atoms, bonding density in the benzene ring bonds, and appreciable asymmetry around the chromium atom. As predicted the minimal basis set (MBS) calculations underestimate the density in the CO and C—C bonds.

The agreement in these and other studies has been often described as "good," yet quite significant disagreements (such as the density around the nitrogen atom in glycine) are often left unexplained and may be due to either basis-set limitations or experimental inadequacies.

A comparison with a more accurate calculation is available for the azide ion (Fig. 9). The calculation uses a (11s 5p 1d/5s 3p 1d) Gaussian bases and gives a better energy than calculations reported by Archibald and Sabin (1971). Both theory and experiment agree on the appearance of lone-pair and bond peaks, their approximate height, and especially the displacement of the bond density from the midpoint of the bond toward the central nitrogen atom (Stevens *et al*., 1977).

For a really quantitative comparison, the thermal averaging in the density maps has to be allowed for. We will argue here that attempts to extrapolate the experimental deformation densities to zero motion, will not lead to true high-resolution rest densities because the required assumptions regarding the atomic deformation functions will bias the final results. Alternatively, the thermally averaged theoretical density $\langle\rho\rangle$ can be obtained within the Born–Oppenheimer approximation as

$$\langle\rho\rangle = \int \phi^2(\mathbf{R}_1, \mathbf{R}_2 \cdots \mathbf{R}_N)\rho(\mathbf{R}_1, \mathbf{R}_2 \cdots \mathbf{R}_N)\, d(\mathbf{R}_1, \mathbf{R}_2 \cdots \mathbf{R}_N), \qquad (13)$$

where ϕ^2 is the nuclear distribution function, and $\mathbf{R}_1 \cdots \mathbf{R}_N$ are the nuclear position vectors. Rigid evaluation of this function requires calculation of the molecule and its electron density at several nuclear configurations to account for internal vibrations.

The only result based on such a procedure is from a recent calculation by P. Becker (private communication) on BeH. Ermler and Kern (1971) have calculated the water molecule at 45 geometries along the normal mode coordinates and have obtained zero-point vibration-averaged, one-electron density functions (but not the total charge distribution) by fitting to a polynomial expansion in the displacement coordinates. Their results show, for example, that if only internal modes are considered, the "smeared" density at the oxygen nucleus is even higher than the rest density. Such a result could never be observed in crystallographic studies, because the important contributions of low-frequency lattice modes would always reduce the electron density. Nevertheless, it adds some urgency to a full quantitative study of thermal smearing due to both internal and external vibrations.

A more approximate method has been discussed by Coulson and Thomas (1971; Thomas, 1971). It is based on a simple convolution which, for isolated atoms, may be written

$$\langle \rho \rangle = \int \rho(r - t)\phi^2(t)\, dt. \tag{14}$$

Calculations on H_2 indicate that near the nucleus thermal smearing is well represented by Eq. (14). The treatment of two-center terms in the theoretical one-electron density functions is less obvious in this approximation. Ruysink and Vos (1974), assuming that all two-center coefficients in the expansion,

$$\rho = \sum_{\mu} \sum_{\nu} P_{\mu\nu}\psi_{\mu}\psi_{\nu}, \tag{15}$$

are only affected by the need to renormalize for different nuclear configurations, have evaluated the thermally smeared density for acetylene. Although their results appear quite reasonable, there seems to be little *a priori* justification for the assumed invariance of the two-center coefficients.

Although an experimental density of acetylene is not available, the theoretical results are in reasonable agreement with experimental densities in other molecules containing C≡C bonds (Ruysink, 1973). A comparison by Becker of the smearing of the BeH density in the convolution approximation with the more rigorous Born–Oppenheimer treatment indicates the discrepancies to be small (Becker, 1975). Although this result is encouraging it should be verified for molecules undergoing more complicated internal vibrations (such as the π-type vibrations in acetylene).

A different procedure has been followed in the comparison with theory of the bond and lone-pair density in tetracyanoethylene (Coppens, 1974). By approximating the bond and lone pairs in the RHF density on dicyanogen (Hirshfeld, 1971) by single Gaussians, (isotropic) density functions are obtained that can be easily convoluted with a harmonic thermal smearing function. The effect of thermal smearing is pronounced and inverts the relative heights of the bond and lone-pair peaks. The thermally corrected theoretical values agree in relative peak height with the experimental results on tetracyanoethylene, but the latter are 10–15% higher. Further work should indicate whether such discrepancies are due to remaining systematic errors in the experiment, inadequate thermal averaging, or shortcoming of the theoretical model.

In the study of the azide ion thermal motion was allowed for as if it consisted completely of rigid body motion. For a simple entity, such as N_3^-, it is impossible to distinguish from the diffraction data alone between motion due to *internal* and *external* modes, so the rigid body treatment may be considered an opportune first approximation, which is reasonably valid because the low-frequency external modes are usually dominant. From the theoretical density at rest and the known crystal structure of NaN_3, structure factors were calculated. By truncation of the thermally attenuated theoretical structure factors at the limit of experimental observation and Fourier transformation of the truncated set back to crystal space, a theoretical density is obtained which is not only thermally smeared but also has the same limited resolution of the experimental map. A quantitative comparison is thus possible, especially with the aid of the error distribution map derived as described in Section IV and reproduced in Fig. 9d. The main features such as peak heights agree remarkably well, the experimental heights being 0.52(5) and 0.23(3)e Å^{-3} in the bond and lone pair, respectively, whereas the corresponding theoretical values are 0.44 and 0.26e Å^{-3}. But the bond peak is more extended along the molecular axis in the experimental map which also shows significantly more density in the plane through the nitrogen atoms perpendicular to the molecular axis. It is most interesting that this last discrepancy is similar to the differences between HF and CI densities obtained for CO and N_2 (see Section VI,A). This may be the first experimental confirmation of the errors in the Hartree-Fock electron density due to neglect of electron correlation.

VIII. Functions Based on Experimental Densities

A. Discussion of Methods

Although the electron density maps are the most detailed functions available, one may argue that they are cumbersome to use and that other one-electron properties, defined by

$$\langle \hat{O} \rangle = \int \hat{O}\rho(\mathbf{r})\, d\tau, \tag{16}$$

where $\hat{O}$ is any operator, are more convenient if not necessarily equally sensitive criteria for comparison with theory or with other experimental techniques.

Experimental values of Eq. (16) can be derived by alternative methods. The first is a two-step process in which the electron density ρ is evaluated as usual [Eq. (9)], and the integration is performed *numerically* by summation over a sufficiently fine grid of points covering the volume under consideration. A second method may be applied if the Fourier transform

$$\Phi = \int_{\substack{\text{volume of} \\ \text{interest}}} \hat{O} \exp(-2\pi i\mathbf{H} \cdot \mathbf{r})\, d\tau \tag{17}$$

can be evaluated *analytically*. In this case the desired quantity is given by

$$\langle \hat{O} \rangle = \frac{1}{V} \sum_{\substack{\text{all} \\ \text{observations}}} \Phi(\mathbf{H})F(\mathbf{H}) \exp(-2\pi i\mathbf{H} \cdot \mathbf{r}). \tag{18}$$

As this summation is very similar to Eq. (9), it can be performed routinely with existing programs (Kurki-Suonio, 1959; Coppens and Hamilton, 1968).

In addition to Fourier methods, least-squares fitting of density functions may be accomplished directly in scattering space analogous to least-squares adjustment of atomic parameters in conventional crystallography. Atomic or molecular parameters adjusted may range from the occupancy of a spherical valence shell and atomic dipole and higher multipole functions (Stewart, 1970; Hirshfeld, 1971) to atomic radial distribution parameters. Other properties can then be derived analytically from the density functions. An advantage of this reciprocal space analysis is that thermal motion is allowed for explicitly, so that, within the limits of the vibration formalism and the density functions used, a deconvolution of thermal motion is achieved. On the other hand, the results of the analysis may be biased by incompleteness of the set of density functions, so that it is in general desirable to perform both reciprocal space and density space analysis of the experimental results. We will discuss some of the most interesting examples and present a few comparisons of the different methods.

B. Net Atomic and Molecular Charges

The difficulty of defining a net atomic charge in a molecule in which the charge density is by definition continuous has not prevented its widespread application as a useful if somewhat ambiguous concept. The Mulliken population analysis or similar methods are not applicable to diffraction densities unless the population parameters $P_{\mu\nu}$ of Eq. (15) have been derived through a fit to the experimental density. Because of appreciable correlations between the experimental parameters, this is a still controversial procedure (Jones *et al.*, 1972; Stewart, 1973a). Better defined, but less directly related to the LCAO expansion of molecular orbitals are formalisms containing only one-center density functions. Methods using spherical valence shells will be referred to here as *valence shell projection methods* because the experimental distribution is projected in a valence shell with either the isolated atom radial distribution or modified to correspond, for example, to the standard-exponent Slater orbitals proposed by Hehre *et al.* (1969). In *valence shell deformation methods*, atomic asymmetry is explicitly introduced, so that not only the electron population but also dipole and higher moments are obtained. Since two-center density is absorbed into the one-center terms, the total molecular density is divided into "pseudoatoms" (Stewart, 1975).

In general, valence shell projection and deformation analysis lead to results that are essentially identical if the conceptual ambiguity is allowed for, and which are also in reasonable agreement with results of the Mulliken population analysis. The following two examples illustrate the results.

1. In glycylglycine, theoretical charges from an INDO calculation (Momany *et al.*, 1974) are in qualitative agreement with those from a valence shell projection (Griffin and Coppens, 1975). Although discrepancies up to 0.2 electron are observed, both methods agree on oxygen and nitrogen atoms having net negative charges of about 0.3–0.4 electron, hydrogen atoms being positive by 0.1–0.3 electron units and carbon atoms being positive in a polar and almost neutral or slightly negative in an aliphatic environment (Fig. 11). Similar experimental results have been obtained in several

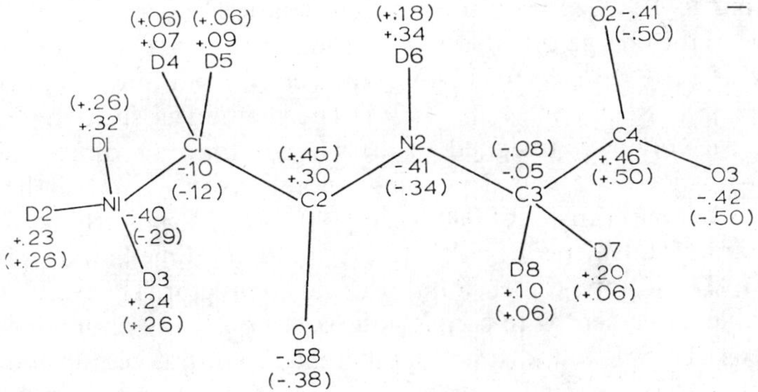

Fig. 11. Experimental net atomic charges for glycylglycine (Griffin and Coppens, 1975). Values in parenthesis are from theoretical calculations of Momany *et al.*, (1974).

other compounds (Coppens, 1975a,b, and references cited therein; Allen-Williams *et al.*, 1975; Harel and Hirshfeld, 1975).

2. In benzene chromiumtricarbonyl (Rees and Coppens, 1973), the chromium atom is found to be positive by about 1.2 electrons, whereas carbonyl groups and the benzene ring are negative by about -0.3 electron units. This contrasts with self-consistent charge and configuration (SCCC) calculations that predict a small positive charge on Cr (of 0.1–0.5 electron) and on the benzene ring (0.04–0.40 electron), but agrees somewhat better with a recent MBS *ab initio* calculation (Guest *et al.*, 1975) according to which Cr is positive by $2.01e$ and the benzene and carbonyl groups negative by -0.84 and $-0.41e$, respectively. However, this result seems not uncontroversial as

relative acidities of phenol and phenolchromiumtricarbonyl (Wu *et al.*, 1972) and NMR chemical shifts (Brill and Kotlar, 1974) suggest a net positive charge on the benzene ring.

The valence shell projection methods are undoubtedly more difficult in the latter example, because of uncertainties about the appropriate chromium valence density function. This ambiguity is avoided in the numerical integration of the electron density if a well-defined volume of integration can be selected. This method is therefore most suitable for integration over the molecular volume in mixed molecular crystals such as hydrates or charge transfer complexes, but it has also been applied to the superconducting alloy SiV_3 which has the β-tungsten structure.

The expression used is

$$\langle \hat{O} \rangle = \frac{1}{V} \sum_{\text{cube}} \rho_{\text{integrated}}, \tag{19}$$

where the sum is over a number of elementary cubes defining the total volume of integration, and the integration over each cube is performed according to Eq. (18). For the low-temperature superconductor SiV_3, an analysis of the charge transfer as a function of volume of integration leads to a value of 2.2 ± 0.3 electrons being transferred from Si to V_3 at room temperature (Staudenmann *et al.*, 1976). The method is such that a roughly polyhedral volume is assigned to the silicon atom, the dimensions of this pseudo-Wigner–Seitz cell depending on the *relative* size of the adjacent atoms. A charge transfer of 0.48–0.60 $(\pm 0.15)e$ has similarly been obtained from tetrathiofulvalene(TTF) to tetracyanoquinodimethanide (TCNQ) in the TTF–TCNQ one-dimensional organic conductor. This value is in quite reasonable agreement with conclusions based on X-ray photoemission data (Coppens, 1975c), but distinctly different from the tetracyanoethylene–pyrene "charge transfer" complex, in which charge transfer was found to be absent or less than $0.15e$ (Larsen *et al.*, 1975).

An unexpected result obtained for several crystalline hydrates is the positive charge of about $0.2e$ on the water molecule, observed first in valence shell projection studies of the trihydrate of an organometallic nickel complex and in oxalic acid dihydrate (Coppens *et al.*, 1971). Because it could have been an artifact of the valence density functions used, it is most important that the numerical integration method [Eq. (19)] has recently led to an identical result for $NaCN \cdot 2H_2O$. For the 2 water molecules in this complex, net charges of $+0.22$ and $+0.24e$ are obtained, compared with $+0.13$ and $+0.26e$ with the valence shell projection method (Bats, 1977). The differences are within the experimental errors. The positive charge on the water molecule should be observable with other physical methods. A theoretical analysis of a crystalline hydrate seems also indicated.

C. Molecular Dipole Moments

The integrations represented by Eqs. (18) and (19) can be used to obtain molecular dipole and higher moments. However, since the integration,

$$\mu_n = \int_{\text{molecule}} \mathbf{r}^n \rho(\mathbf{r})\, d\tau, \tag{20}$$

places increasing weight on the peripheral areas of the molecule where relative experimental uncertainties are large, prospects for obtaining higher moments are limited. For the dipole moments, recent applications to sulfamic acid (SO_3NH_3) and to the cyanide ion and the water molecules in $NaCN \cdot 2H_2O$ indicate fair agreement with solution dipole moments, although the solid state results tend to be smaller in magnitude.

D. Orbital Exponents

Within the "atoms in molecule" approximation one may use the experimental data to examine the radial dependence of the atomic density for comparison with optimized molecular exponents frequently used in minimal basis set calculations (Hehre *et al.*, 1969; Stevens *et al.*, 1971). The procedure used involves a least-squares fit of the experimental structure factors including as variable the orbital exponent ζ. This may be done with a spherical density function describing the atomic valence shell in the spirit of the valence shell projection method or with a formalism including angular distortion functions. However, in the latter case the density in the bonds will have a pronounced effect on the radial distribution which will tend to lower the ζ-value as the two-center bond density represents a migration of density away from the nuclear region. This is confirmed by the results of Stewart utilizing powder data on diamond, which show a decrease in ζ when higher order multipoles are included (spherical, $1.68(3)a_0^{-1}$; nonspherical, 1.56–$1.58a_0^{-1}$, compared with $1.72a_0^{-1}$ for the "standard" molecular value) (Stewart, 1973b). The contraction of the carbon L shell compared with the isolated atom value is also found in spherical analyses of glycylglycine, oxalic acid, and *p*-nitropyridine N-oxide, whereas oxygen atoms remain unchanged or expand by about 2% (P. Coppens, Y. W. Yang, and P. Becker, unpublished results).

A quite informative example is provided by orthorhombic sulfur. As S_8 contains only one type of atom, the radial dependence can be derived directly, without resorting to least-squares techniques, by combining X-ray structure factors and geometrical structure factors G, based on neutron atomic parameters. If

$$G = \sum_{\substack{\text{all} \\ \text{atoms}}} T_i \exp(2\pi i \mathbf{H} \cdot \mathbf{r}),$$

where T_i is the temperature factor of atom i, the experimental scattering factor f is obtained as

$$F(S) = \langle F(\text{obs})/G \rangle$$

in which the average is over the reflections in small ranges of $|S| = 2 \sin \theta/\lambda$. In Fig. 12, the experimental f curve is compared with an isolated atom-scattering factor and with the results of the orbital exponent refinement. Although the latter represents a significant improvement, it is clear that a single orbital exponent cannot fully represent the modification of the radial distribution upon bond formation. Specifically, the experimental curve is higher near the origin and lower at large S values, indicating an expansion of the charge density near the nucleus and a contraction at larger distance.

The most pronounced contraction is found for the hydrogen atom, in agreement with theoretical orbital exponents and an analysis of the density in the H_2 molecule (Stewart *et al.*, 1965).

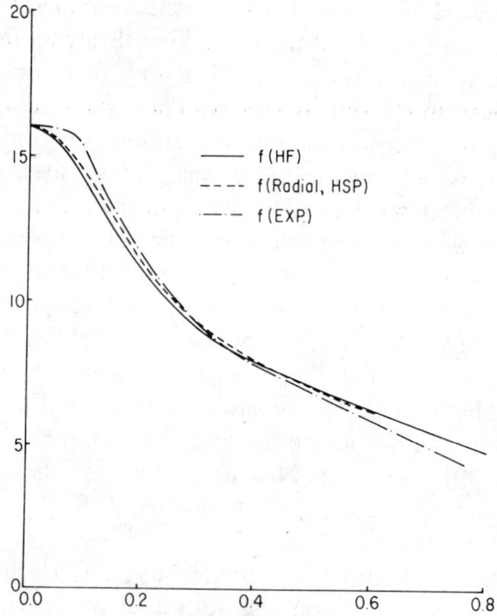

Fig. 12. Experimental scattering factor f(EXP) of orthorhombic sulfur compared with the isolated atom scattering factor f(HF) and the result of an orbital exponent refinement f(Radial, HSP) as a function of sin θ/λ (Å^{-1}). (From Coppens *et al.*, 1977.)

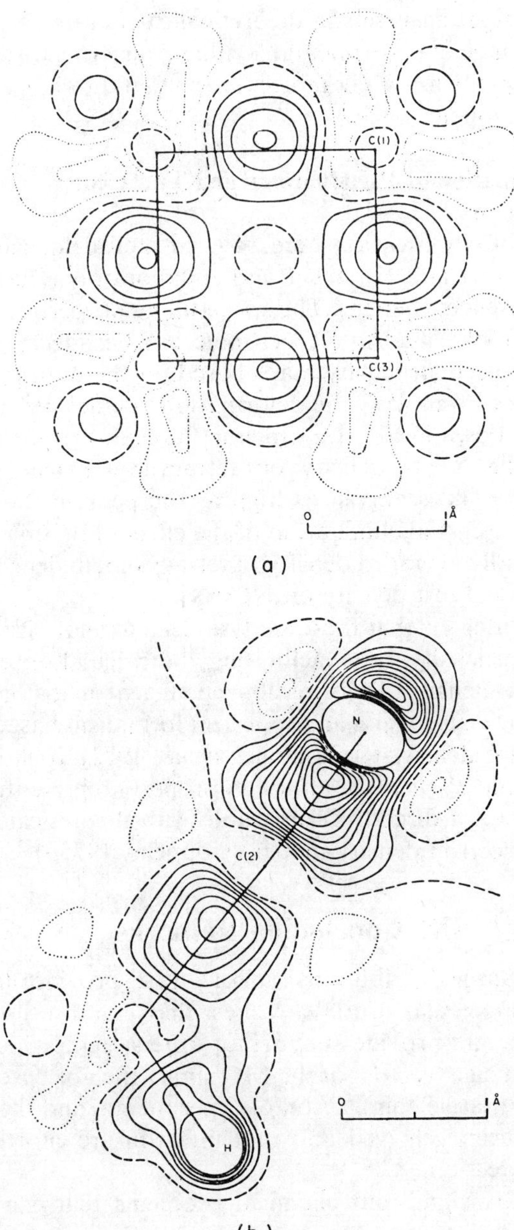

Fig. 13. Deformation density of tetracyanocyclobutane from least-squares refinement of deformation coefficients. (a) Cyclobutane ring; (b) HCCN plane. Contours at $0.10e\ \text{Å}^{-3}$. Very negative contours near atoms are omitted. Noncrystallographic symmetry constraints have been imposed on the deformation coefficients. (From Harel and Hirshfeld, 1975.)

The trend to larger basis sets in theoretical calculations has reduced the immediate importance of experimental ζ-values. Nevertheless, MBS calculations are still a necessity for larger molecules, so that the choice of the "best" ζ-value remains pertinent.

E. Angular Deformation of Pseudoatoms and Two-Center Population Parameters

We will only briefly mention here valence shell deformation methods introduced by Stewart and Hirshfeld and collaborators (Harel and Hirshfeld, 1975, and references therein). The space-filling properties of formalisms including atomic dipole, quadrupole, octopole, and hexadecapole terms may be judged from molecular density maps based on the multipole expansion coefficient. A recent example is the deformation coefficients map of tetracyanocyclobutane, shown in Fig. 13, which in the cyano group bears a strong resemblance to the X—N density on tetracyanoethylene (Fig. 8). The nitrogen lone-pair peak is almost as high as the peak in the C≡N bond, indicating at least a partial elimination of the effect of thermal motion [it is lower in the thermally averaged density of tetracyanoethylene (Fig. 8b), but higher in a theoretical rest density on NCCN].

An added advantage is that these analyses lead to analytical descriptions of the deformation density, thus facilitating the calculation of other functions based on the density. Deformation coefficient maps have also been obtained for cyanuric acid using an aspherical formalism based on Eq. (15), including two-center density terms (Jones *et al.*, 1972). It should be noted that the experimental $P_{\mu\nu}$ values, although properly representing the deformation density, are not directly comparable with theoretical diagonal elements of the one-electron density matrix (Coppens, 1975b).

IX. Concluding Remarks

Experimental charge distributions can, of course, provide no information on the individual molecular orbitals. At best they can discriminate against certain calculations and provide support for others, upon which analysis of the eigenvalues and eigenvectors of the MO's may then be based. It is fair to say that truly quantitative comparison of experimental and theoretical densities has not yet been achieved and that much future effort will be concentrated in this direction.

Among the many significant chemical problems that can fruitfully be studied in a qualitative way, we mention here the nature of the short hydrogen bond (ionic or covalent?), electronic structure of propellanes (to complement existing theoretical work), the nature of bonding in S_4N_4 (are there S—S bonds across the ring?), and metal–metal bonding in 3d metal-containing cluster compounds. The field is quite generally applicable at least to light-atom systems and should be of considerable value in elucidating poorly understood and controversial bonding situations.

ACKNOWLEDGMENTS

The support of the National Science Foundation for our part of the research surveyed here is gratefully acknowledged as is the willingness of several colleagues to communicate results prior to publication. The authors would also like to thank Professor R. Daudel for encouraging the writing of this manuscript.

REFERENCES

Aldred, P. J. E., and Hart, M. (1973). *Proc. R. Soc. London, Ser. A* **332**, 223.

Allen-Williams, A. J., Delaney, W. T., Farina, R., Maslen, E. N., O'Connor, B. H., Varghese, J. N., and Yung, F. H. (1975). *Acta Crystallogr., Sect. A* **31**, 101.

Almlöf, J., Kvick, A., and Thomas, J. O. (1973). *J. Chem. Phys.* **59**, 3901.

Archibald, T. W., and Sabin, J. R. (1971). *J. Chem. Phys.* **55**, 1821.

Bader, R. F. W., Henneker, W. H., and Cade, P. E. (1967a). *J. Chem. Phys.* **46**, 3341.

Bader, R. F. W., Keaveny, I., and Cade, P. E. (1967b). *J. Chem. Phys.* **47**, 3381.

Bats, J. W. (1977). *Acta Crystallogr. Sect. B* **33**, 466.

Bats, J. W., and Coppens, P. (1977). *Acta Crystallogr.* (in press).

Becker, P., Coppens, P., and Ross, F. K. (1973). *J. Am. Chem. Soc.* **95**, 7604.

Becker, P. J. (1975). To be published.

Bent, H. A. (1961). *Chem. Rev.* **61**, 275.

Bentley, J., and Stewart, R. F. (1974). *Acta Crystallogr., Sect. A* **30**, 60.

Berkovitch-Yellin, Z., and Leiserowitz, L. (1976). *J. Am. Chem. Soc.* **97**, 5627.

Berlin, T. (1951). *J. Chem. Phys.* **19**, 208.

Boyd, D. B. (1972). *J. Am. Chem. Soc.* **94**, 64.

Brill, T. B., and Kotlar, A. J. (1974). *Inorg. Chem.* **13**, 470.

Cade, P. E. (1972). *Trans. Am. Crystallogr. Assoc.* **8**, 1.

Chelikowsky, J. R., and Cohen, M. L. (1974). *Phys. Rev. Lett.* **33**, 1339.

Coppens, P. (1974). *Acta Crystallogr., Sect. B* **30**, 255.

Coppens, P. (1975a). *MTP Inter. Rev. Sci., Ser. Two*, **11**, 21.

Coppens, P. (1975b). *In* "Electronic Structure of Polymers and Molecular Solids" (J. M. Andre and J. Ladik, eds.), p. 227. Plenum, New York.

Coppens, P. (1975c). *Phys. Rev. Lett.* **35**, 98.

Coppens, P., and Hamilton, W. C. (1968). *Acta Crystallogr., Sect. B* **24**, 925.

Coppens, P., and Lehmann, M. S. (1976). *Acta Crystallogr., Sect. B* **32**, 1777.

Coppens, P., Sabine, T. M., Delaplane, R. G., and Ibers, J. A. (1969). *Acta Crystallogr., Sect. B* **25**, 2451.

Coppens, P., Pautler, D., and Griffin, J. F. (1971). *J. Am. Chem. Soc.* **93**, 1051.

Coppens, P., Yang, Y. W., Blessing, R. H., Cooper, W. F., and Larsen, F. K. (1977). *J. Am. Chem. Soc.* **99**, 760.

Coulson, C. A., and Moffit, W. A. (1949). *Philos. Mag.* **40**, 1.
Coulson, C. A., and Thomas, M. W. (1971). *Acta Crystallogr., Sect. B* **27**, 1354.
Cruickshank, D. W. J. (1949). *Acta Crystallogr.* **2**, 65.
Cruickshank, D. W. J., and Rollett, J. S. (1953). *Acta Crystallogr.* **6**, 705.
DeWith, G., and Feil, D. (1975). *Chem. Phys. Lett.* **30**, 279.
DeWith, G., Harkema, S., and Feil, D. (1975). *Acta Crystallogr., Sect. A* **31**, S227.
Dixon, R. N., and Hugo, J. M. V. (1975). *Mol. Phys.* **29**, 953.
Dunning, T. H., and Winter, N. W. (1971). *Chem. Phys. Lett.* **11**, 194.
Ermler, W. C., and Kern, C. W. (1971). *J. Chem. Phys.* **55**, 4851.
Goodisman, J., and Klemperer, W. (1963). *J. Chem. Phys.* **38**, 721.
Griffin, J. F., and Coppens, P. (1975). *J. Am. Chem. Soc.* **97**, 3496.
Grimaldi, F., Lecourt, A., and Moser, C. (1967). *Int. J. Quantum Chem., Symp.* **1**, 153.
Groenewegen, P. P. M., Zeevalkink, J., and Feil, D. (1971). *Acta Crystallogr., Sect. A* **27**, 487.
Guest, M. F., Hillier, I. H., Higginson, B. R., and Lloyd, D. R. (1975). *Mol. Phys.* **29**, 113.
Harel, M., and Hirshfeld, F. L. (1975). *Acta Crystallogr., Sect. B* **31**, 162.
Hartmann, A., and Hirshfeld, F. L. (1966). *Acta Crystallogr.* **20**, 80.
Hehre, W. J., Stewart, R. F., and Pople, J. A. (1969). *J. Chem. Phys.* **51**, 2657.
Helmholdt, R. B. (1975). Thesis, University of Groningen, The Netherlands.
Helmholdt, R. B., Ruysink, A. J. F., Reynaers, H., and Kemper, G. (1972). *Acta Crystallogr., Sect. B* **28**, 318.
Hirshfeld, F. L. (1971). *Acta Crystallogr., Sect. B* **27**, 769.
Hirshfeld, F. L., and Rzotkiewicz, S. (1974). *Mol. Phys.* **27**, 1319.
Ito, T., and Sakurai, T. (1974). *Acta Crystallogr., Sect. B* **29**, 1594.
Iwata, M., and Saito, Y. (1973). *Acta Crystallogr., Sect. B* **29**, 822.
Jones, D. S., Pautler, D., and Coppens, P. (1972). *Acta Crystallogr., Sect. A* **28**, 635.
Kurki-Suonio, K. (1959). *Ann. Acad. Sci. Fenn., Ser. A* **6**, 31.
Larsen, F. K., Little, R. G., and Coppens, P. (1975). *Acta Crystallogr., Sect. B* **31**, 430.
Laws, E. A., and Lipscomb, W. N. (1972). *Isr. J. Chem.* **10**, 77.
McIver, J. W., Coppens, P., and Nowak, D. (1971). *Chem. Phys. Lett.* **11**, 82.
Martenson, O., and Sperber, G. (1970). *Acta Chem. Scand.* **24**, 1749.
Marumo, F., Isobe, M., Saito, Y., Yagi, T., and Akimoto, S. (1974). *Acta Crystallogr., Sect. B* **30**, 1904.
Matthews, D. A., and Stucky, G. D. (1971). *J. Am. Chem. Soc.* **93**, 5954.
Momany, F. A., Carruthers, L. M., and Sheraga, H. A. (1974). *J. Phys. Chem.* **78**, 1621.
Moon, R. M. (1972). *Trans. Am. Crystallogr. Assoc.* **8**, 59.
Newton, M. D. (1976). *In* "Modern Theoretical Chemistry" (H. F. Schaefer, III, ed.), Vol. IV. Plenum, New York.
Ransil, B. J., and Sinai, J. J. (1967). *J. Chem. Phys.* **46**, 4050.
Rees, B. (1976). *Acta Crystallogr., Sect. A* **32**, 483.
Rees, B., and Coppens, P. (1973). *Acta Crystallogr., Sect. B* **29**, 2516.
Rees, B., and Mitschler, A. (1976). *J. Am. Chem. Soc.* **98**, 7918.
Rosenfeld, J. L. (1964). *Acta Chem. Scand.* **18**, 1719.
Roux, M., and Daudel, R. (1955). *C. R. Helvd. Seances Acad. Sci.* **240**, 90.
Ruysink, A. F., and Vos, A. (1974). *Acta Crystallogr., Sect. A* **30**, 497.
Ruysink, A. F. J. (1973). Thesis, University of Groningen, The Netherlands.
Rys, J., Dupuis, M., and Stevens, E. D. (1975). To be published.
Smith, P. R., and Richardson, J. W. (1967). *J. Chem. Phys.* **71**, 924.
Staudenmann, J. L., Coppens, P., and Muller, R. (1976). *Solid State Commun.* **19**, 29.
Stevens, E. D. (1973). Thesis, University of California at Davis.
Stevens, E. D. (1977). To be published.

Stevens, E. D., and Coppens, P. (1975). *Acta Crystallogr., Sect. A* **31**, 612.
Stevens, E. D., and Hope, H. (1975). *Acta Crystallogr., Sect. A* **31**, 494.
Stevens, E. D., and Hope, H. (1977). *Acta Crystallogr.* (in press).
Stevens, E. D., Rys, J., and Coppens, P. (1977). *J. Am. Chem. Soc.* **99**, 265.
Stevens, R. M., Switkes, E., Laws, E. A., and Lipscomb, W. N. (1971). *J. Am. Chem. Soc.* **93**, 2603.
Stewart, R. F. (1968). *Acta Crystallogr., Sect. A* **24**, 497.
Stewart, R. F. (1970). *J. Chem. Phys.* **53**, 205.
Stewart, R. F. (1973a). *J. Chem. Phys.* **58**, 1668.
Stewart, R. F. (1973b). *J. Chem. Phys.* **58**, 4430.
Stewart, R. F. (1975). *Acta Crystallogr., Sect. A* **31**, S218.
Stewart, R. F., and Jensen, L. H. (1969). *Z. Kristallogr., Kristallgeom., Kristallphys., Kristallchem.* **128**, 133.
Stewart, R. F., Davidson, E. R., and Simpson, W. T. (1965). *J. Chem. Phys.* **42**, 3175.
Tanemura, S., and Kato, N. (1972). *Acta Crystallogr., Sect. A* **28**, 69.
Thomas, M. W. (1971). *Acta Crystallogr., Sect. B* **27**, 1760.
Tofield, B. C. (1975). *Struct. Bonding (Berlin)* **21**, 1.
Waller, I., and Hartree, D. R. (1929). *Proc. R. Soc., London, Ser. A* **124**, 119.
Walter, J. P., and Cohen, M. L. (1971). *Phys. Rev. B* **4**, 1877.
Wang, Y. W., and Coppens, P. (1976). *Inorg. Chem.* **15**, 1122.
Wang, Y. W., Blessing, R. B., Ross, F. K., and Coppens, P. (1976). *Acta Crystallogr., Sect. B* **32**, 572.
Wu, A., Biehl, E. R., and Reeves, P. C. (1972). *J. Chem. Soc., Perkin Trans. 2* p. 449.
Yang, Y. W. (1976). Thesis, State University of New York at Buffalo.
Yang, Y. W., and Coppens, P. (1974). *Solid State Commun.* **15**, 1555.

Evaluation of Momentum Distributions and Compton Profiles for Atomic and Molecular Systems

PER KAIJSER* and VEDENE H. SMITH, JR.

Department of Chemistry
Queen's University
Kingston, Ontario, Canada

I. Introduction 37
II. Dirac–Fourier Transformation of Position Space Wavefunctions and One-Particle Density Matrices 40
III. Dirac–Fourier Transformation of Atomic Orbitals 44
A. Spherical Harmonic Orbitals 44
B. Cartesian and Ellipsoidal Gaussian-Type Orbitals 55
IV. Evaluation of the Momentum Density $\rho(\mathbf{p})$ 58
A. Spherical Harmonic Basis Functions 58
B. Cartesian Gaussian-Type Functions 60
V. Evaluation of the Directional Compton Profile $J(\mathbf{q})$ 62
A. From Momentum Densities 63
B. From Basis Overlap Integrals 69
VI. The Spherically Averaged Momentum Density $\overline{\rho(p)}$ and the Isotropic Compton Profile $\overline{J(q)}$ 71
A. Spherical Harmonic Basis Functions 71
B. Cartesian Gaussian Basis Functions 72
VII. Concluding Remarks 73
References 74

I. Introduction

In recent years, the improved accuracy of existing experiments and the development of new experimental probes have given great promise of access to the electron momentum distribution $\rho(\mathbf{p})$ or to various integrals of $\rho(\mathbf{p})$. At the present time, these experiments include X-ray and γ-ray Compton scattering, high-energy electron Compton scattering, the so-called (e, 2e) reaction experiment, and positron annihilation. It is not our purpose to review in detail here these experiments and the underlying fundamental

* *Present address:* Quantum Theory Project, University of Florida, Gainesville, Florida 32611.

assumptions, approximations, and corrections that have been extensively reviewed in the recent literature (Williams, 1977; Reed, 1976), but rather to point out those double integrals of $\rho(\mathbf{p})$ that are related to the experimentally measured quantities and whose calculation together with that of $\rho(\mathbf{p})$ itself is the subject of our investigation.

In the case of photon scattering, one may show that within the context of the impulse approximation, the shape of the Compton line is related to the directional Compton profile (DCP):

$$J(\mathbf{q}) = \int \rho(\mathbf{p})\, \delta(\mathbf{p} \cdot \hat{\mathbf{q}} - q)\, d\mathbf{p}, \tag{1}$$

where $\hat{\mathbf{q}}$ is a unit vector along the scattering vector $\mathbf{q}$, and $q = |\mathbf{q}|$. It is apparent that $J(\mathbf{q})$ is a double integral over the momentum distribution which is obtained by the integration of $\rho(\mathbf{p})$ over planes orthogonal to $\mathbf{q}$. The directional Compton profile is sometimes described as a one-dimensional momentum distribution. This identification becomes apparent if one chooses the p_z axis in the $\mathbf{q}$ direction and writes Eq. (1) in the familiar form

$$J(p_z) = \iint \rho(\mathbf{p})\, dp_x\, dp_y\,. \tag{2}$$

When the target system is randomly oriented, as in the gas or liquid phase, the quantity measured is the isotropic Compton profile (ICP) or the spherical average of $J(\mathbf{q})$, namely

$$\overline{J(q)} = (4\pi)^{-1} \int_{|\mathbf{q}|=q} J(\mathbf{q})\, d\Omega_{\hat{\mathbf{q}}} \tag{3}$$

or

$$\overline{J(q)} = 2\pi \int_{|q|}^{\infty} p\overline{\rho(p)}\, dp \tag{4}$$

where $\overline{\rho(p)}$ is the spherical average of $\rho(\mathbf{p})$, that is,

$$\overline{\rho(p)} = (4\pi)^{-1} \int_{|\mathbf{p}|=p} \rho(\mathbf{p})\, d\Omega_{\hat{\mathbf{p}}}\,. \tag{5}$$

Single crystals and other largely aligned or layered targets permit one to measure the directional Compton profile for certain directions. These give us more information about the electronic structure and momentum density that $\overline{J(p)}$ does. In fact, much research (Mijnarends, 1977) has been devoted to the reconstruction of $\rho(\mathbf{p})$ from the knowledge of $J(\mathbf{q})$.

The employment of positrons as a probe of the electron momentum distribution is based on the fact that the probability $P(\mathbf{p})$ of annihilation of the positron with an electron of momentum $\mathbf{p}$ reduces to $\rho(\mathbf{p})$ if it is assumed that the positron wavefunction is a constant. In the conventional "long-slit" geometry, the measurements are proportional to $\iint P(\mathbf{p})\, dp_x\, dp_y$ and thus to $J(\mathbf{q})$ under the foregoing assumption about the positron wavefunction.

It is possible to use electrons to investigate the momentum distribution of the target system in two different experiments, both of which have been implemented only in the gas phase. In the first of these, high-energy electron Compton scattering, the isotropic Compton profile $\overline{J(q)}$ is obtained. In the second, the (e, 2e) reaction experiment (Williams, 1977; McCarthy and Weigold, 1976), the differential cross section is related to the momentum distribution of the ejected electron defined by the square of the Fourier transform of the overlap integral between the wavefunction of the N-electron target and the residual $(N - 1)$-electron ion. When the target is described by the independent-particle model and the ion by the frozen orbital approximation, this is just the momentum distribution $\rho_i(\mathbf{p})$ of the orbital corresponding to the ejected electron in the ground state of the target. Since the target is in the gas phase, the measured quantity is $\overline{\rho(p)}$. There is a one-to-one correspondence between the two quantities $\overline{\rho(p)}$ and $\overline{J(q)}$. However, for comparison with theory, as we will see later, the former is one step closer to the wavefunction.

Use of these experiments to test the quality of theoretical wavefunctions and models requires calculations of $\rho(\mathbf{p})$ and of the Compton profiles $J(\mathbf{q})$ and $\overline{J(q)}$. Such calculations are based on the knowledge of the electronic wavefunction for the target system. For atoms and molecules, these are with a few exceptions (McWeeny and Coulson, 1949; Henderson and Scherr, 1960) expressed in the position space representation. It is for this reason that the present article is directed toward the calculation of the momentum density as well as the isotropic and directional Compton profiles from position space wavefunctions usually employed to describe atoms and molecules (Richards *et al.*, 1971, 1974).

In the next section we introduce the one-electron density matrix as a convenient tool for determining $\rho(\mathbf{p})$ and consider the Dirac–Fourier transformation of the wavefunction and the density matrix. The transformation of single-particle functions required for the density matrix procedure is described in Section III for typical atomic orbital or basis functions. The momentum distributions for atoms and molecules are derived in Section IV, whereas the determination of the directional Compton profiles is dealt with in Section V. The isotropic profiles and spherically averaged momentum densities are discussed in Section VI. The presentation in these last three sections is restricted to atomic and molecular wavefunctions which can be written in terms of atomic orbital basis functions, that is as the so-called linear combination of atomic orbitals (LCAO) type. The treatment of wavefunctions with explicit r_{12} dependence (Benesch, 1976a) and spheroidal wavefunctions (Liu and Smith, 1977) may be found in these references.

The reader may find the treatises of Erdélyi *et al.* (1953, 1954) and Abramowitz and Stegun (1964) useful as references for the various mathematical functions and integral transforms involved in our study. We have followed the notation and definitions of the former except when otherwise stated.

Although a number of calculations of momentum distributions have been made during the years since the pioneering investigations of Podolsky and Pauling (1929) for atoms and Coulson and Duncanson (1941) for molecules, a comprehensive study of the methods has not yet been presented. It is hoped that our discussion will fill this need.

II. Dirac–Fourier Transformation of Position Space Wavefunctions and One-Particle Density Matrices

Dirac (1958) has shown that the Fourier transform

$$\hat{\Psi}(\mathbf{P}, \sigma) = (2\pi)^{-3N/2} \int e^{-i\mathbf{P}\cdot\mathbf{R}} \Psi(\mathbf{R}, \sigma)\, d\mathbf{R} \tag{6}$$

is the bridge between the position and momentum space representations of the N-electron wavefunction. Here, $\mathbf{P}$, $\mathbf{R}$, and σ stand for the collection of momentum coordinates $(\mathbf{p}_1, \mathbf{p}_2, \ldots, \mathbf{p}_N)$, position coordinates $(\mathbf{r}_1, \mathbf{r}_2, \ldots, \mathbf{r}_N)$, and spin coordinates $(\sigma_1, \sigma_2, \ldots, \sigma_N)$, respectively, whereas $d\mathbf{R} = d\mathbf{r}_1, d\mathbf{r}_2, \ldots, d\mathbf{r}_N$.

In order to obtain $\rho(\mathbf{p})$, one could proceed by taking the total wavefunction $\Psi(\mathbf{R}, \sigma)$, compute its $3N$-dimensional Dirac–Fourier transform $\hat{\Psi}(\mathbf{P}, \sigma)$ by means of Eq. (6), then integrate out the momentum space coordinates of $N - 1$ electrons and trace the spin coordinates of N electrons from $|\hat{\Psi}(\mathbf{P}, \sigma)|^2$:

$$\rho(\mathbf{p}) = N \int |\hat{\Psi}(\mathbf{P}, \sigma)|^2\, d\sigma\, d\mathbf{p}_2 \cdots d\mathbf{p}_N . \tag{7}$$

In the case of an independent-particle model wavefunction, that is, one which is representable by a single Slater determinant,

$$\Psi(\mathbf{R}, \sigma) = (N!)^{-1/2} \det|\psi_a(\mathbf{r}_1, \sigma_1)\psi_b(\mathbf{r}_2, \sigma_2) \cdots \psi_c(\mathbf{r}_N, \sigma_N)|, \tag{8}$$

one needs only to transform each individual spin orbital by means of the single-particle version of Eq. (6). Similarly, in the case of a wavefunction $\Psi(\mathbf{R}, \sigma)$ which is expanded in terms of Slater determinants over a set of position space spin orbitals, the transformation of each individual spin orbital leads to $\hat{\Psi}(\mathbf{P}, \boldsymbol{\sigma})$ expressed as the same expansion of Slater determinants but over the transformed set of spin orbitals.

Since we are interested in $\rho(\mathbf{p})$ and not $\hat{\Psi}(\mathbf{P}, \boldsymbol{\sigma})$ itself, a convenient procedure (Benesch and Smith, 1970, 1971, 1973) is to transform the one-electron density matrix $\gamma(\mathbf{r}, \sigma \,|\, \mathbf{r}', \sigma')$ to $\hat{\gamma}(\mathbf{p}, \sigma \,|\, \mathbf{p}', \sigma')$ via the analogous Dirac–Fourier transformation to Eq. (6). In fact, since the spin variables are not involved in the Dirac–Fourier transformation, we can restrict our study to the transformation of the spin-traced one-electron density matrix (or charge-density matrix),

$$\gamma(\mathbf{r} \,|\, \mathbf{r}') = N \int \Psi^*(\mathbf{r}', \mathbf{r}_2, \ldots, \mathbf{r}_N; \boldsymbol{\sigma})\Psi(\mathbf{r}, \mathbf{r}_2, \ldots, \mathbf{r}_N; \boldsymbol{\sigma}) \, d\boldsymbol{\sigma} \, d\mathbf{r}_2 \cdots d\mathbf{r}_N, \quad (9)$$

to obtain

$$\hat{\gamma}(\mathbf{p} \,|\, \mathbf{p}') = (2\pi)^{-3} \int \gamma(\mathbf{r} \,|\, \mathbf{r}') e^{-i\mathbf{p} \cdot \mathbf{r} + i\mathbf{p}' \cdot \mathbf{r}'} \, d\mathbf{r} \, d\mathbf{r}'. \quad (10)$$

By comparison with the definition in Eq. (7), we find that $\rho(\mathbf{p})$ is just the diagonal part of $\gamma(\mathbf{p} \,|\, \mathbf{p}')$,

$$\rho(\mathbf{p}) = \hat{\gamma}(\mathbf{p} \,|\, \mathbf{p}) = (2\pi)^{-3} \int \gamma(\mathbf{r} \,|\, \mathbf{r}') e^{-i\mathbf{p} \cdot (\mathbf{r} - \mathbf{r}')} \, d\mathbf{r} \, d\mathbf{r}'. \quad (11)$$

It is important to note that $\rho(\mathbf{p})$ is different from the familiar scattering factor $F(\mathbf{s})$ which is the Fourier transform of the position space charge density $\rho(\mathbf{r}) = \gamma(\mathbf{r} \,|\, \mathbf{r})$, that is,

$$F(\mathbf{s}) = \int \gamma(\mathbf{r} \,|\, \mathbf{r}) e^{+i\mathbf{s} \cdot \mathbf{r}} \, d\mathbf{r}. \quad (12)$$

The determination of $\rho(\mathbf{p})$ involves the entire one-particle charge density matrix $\gamma(\mathbf{r} \,|\, \mathbf{r}')$, whereas for $F(\mathbf{s})$ only the diagonal component $\gamma(\mathbf{r} \,|\, \mathbf{r})$ is required. Another expression of this difference may be obtained (Benesch *et al.*, 1971) by rewriting $F(\mathbf{s})$ as

$$F(\mathbf{s}) = \int \hat{\gamma}(\mathbf{p} \,|\, \mathbf{p} + \mathbf{s}) \, d\mathbf{p}, \quad (13)$$

that is as a convolution of the one-particle momentum density matrix $\hat{\gamma}(\mathbf{p} \,|\, \mathbf{p}')$.

As indicated in the foregoing, Eq. (11) is the logical starting point for the calculation of $\rho(\mathbf{p})$. We take advantage of the fact that $\gamma(\mathbf{r} \,|\, \mathbf{r}')$ may be expressed in terms of its eigenfunctions, the natural orbitals $\boldsymbol{\chi}(\mathbf{r}) = [\chi_1(\mathbf{r}), \chi_2(\mathbf{r}), \ldots,]$. These, in turn, are assumed to be linear combinations of a set of linearly independent basis orbitals $\boldsymbol{\phi}(\mathbf{r}) = [\phi_1(\mathbf{r}), \phi_2(\mathbf{r}), \ldots, \phi_m(\mathbf{r})]$ with the coefficients collected in the matrix C. We may then write

$$\gamma(\mathbf{r}\,|\,\mathbf{r}') = \boldsymbol{\chi}(\mathbf{r})\mathsf{N}\boldsymbol{\chi}^\dagger(\mathbf{r}') \tag{14}$$

$$= \boldsymbol{\phi}(\mathbf{r})\mathsf{CNC}^\dagger\boldsymbol{\phi}^\dagger(\mathbf{r}') \tag{15}$$

$$= \boldsymbol{\phi}(\mathbf{r})\mathsf{D}\boldsymbol{\phi}^\dagger(\mathbf{r}'). \tag{16}$$

The diagonal occupation number matrix N together with C define the Hermitian matrix D. The latter contains all the mixing coefficients between the basis orbitals $\boldsymbol{\phi}(\mathbf{r})$ in $\gamma(\mathbf{r}\,|\,\mathbf{r}')$ and will thus logically be referred to as the coupling matrix. We shall assume for the remainder of this article that $\gamma(\mathbf{r}\,|\,\mathbf{r}')$ is of finite one-rank, that is, the number of nonzero elements in the diagonal occupation number or eigenvalue matrix N is finite. In addition, we assume that the basis is finite as well, so that C and D are of finite dimension.

Insertion of Eqs. (14)–(16) into Eq. (11) leads to the respective expressions

$$\rho(\mathbf{p}) = \hat{\boldsymbol{\chi}}(\mathbf{p})\mathsf{N}\hat{\boldsymbol{\chi}}^\dagger(\mathbf{p}), \tag{17}$$

$$= \hat{\boldsymbol{\phi}}(\mathbf{p})\mathsf{CNC}^\dagger\hat{\boldsymbol{\phi}}^\dagger(\mathbf{p}), \tag{18}$$

$$= \hat{\boldsymbol{\phi}}(\mathbf{p})\mathsf{D}\hat{\boldsymbol{\phi}}^\dagger(\mathbf{p}), \tag{19}$$

where $\hat{\boldsymbol{\chi}}(\mathbf{p})$ and $\hat{\boldsymbol{\phi}}(\mathbf{p})$ are the Fourier transforms defined by Eq. (6) of the natural orbitals $\boldsymbol{\chi}(\mathbf{r})$ and the basis orbitals $\boldsymbol{\phi}(\mathbf{r})$, that is,

$$\hat{\phi}_i(\mathbf{p}) = (2\pi)^{-3/2} \int e^{-i\mathbf{p}\cdot\mathbf{r}}\phi_i(\mathbf{r})\, d\mathbf{r} \tag{20}$$

$$= (2\pi)^{-3/2} e^{-i\mathbf{p}\cdot\mathbf{A}_i} \int e^{-i\mathbf{p}\cdot\mathbf{r}_i}\phi_i(\mathbf{r}_i)\, d\mathbf{r}_i \tag{21}$$

and similarly for $\hat{\chi}_i$. In the above, $\mathbf{A}_i$ is the position of the center for ϕ_i from which $\mathbf{r}_i$ is defined, that is, $\mathbf{r} = \mathbf{A}_i + \mathbf{r}_i$. Equation (18) reflects the linearity of the Fourier transformation for which operation we use the symbol ^ (caret).

We observe that the evaluation of $\rho(\mathbf{p})$ has been reduced to the determination of only two quantities, namely the set of **p**-space basis orbitals $\hat{\boldsymbol{\phi}}(\mathbf{p})$ and their coupling matrix D. Transformation (20) deals with the former and a detailed description of the procedure for each of the most common types of basis orbitals will be given in the next section. For the D matrix, we note that it is the same in both the **r**- and **p**-space representations. From its definition,

$$\mathsf{D} = \mathsf{CNC}^\dagger, \tag{22}$$

and the fact that the C and N matrices are part of the output from any molecular orbital (MO) calculation, it may be easily evaluated for such wavefunctions. A prescription for the evaluation of D in the case of configuration interaction (CI) wavefunctions (those which are expressed as an expansion in Slater determinants) has been given by Löwdin (1955). Of course, if a natural (spin) orbital analysis of a wavefunction has been performed, the matrices C and N are available. This permits the determination of the contributions from the individual natural orbitals χ_i to $\rho(\mathbf{p})$:

$$\rho(\mathbf{p}) = \sum_i n_i \rho_i(\mathbf{p}), \tag{23}$$

where

$$\rho_i(\mathbf{p}) = |\hat{\chi}_i(\mathbf{p})|^2. \tag{24}$$

The individual natural contributions to $J(\mathbf{q})$ and $\overline{J(q)}$ are defined analogously. They are useful in analyzing the effects of electron correlation especially by comparison with the Best-density independent-particle model (Kutzelnigg and Smith, 1964; Larsson and Smith, 1969) for which

$$\rho_{\mathrm{BD}}(\mathbf{p}) = \sum_{i=1}^{N} \rho_i(\mathbf{p}). \tag{25}$$

This reference model is sometimes called the first natural configuration (Bender and Davidson, 1968).

A similar component analysis of $\rho(\mathbf{p})$ may be made with respect to the nondiagonal basis orbital representation, $\hat{\boldsymbol{\phi}}(\mathbf{p})$. We write

$$\begin{aligned} \rho(\mathbf{p}) &= \hat{\boldsymbol{\phi}}(\mathbf{p})\mathsf{D}\hat{\boldsymbol{\phi}}^\dagger(\mathbf{p}) \\ &= \sum_{i,j} \hat{\phi}_i(\mathbf{p}) D_{ij} \hat{\phi}_j^*(\mathbf{p}) \qquad (26) \\ &= \sum_i \rho_{ii}(\mathbf{p}) + \sum_{i<j} \rho_{ij}(\mathbf{p}), \qquad (27) \end{aligned}$$

where

$$\rho_{ii}(\mathbf{p}) = D_{ii} |\hat{\phi}_i(\mathbf{p})|^2 \tag{28}$$

and

$$\rho_{ij}(\mathbf{p}) = \hat{\phi}_i(\mathbf{p}) D_{ij} \hat{\phi}_j^*(\mathbf{p}) + \hat{\phi}_j(\mathbf{p}) D_{ji} \hat{\phi}_i^*(\mathbf{p}). \tag{29}$$

By analogy with the "population analysis" of position space charge densities (Mulliken, 1955), we define normalized charge and overlap distributions

$$a_i(\mathbf{p}) = |\hat{\phi}_i(\mathbf{p})|^2, \tag{30}$$

$$b_{ij}(\mathbf{p}) = \hat{\phi}_i(\mathbf{p}) \hat{\phi}_j^*(\mathbf{p}) S_{ij}^{-1}, \qquad S_{ij} \neq 0 \tag{31}$$

and populations

$$Q_i = D_{ii}; \qquad Q_{ij} = S_{ij} D_{ij}, \tag{32}$$

where

$$S_{ij} = \int \hat{\phi}_i(\mathbf{p}) \hat{\phi}_j^*(\mathbf{p}) \, d\mathbf{p}. \tag{33}$$

Since

$$\int \rho(\mathbf{p}) \, d\mathbf{p} = N, \tag{34}$$

it follows that

$$N = \sum_i Q_i + \sum_{i<j} (Q_{ij} + Q_{ji}), \tag{35}$$

that is, the total charge is divided into contributions from each orbital density a_i and each overlap density b_{ij} in agreement with the familiar population analysis.

Furthermore, we can discuss orbital and overlap contributions to $J(\mathbf{q})$ and $\overline{J(q)}$ which are calculated by the fundamental definitions for $a_i(\mathbf{p})$ and $b_{ij}(\mathbf{p})$.

III. Dirac–Fourier Transformation of Atomic Orbitals

As we have discussed in the last section, the calculation of $\rho(\mathbf{p})$ from an N-electron wavefunction $\Psi(\mathbf{R}, \boldsymbol{\sigma})$ or a one-particle density matrix $\gamma(\mathbf{r} \mid \mathbf{r}')$ ultimately rests on the transformation of single-particle functions (orbitals) from the position to the momentum space representation. In the present section we discuss the transformation for several types of orbital basis functions which are more commonly used for the construction of atomic and molecular wavefunctions.

A. Spherical Harmonic Orbitals

Due to the separability in spherical polar coordinates of Schrödinger's equation for an atomic central field, the position space solutions may be written in the familiar form

$$\phi_{nlm}(\mathbf{r}) = f_{nl}(r) Y_{lm}(\theta, \phi), \tag{36}$$

where $Y_{lm}(\theta, \phi)$ is the usual orthonormalized spherical harmonic function (Messiah 1967), defined for $m \geq 0$ by

$$Y_{lm}(\theta, \phi) = k_{lm} P_l^m(\cos\theta) e^{im\phi}, \tag{37}$$

$$k_{lm} = (-1)^m \left[\frac{2l+1}{4\pi} \cdot \frac{(l-m)!}{(l+m)!} \right]^{1/2}, \tag{38}$$

and for $m < 0$ via the complex conjugate relation,

$$Y^*_{lm}(\theta, \phi) = (-1)^m Y_{l-m}(\theta, \phi). \tag{39}$$

Here $P_l^m(\cos\theta)$ is the (unnormalized) associated Legendre function.

The radial function $f_{nl}(r)$ is normalized so that

$$\int_0^\infty r^2 f^2_{nl}(r)\, dr = 1 \tag{40}$$

and is unspecified except that $\lim_{r\to 0} r^{-l} f_{nl}(r)$ exists.

In order to transform $\phi(\mathbf{r})$, we first change variables to its center, $\mathbf{A}$, and use in Eq. (21) the expansion (Messiah, 1967),

$$e^{-i\mathbf{p}\cdot\mathbf{r}} = 4\pi \sum_{l=0}^{\infty} \sum_{m=-l}^{l} (-i)^l j_l(pr) Y^*_{lm}(\theta, \phi) Y_{lm}(\hat{\theta}, \hat{\phi}), \tag{41}$$

where $\hat{\theta}$ and $\hat{\phi}$ are the angular variables in momentum space and j_l is a spherical Bessel function. Integration over θ and ϕ and the use of the orthonormality properties of the $Y_{lm}(\theta, \phi)$ yields

$$\hat{\phi}_{nlm}(\mathbf{p}) = e^{-i\mathbf{p}\cdot\mathbf{A}}(-i)^l Y_{lm}(\hat{\theta}, \hat{\phi}) \left(\frac{2}{\pi}\right)^{1/2} \int_0^\infty j_l(pr) f_{nl}(r) r^2\, dr. \tag{42}$$

It should be noted that the angular dependence of $\hat{\phi}(\mathbf{p})$ is identical to that of the original $\phi(\mathbf{r})$ except for the phase factor $e^{-i\mathbf{p}\cdot\mathbf{A}}$. The latter can be disregarded in atomic problems by choosing the origin at the nucleus. In this case, we may write

$$\hat{\phi}_{nlm}(\mathbf{p}) = u_{nl}(p) Y_{lm}(\hat{\theta}, \hat{\phi}). \tag{43}$$

The radial part of our momentum space orbital, $u_{nl}(p)$, is a symmetry-dependent transformation of the radial part of $\phi_{nlm}(\mathbf{r})$ involving a spherical Bessel function of the same order as the orbital quantum number l,

$$\begin{aligned} u_{nl}(p) &= (-i)^l \left(\frac{2}{\pi}\right)^{1/2} \int_0^\infty j_l(pr) f_{nl}(r) r^2\, dr \\ &= (-i)^l p^{-1} h_{nl}(p), \end{aligned} \tag{44}$$

where $h_{nl}(p)$ is called (Erdélyi *et al.*, 1954) the Hankel transform of order $l + \frac{1}{2}$ of the function $r f_{nl}(r)$ and is defined as follows:

$$h_{nl}(p) = \int_0^\infty [r f_{nl}(r)] J_{l+1/2}(pr)(pr)^{1/2}\, dr. \tag{45}$$

In the above $J_n(x)$ is the ordinary Bessel function of order n.

Since $u_{nl}(p)$ is given by this symmetry-dependent transformation and not by the familiar one-dimensional Fourier transformation, one must be careful in making statements about $u_{nl}(p)$ from knowledge of $f_{nl}(r)$, especially when comparing orbitals with different l.

Before we turn to the transformation of some of the more common forms of $f_{nl}(r)$, we note some of the general properties of $u_{nl}(p)$ (Benesch and Smith, 1973). The radial momentum functions are normalized according to

$$\int_0^\infty p^2 |u_{nl}(p)|^2 \, dp = 1. \tag{46}$$

For $p = 0$, only functions with $l = 0$ are nonvanishing,

$$u_{n0}(0) = \left(\frac{2}{\pi}\right)^{1/2} \int_0^\infty r^2 f_{n0}(r) \, dr, \tag{47}$$

whereas for large p, the asymptotic behavior is

$$u_{n0}(p) = -\left(\frac{8}{\pi}\right)^{1/2} \left[\left(\frac{df_{n0}(r)}{dr}\right)_{r=0} p^{-4} + O(p^{-6})\right]. \tag{48}$$

The leading contribution for nonzero l may be shown to be of higher order, namely $O(p^{-l-4})$. For those radial functions that satisfy the electron-nuclear cusp condition (Löwdin, 1953; Kato, 1957; Smith, 1971),

$$\left(\frac{df(r)}{dr}\right)_{r=0} = -Zf(0), \tag{49}$$

Eq. (47) reduces to

$$u_{n0}(p) = \left(\frac{8}{\pi}\right)^{1/2} [Zf_{n0}(0)p^{-4} + O(p^{-6})]. \tag{50}$$

1. *Slater-Type (Exponential) Atomic Orbitals*

These orbitals, introduced by Slater (1932), are extensively used as basis functions for atomic (Clementi, 1965) and molecular (Cade and Wahl, 1974) wavefunctions. They may be written in the form of Eq. (36) with

$$f_{nl}(r) = N(n, \alpha) r^{n-1} e^{-\alpha r}, \qquad \alpha > 0, \tag{51}$$

where the normalization factor is

$$N(n, \alpha) = [(2\alpha)^{2n+1}/\Gamma(2n+1)]^{1/2} \tag{52}$$

and $n - 1 \geq l \geq 0$.

For these functions, the general expression for the Hankel transform is

$$h_{nl}(p) = N(n, \alpha) \int_0^\infty r^n e^{-\alpha r} J_{l+1/2}(pr)(pr)^{1/2}\, dr$$
$$= N(n, \alpha) \frac{\Gamma(n + l + 2)p^{l+1}}{2^{l+2}\alpha^{n+l+2}\Gamma(l + \frac{3}{2})}$$
$$\times {}_2F_1\left(\frac{n + l + 2}{2}, \frac{n + l + 3}{2}; l + \tfrac{3}{2}; -\frac{p^2}{\alpha^2}\right), \tag{53}$$

where ${}_2F_1$ is a hypergeometric function (Erdélyi *et al.*, 1953).

$${}_2F_1(a, b; c; z) = \sum_{k=0} \frac{(a)_k(b)_k}{(c)_k} \frac{z^k}{k!} \tag{54}$$

with

$$(a)_k = a \cdot (a + 1)(a + 2) \cdots (a + k - 1). \tag{55}$$

By use of a standard transformation among hypergeometric functions, Eq. (53) may be written (Watson, 1944) in the form

$$h_{nl}(p) = N(n, \alpha) \frac{\Gamma(n + l + 2)p^{l+1}\alpha^{n-l}}{2^{l+2}\Gamma(l + \frac{3}{2})(\alpha^2 + p^2)^{n+1}}$$
$$\times {}_2F_1\left(\frac{l - n}{2}, \frac{l - n + 1}{2}; l + \tfrac{3}{2}; \frac{-p^2}{\alpha^2}\right). \tag{56}$$

The series for the hypergeometric function [Eq. (54)] reduces to a polynomial of degree m in z when either a or b is equal to $-m$ (m, a nonnegative integer). As a result, we observe that Eq. (56) for $h_{nl}(p)$ involves a polynomial in $(-p^2/\alpha^2)$ of degree $(n - l)/2$ or $(n - l - 1)/2$ according to whether $(n - l)$ is even or odd.

Similar transformations lead to two other closed form expressions which may prove useful:

$$h_{nl}(p) = N(n, \alpha) \frac{\Gamma(n + l + 2)p^{l+1}}{2^{l+2}\Gamma(l + \frac{3}{2})(p^2 + \alpha^2)^{(n+l+2)/2}}$$
$$\times {}_2F_1\left(\frac{n + l + 2}{2}, \frac{l - n}{2}; l + \tfrac{3}{2}; \frac{p^2}{p^2 + \alpha^2}\right) \tag{57}$$

and

$$h_{nl}(p) = N(n, \alpha) \frac{\Gamma(n + l + 2)p^{l+1}\alpha}{2^{l+2}\Gamma(l + \frac{3}{2})(p^2 + \alpha^2)^{(n+l+3)/2}}$$
$$\times {}_2F_1\left(\frac{n + l + 3}{2}, \frac{l - n + 1}{2}; l + \tfrac{3}{2}; \frac{p^2}{p^2 + \alpha^2}\right). \tag{58}$$

When $(n - l)$ is even, the hypergeometric function in the first of these is a polynomial in $p^2/(p^2 + \alpha^2)$ of degree $(n - l)/2$. When $(n - l)$ is odd, the hypergeometric function in the second expression is a polynomial in $p^2/(p^2 + \alpha^2)$ of degree $(n - l - 1)/2$.

An explicit expression may be obtained as well by repeated differentiation:

$$\begin{aligned} p^{-1}h_{nl}(p) &= N(n, \alpha)\left(\frac{2}{\pi}\right)^{1/2}(-1)^{n-l}l! \\ &\quad \times (2p)^l \frac{d^{n-l}}{d\alpha^{n-l}}(\alpha^2 + p^2)^{-(l+1)} \\ &= (-1)^{n-l}\left[\frac{4\alpha}{\pi\Gamma(2n+1)}\right]^{1/2}\Gamma(l+1)(2\alpha)^n \\ &\quad \times (2p)^l \frac{d^{n-l}}{d\alpha^{n-l}}(\alpha^2 + p^2)^{-(l+1)}. \end{aligned} \tag{59}$$

A useful recursion relation for the numerical evaluation of $h_{nl}(p)$ for Slater-type orbitals is one for fixed l (Linderberg and Bystrand, 1964). In terms of the auxiliary function $K_n^l(p)$ defined by

$$h_{nl}(p) = [4\pi\alpha\Gamma(2n+1)]^{-1/2}K_n^l(p), \tag{60}$$

it is

$$K_n^l(p) = [4\alpha^2/(\alpha^2 + p^2)][nK_{n-1}^l(p) - (n + l)(n - l - 1)K_{n-2}^l(p)] \tag{61}$$

with

$$K_l^l(p) = \Gamma(l+1)[4\alpha p/(\alpha^2 + p^2)]^{l+1} \tag{62}$$

and

$$K_n^l = 0 \qquad \text{for } n < l. \tag{63}$$

Specific expressions for the $\hat{\phi}_{nlm}(\mathbf{p})$ are presented in Table I for $1 \leq n \leq 4$.

2. *Hydrogen-Like and Other Laguerre Function Atomic Orbitals*

The exactly solvable atomic one-electron problem for the Coulomb potential results in a discrete set of orbitals of the general form of Eq. (36) with the radial part $f_{nl}(r)$ involving Laguerre functions[1]:

$$f_{nl}^H(r) = 2\alpha\left[\frac{\alpha(n-l-1)!}{n(n+l)!}\right]^{1/2} e^{-\alpha r}(2\alpha r)^l L_{n-l-1}^{2l+1}(2\alpha r), \tag{64}$$

[1] The reader should note that there are different conventions in the literature for the definition of Laguerre functions. We follow the notation of the Bateman Manuscript Project (Erdélyi *et al.*, 1953). Their notation (left superscript B) is related to those of Shull and Löwdin (1955) and Podolsky and Pauling (1929) (left superscript P) and those of Messiah (1967) (left superscript M) by

$$(\nu + k)!\,{}^{B}L_k^\nu(x) = (-1)^\nu\,{}^{P}L_{\nu+k}^\nu(x) = {}^{M}L_k^\nu(x).$$

TABLE I

FOURIER TRANSFORMS OF NORMALIZED SLATER-TYPE ORBITALS WITH ORBITAL EXPONENT α

n quantum number	Position space representation	Momentum space representation				
		Common factor	Factor dependent on the l quantum number			
			$l = 0$	$l = 1$	$l = 2$	$l = 3$
1	$2\alpha^{3/2}e^{-\alpha r}Y_{00}(\theta, \phi)$	$(2\alpha/\pi)^{1/2} \times (\alpha^2 + p^2)^{-2}$	$4\alpha^2 Y_{00}(\hat{\theta}, \hat{\phi})$	—	—	—
2	$(4/3)^{1/2}\alpha^{5/2}re^{-\alpha r} \times Y_{lm}(\theta, \phi)$	$(2\alpha/3\pi)^{1/2} \times (\alpha^2 + p^2)^{-3}$	$4\alpha^2(3\alpha^2 - p^2)Y_{00}(\hat{\theta}, \hat{\phi})$	$-i16p\alpha^3 Y_{1m}(\hat{\theta}, \hat{\phi})$	—	—
3	$(2/3)(2/5)^{1/2}\alpha^{7/2}r^2e^{-\alpha r} \times Y_{lm}(\theta, \phi)$	$(\alpha/5\pi)^{1/2} \times (\alpha^2 + p^2)^{-4}$	$32\alpha^4(\alpha^2 - p^2)Y_{00}(\hat{\theta}, \hat{\phi})$	$-i(32/3)p\alpha^3 \times (5\alpha^2 - p^2)Y_{1m}(\hat{\theta}, \hat{\phi})$	$-64p^2\alpha^4 Y_{2m}(\hat{\theta}, \hat{\phi})$	—
4	$(2/3)(35)^{-1/2}\alpha^{9/2}r^3e^{-\alpha r} \times Y_{lm}(\theta, \phi)$	$(2\alpha/35\pi)^{1/2} \times (\alpha^2 + p^2)^{-5}$	$16\alpha^4(p^4 - 10p^2\alpha^2 + 5\alpha^4) \times Y_{00}(\hat{\theta}, \hat{\phi})$	$-i32p\alpha^5 \times (5\alpha^2 - 3p^2)Y_{1m}(\hat{\theta}, \hat{\phi})$	$-32p^2\alpha^4(7\alpha^2 - p^2) \times Y_{2m}(\hat{\theta}, \hat{\phi})$	$i256p^3\alpha^5 \times Y_{3m}(\hat{\theta}, \hat{\phi})$

where the parameter α is equal to Z/n and Z is the effective nuclear charge. This discrete set is orthogonal but not a complete set (Schrödinger, 1926) and the continuum must be included for completeness. A discrete, complete but not orthogonal set may be obtained (Schrödinger, 1926) by setting $\alpha = Z$ in Eq. (63), that is, by omitting the principal quantum number n. Hylleraas (1929), Shull and Löwdin (1955), and Holøien (1956) have suggested the use of the complete, discrete orthogonal set,

$$f_{nl}^{SL}(r) = (2\alpha)^{3/2}\left[\frac{(n-l-1)!}{(n+l+1)!}\right]^{1/2} e^{-\alpha r}(2\alpha r)^l L_{n-l-1}^{2l+2}(2\alpha r) \tag{65}$$

with α common to all members of the set, that is, independent of n and l. The Hankel transform of the radial functions $f_{nl}^{H}(r)$ and $f_{nl}^{SL}(r)$ defined respectively by Eqs. (64) and (65) are

$$h_{nl}^{H}(p) = \left[\frac{2\alpha n(n-l-1)!}{\pi(n+l)!}\right]^{1/2} \frac{l!\,\alpha^{l+2}(4p)^{l+1}}{(\alpha^2+p^2)^{l+2}} C_{n-l-1}^{l+1}(t) \tag{66}$$

and

$$h_{nl}^{SL}(p) = \left[\frac{4\alpha(n-l-1)!}{\pi(n+l+1)!}\right]^{1/2} \frac{(l+1)!\,\alpha^{l+2}(4p)^{l+1}}{(\alpha^2+p^2)^{l+2}} \times \{C_{n-l-1}^{l+2}(t) + C_{n-l-2}^{l+2}(t)\}, \tag{67}$$

where

$$t = (p^2-\alpha^2)/(p^2+\alpha^2) \tag{68}$$

and the C_k^{ν} are the Gegenbauer polynomials (Erdélyi *et al.*, 1954) defined by the generating function

$$(1 - 2ut + u^2)^{-\nu} \equiv \sum_{k=0}^{\infty} C_k^{\nu}(t)u^k. \tag{69}$$

Expression (67) for $h_{nl}^{SL}(p)$ may be rewritten in terms of Jacobi polynomials $P_k^{\alpha,\beta}(t)$:

$$h_{nl}^{SL}(p) = \left[\frac{16\alpha(n-l-1)!}{\pi(n+l+1)!}\right]^{1/2} \times \frac{(l+1)!\,\alpha^{l+2}(4p)^{l+1}\Gamma(n+l+2)\Gamma(l+\frac{5}{2})}{(\alpha^2+p^2)^{l+2}\Gamma(2l+4)\Gamma(n+\frac{1}{2})} P_{n-l-1}^{(l+3/2,\,l+1/2)}(t), \tag{70}$$

where

$$P_k^{\alpha,\beta}(t) = \binom{k+\alpha}{k} {}_2F_1(-k, k+\alpha+\beta+1; \alpha+1; (1-t)/2) \tag{71}$$

and

$$\binom{k+\alpha}{k} = \frac{(\alpha+1)_k}{k!}. \tag{72}$$

Equation (66) was derived by Podolsky and Pauling (1929). We follow their approach to derive Eq. (67) for the Hankel transform of the Shull–Löwdin radial function. From the definitions of the Hankel transform [Eq. (45)] and of the $f_{nl}^{SL}(r)$ [Eq. (65)], we write

$$h_{nl}^{SL}(p) = \left[\frac{(n-l-1)!}{2\alpha(n+l+1)!}\right]^{1/2} g_{nl}^{SL}(y), \tag{73}$$

where

$$g_{nl}^{SL}(y) = \int_0^\infty e^{-x/2} J_{l+1/2}\left(\frac{xy}{2}\right) x^{l+1} L_{n-l-1}^{2l+2}(x)\left(\frac{xy}{2}\right)^{1/2} dx, \tag{74}$$

$y = p/\alpha$, and $x = 2\alpha r$.

To evaluate $g_{nl}^{SL}(y)$, we introduce the generating function

$$U_l(y, u) = \sum_{n=l+1}^{\infty} g_{nl}^{SL}(y) u^{n-l-1}, \tag{75}$$

interchange the order of summation and integration, and utilize the generating function for the Laguerre polynomials,

$$\sum_{n=l+1}^{\infty} L_{n-l-1}^{2l+2}(x) u^{n-l-1} \equiv \frac{e^{-xu/(1-u)}}{(1-u)^{2l+3}}, \tag{76}$$

to obtain

$$U_l(y, u) = (1-u)^{-(2l+3)} \int_0^\infty e^{-x(1+u)/[2(1-u)]} x^{l+1} J_{l+1/2}\left(\frac{xy}{2}\right)\left(\frac{xy}{2}\right)^{1/2} dx. \tag{77}$$

Since this is a known integral (Erdélyi *et al.*, 1954), we have

$$U_l(y, u) = \frac{2^{l+1/2}\Gamma(l+2)(1+u)}{\pi^{1/2}(1-u)^{2l+4}} \cdot \left\{\left[\frac{1+u}{2(1-u)}\right]^2 + \left(\frac{y}{2}\right)^2\right\}^{-l-2} \tag{78}$$

$$= \frac{2^{3/2}\alpha^{l+3}(l+1)!\,(4p)^{l+1}}{\pi^{1/2}(\alpha^2+p^2)^{l+2}} \cdot \left[\frac{1+u}{(1-2tu+u^2)^{l+2}}\right], \tag{79}$$

where

$$t = \frac{y^2-1}{y^2+1} = \frac{p^2-\alpha^2}{p^2+\alpha^2}. \tag{80}$$

Use of the generating function for the Gegenbauer polynomials [Eq. (69)] enables us to write the factor in square brackets in Eq. (79) as a power series in u:

$$\left[\frac{1+u}{(1-2tu+u^2)^{l+2}}\right] = \sum_{k=0}^{\infty} [C_k^{l+2}(t) + C_{k-1}^{l+2}(t)]u^k. \tag{81}$$

By a term-by-term comparison of the two power series for $U(y, u)$, Eq. (75) and Eqs. (79) and (81), we obtain an explicit expression for $g_{nl}^{SL}(y)$ and hence via Eq. (73) the desired Hankel transform [Eq. (67)] of the Shull–Löwdin radial function.

Since

$$C_k^{\nu}(t) = \frac{(2\nu)_k}{(\nu+\frac{1}{2})_k} P_k^{(\nu-1/2,\,\nu-1/2)}(t) \tag{82}$$

and

$$(2k+\alpha+\beta)P_k^{(\alpha,\,\beta-1)}(t) = (k+\alpha+\beta)P_k^{\alpha,\,\beta}(t) + (k+\alpha)P_{k-1}^{\alpha,\,\beta}(t). \tag{83}$$

Equation (70) may be otained immediately from Eq. (67).

To complete our discussion, we note the following recursion formulas for the Gegenbauer polynomials:

$$C_k^{\nu}(t) = 2\nu k^{-1}[tC_{k-1}^{\nu+1}(t) - C_{k-2}^{\nu+1}(t)], \tag{84}$$

$$C_{k+1}^{\nu}(t) = (k+1)^{-1}[2(k+\nu)tC_k^{\nu}(t) - (2\nu+k-1)C_{k-1}^{\nu}(t)] \tag{85}$$

with

$$C_0^{\nu}(t) = 1 \quad \text{and} \quad C_1^{\nu}(t) = 2\nu t. \tag{86}$$

3. *Gaussian-Type Orbitals*

Gaussian-type orbitals are defined as the class of functions with the dominating long-range behavior of a Gaussian exponential (i.e., $e^{-\alpha r^2}$) and were introduced for atomic and molecular calculations by Boys (1950) and McWeeny (1950). In this section, we consider the spherical harmonic Gaussian-type orbitals (Harris, 1963), namely those of the form defined by Eq. (36) with the radial function

$$f_{nl}(r) = B(n, \alpha)r^{n-1}e^{-\alpha r^2} \tag{87}$$

with the normalization constant

$$B(n, \alpha) = \left[\frac{2(2\alpha)^{n+1/2}}{\Gamma(n+\frac{1}{2})}\right]^{1/2} \tag{88}$$

and $n-1 \geq l \geq 0$. Other Gaussian-type orbitals will be discussed in Section III,B.

The Hankel transform for these orbitals is (Erdélyi *et al.*, 1954)

$$h_{nl}(p) = \frac{B(n, \alpha)\Gamma(a)}{2^c \alpha^a \Gamma(c)} p^{l+1} {}_1F_1(a; c; -p^2/4\alpha), \tag{89}$$

where $a = (n + l + 2)/2$ and $c = l + \frac{3}{2}$. Kummer's confluent hypergeometric series

$${}_1F_1(a; c; z) = \sum_{k=0}^{\infty} \frac{(a)_k}{(c)_k} \cdot \frac{z^k}{k!} \tag{90}$$

satisfies the recursion relation

$${}_1F_1(a; c; z) = \frac{2(a-1) - c + 2z}{a-1} {}_1F_1(a-1; c; z) - \frac{a-c-1}{a-c} {}_1F_1(a-2; c; z). \tag{91}$$

Analytical expressions for $h_{nl}(p)$ exist only when $n - l$ is odd, that is, $n - l = 2v + 1$, $v = 0, 1, 2, \ldots$ (Tsapline, 1971). It is interesting to note that this condition is the same as the one for which Gaussian-type orbitals are advantageous in the evaluation of molecular integrals (Harris, 1963). The form of the momentum wavefunctions themselves are in these cases of the Gaussian type as well and involve a generalized Laguerre polynomial $L_v^{l+1/2}(z)$, which may easily be generated by repeated differentiation:

$$h_{nl}(p) = \frac{B(n, \alpha) p^{l+1} v!\, e^{-z}}{2^{l+3/2} \alpha^{(n+l+2)/2}} L_v^{l+1/2}(z) \tag{92}$$

$$= B(n, \alpha) \frac{2^{l-1/2}}{\alpha^{v+1} p^l} \frac{d^v}{dz^v} (e^{-z} z^{(n+l)/2}) \tag{93}$$

where $z = p^2/4\alpha$.

For $l = 0$ and n odd, the solution may be written in terms of a Hermite polynomial $H_n(x)$, that is,

$$h_{n0}(p) = \frac{B(n, \alpha)(-1)^v e^{-z}}{2^{n+1/2} \alpha^{(n+1)/2}} H_n(z^{1/2}), \tag{94}$$

where

$$H_n(x) = (-1)^n e^{x^2} \frac{d^n}{dx^n} (e^{-x^2}). \tag{95}$$

In the special case $v = 0$, which corresponds to $n = l + 1$, that is, 1s, 2p, 3d, etc., orbitals, the Hankel transform simplifies to a single term, namely

$$h_{n\,n-1}(p) = \frac{B(n, \alpha) p^n}{(2\alpha)^{n+1/2}} e^{-p^2/4\alpha}. \tag{96}$$

The $\hat{\phi}_{nlm}(\mathbf{p})$ are presented in Table II for all cases for which $1 \leq n \leq 4$.

TABLE II

FOURIER TRANSFORMS OF NORMALIZED SPHERICAL HARMONIC GAUSSIAN-TYPE ORBITALS WITH ORBITAL EXPONENT α[a]

		Momentum space representation				
			Factor dependent on the l quantum number			
n Quantum number	Position space representation	Common factor	$l = 0$	$l = 1$	$l = 2$	$l = 3$
1	$(2\alpha/\pi)^{1/4}2(2\alpha)^{1/2}e^{-\alpha r^2} \times Y_{00}(\theta, \phi)$	$(2\alpha/\pi)^{1/4}e^{-p^2/4\alpha}$	$\alpha^{-1}Y_{00}(\hat{\theta}, \hat{\phi})$	—	—	—
2	$(2\alpha/\pi)^{1/4}4\alpha(2/3)^{1/2}re^{-\alpha r^2} \times Y_{lm}(\theta, \phi)$	$(2\alpha/\pi)^{1/4}(3\alpha)^{-1/2}e^{-p^2/4\alpha}$	—	$-i(p\alpha^{-1}) \times Y_{1m}(\hat{\theta}, \hat{\phi})$	—	—
3	$(2\alpha/\pi)^{1/4}8\alpha(2\alpha/15)^{1/2}r^2e^{-\alpha r^2} \times Y_{lm}(\theta, \phi)$	$(2\alpha/\pi)^{1/4}(15)^{-1/2}e^{-p^2/4\alpha}$	$(6\alpha^{-1} - p^2\alpha^{-2}) \times Y_{00}(\hat{\theta}, \hat{\phi})$	—	$-p^2\alpha^{-2} \times Y_{2m}(\hat{\theta}, \hat{\phi})$	—
4	$(2\alpha/\pi)^{1/4}16\alpha^2(2/105)^{1/2}r^3e^{-\alpha r^2} \times Y_{lm}(\theta, \phi)$	$(2\alpha/\pi)^{1/4}(105\alpha)^{-1/2}e^{-p^2/4\alpha}$	—	$-i(10p\alpha^{-1} - p^3\alpha^{-2}) \times Y_{1m}(\hat{\theta}, \hat{\phi})$	—	$ip^3\alpha^{-2} \times Y_{3m}(\hat{\theta}, \hat{\phi})$

[a] Only the cases when the Hankel transformed radial wavefunction, Kummer's confluent hypergeometric series, is expressible in a closed analytical form ($n - l$ is odd) are tabulated here [see Eq. (89)].

4. *Tabulated Radial Functions*

In the case of orbitals that are tabulated numerically (Hartree, 1957; Froese-Fischer, 1972; Herman and Skillman, 1963), numerical techniques must be employed in order to obtain the Hankel transform. Two approaches may be considered. In the first method, one could fit the tabulated function to a set of basis functions which are natural to the problem such as Slater-type (exponential) functions in the atomic case (Löwdin and Appel, 1956) and for which the transform is known analytically. Then the desired transform may be obtained as the linear combination of the transforms of the basis functions. One possible difficulty with this procedure could be that if the fitting was carried out according to some criterion in position space, it may not be a satisfactory one in momentum space especially in important regions there.

The other approach is the direct numerical evaluation of the integral in Eq. (44),

$$p^{-1}h_{nl}(p) = \left(\frac{2}{\pi}\right)^{1/2} \int_0^\infty j_l(pr) f_{nl}(r) r^2 \, dr, \tag{97}$$

where difficulties are encountered in accurate calculation due to the oscillatory nature of the integrand. Even for those radial functions $f_{nl}(r)$ that very rapidly approach zero as r approaches infinity, the transform consists of a large number of significant positive and negative contributions of nearly equal size due to the rapid oscillations of $j_l(pr)$ for larger p. It is for this reason that the report of any study involving the numerical transformation of orbitals should include a critical discussion of the numerical methods employed and the accuracy obtained in the calculation. The problem has been considered recently by Thulstrup (1975), by Benesch (1976b), and by Thakkar and Smith (1975, 1976) with the latter authors emphasizing the evaluation to controlled accuracy.

B. Cartesian and Ellipsoidal Gaussian-Type Orbitals

Boys (1950) proposed the use of Cartesian Gaussian-type orbitals in molecular calculations because analytical expressions may be obtained for all one- and two-electron integrals including the difficult multicenter ones. Since these orbitals may be factorized in the coordinates, we write them in the form

$$\phi_{k_1k_2k_3}(\alpha; \mathbf{r} - \mathbf{A}) = N_{k_1k_2k_3} x_{1A}^{k_1} x_{2A}^{k_2} x_{3A}^{k_3} \exp(-\alpha r_A^2) \tag{98}$$

$$= \prod_{i=1}^{3} N_{k_i}(\alpha) x_{iA}^{k_i} \exp(-\alpha x_{iA}^2) \tag{99}$$

$$= \prod_{i=1}^{3} \phi_{k_i}(\alpha; x_{iA}), \tag{100}$$

where

$$N_{k_i}(\alpha) = \left(\frac{(2\alpha)^{k+1/2}}{\Gamma(k+\frac{1}{2})}\right)^{1/2} \tag{101}$$

$$r_A^2 = |\mathbf{r} - \mathbf{A}|^2 = x_{1A}^2 + x_{2A}^2 + x_{3A}^2 \tag{102}$$

and the k_i are nonnegative integers.

Browne and Poshusta (1962) suggested a generalization of the Cartesian Gaussians with different exponents for different coordinates:

$$\phi_{k_1k_2k_3}(\alpha^1, \alpha^2, \alpha^3; \mathbf{r}_A) = \prod_{i=1}^{3} N_{k_i}(\alpha^i)x_{iA}^{k_i} \exp(-\alpha^i x_{iA}^2) = \prod_{i=1}^{3} \phi_{k_i}(\alpha^i; x_{iA}). \tag{103}$$

Although these ellipsoidal Gaussian-type functions are not as commonly used as are the Cartesian Gaussians, it is convenient to consider the transformation of the former and then specialize to the latter. From the definition [Eq. (20)] and Eq. (103), we have

$$\begin{aligned}
\hat{\phi}_{k_1k_2k_3}(\alpha^1, \alpha^2, \alpha^3; \mathbf{p}) &= (2\pi)^{-3/2} \int \exp(-i\mathbf{p}\cdot\mathbf{r}) \\
&\quad \times \phi_{k_1k_2k_3}(\alpha^1, \alpha^2, \alpha^3; \mathbf{r} - \mathbf{A})\, d\mathbf{r} && (104) \\
&= \exp(-i\mathbf{p}\cdot\mathbf{A}) \prod_{i=1}^{3} (2\pi)^{-1/2} \\
&\quad \times \int_{-\infty}^{\infty} \exp(-ip_i x_{iA})\phi_{k_i}(\alpha^i; x_{iA})\, dx_{iA} && (105) \\
&= \exp(-i\mathbf{p}\cdot\mathbf{A}) \prod_{i=1}^{3} (2\pi)^{-1/2} N_{k_i}(\alpha^i) \\
&\quad \times \exp(-p_i^2/4\alpha^i) \int_{-\infty}^{\infty} \left(x_{iA} - \frac{ip_i}{2\alpha^i}\right)^{k_i} \\
&\quad \times \exp(-\alpha_i x_{iA}^2)\, dx_{iA} && (106) \\
&= \exp(-i\mathbf{p}\cdot\mathbf{A}) \prod_{i=1}^{3} \frac{N_{k_i}(\alpha^i)(-i)^{k_i}}{2^{k_i+1/2}(\alpha^i)^{(k_i+1)/2}} \\
&\quad \times \exp(-z_i^2)H_{k_i}(z_i) && (107) \\
&= \exp(-i\mathbf{p}\cdot\mathbf{A}) \prod_{i=1}^{3} (-i)^{k_i} N_{k_i}(\alpha^i) \\
&\quad \times \exp(-z_i^2)K_{k_i}(z_i) && (108) \\
&= \exp(-i\mathbf{p}\cdot\mathbf{A}) \prod_{i=1}^{3} (-i)^{k_i} \\
&\quad \times \exp(-z_i^2)\bar{K}_{k_i}(z_i), && (109)
\end{aligned}$$

where

$$z_i = p_i/2(\alpha^i)^{1/2}. \tag{110}$$

The Hermite polynomial $H_n(x)$ was defined in Eq. (95) and may be evaluated from the recursion relation

$$H_{n+1}(x) - 2xH_n(x) + 2nH_{n-1}(x) = 0 \tag{111}$$

$$H_0(x) = 1 \quad \text{and} \quad H_1(x) = 2x. \tag{112}$$

K_n and $\overline{K_n}$ are two auxilliary functions defined by Eqs. (107)–(109) and fulfill similar recursion schemes, namely

$$2\alpha K_{k+1}(z) = pK_k(z) - kK_{k-1}(z), \tag{113}$$

$$K_0(z) = (2\alpha)^{-1/2} \quad \text{and} \quad K_1(z) = p(2\alpha)^{-3/2} \tag{114}$$

and

$$\bar{K}_{k+1}(z) = \frac{p}{[(2k+1)\alpha]^{1/2}} \bar{K}_k(z) - \frac{2k}{[(2k+1)(2k-1)]^{1/2}} \bar{K}_{k-1}(z), \tag{115}$$

$$\bar{K}_0(z) = (2\alpha\pi)^{-1/4} \quad \text{and} \quad \bar{K}_1(z) = (2\alpha\pi)^{-1/4} p\alpha^{-1/2}, \tag{116}$$

where z, p, and α are related by Eq. (110).

Since these functions factorize in each coordinate p_i, it is sufficient to present the explicit transform for one dimension, $\hat{\phi}_{k_i}(\alpha; p_i)$, as we have done in Table III for $0 \leq k_i \leq 4$. The common factor $e^{-iA_ip_i}$ is omitted from this table.

TABLE III

FOURIER TRANSFORMS OF NORMALIZED CARTESIAN GAUSSIAN-TYPE ORBITALS WITH ORBITAL EXPONENT α (FOR ONE COORDINATE)

Position space representation	Momentum space representation
$\left(\frac{2\alpha}{\pi}\right)^{1/4} e^{-\alpha x^2}$	$(2\alpha\pi)^{-1/4} e^{-p_x{}^2/4x}$
$2\alpha^{1/2} x \left(\frac{2\alpha}{\pi}\right)^{1/4} e^{-\alpha x^2}$	$-i\alpha^{-1/2} p_x (2\alpha\pi)^{-1/4} e^{-p_x{}^2/4x}$
$\frac{4\alpha}{\sqrt{3}} x^2 \left(\frac{2\alpha}{\pi}\right)^{1/4} e^{-\alpha x^2}$	$\frac{-1}{\sqrt{3}}\left(\frac{p_x^2}{\alpha} - 2\right)(2\alpha\pi)^{-1/4} e^{-p_x{}^2/4\alpha}$
$\frac{8\alpha^{3/2}}{\sqrt{15}} x^3 \left(\frac{2\alpha}{\pi}\right)^{1/4} e^{-\alpha x^2}$	$\frac{i\alpha^{-1/2} p_x}{\sqrt{15}}\left(\frac{p_x^2}{\alpha} - 6\right)(2\alpha\pi)^{-1/4} e^{-p_x{}^2/4\alpha}$
$\frac{16\alpha^2}{\sqrt{105}} x^4 \left(\frac{2\alpha}{\pi}\right)^{1/4} e^{-\alpha x^2}$	$\frac{1}{\sqrt{105}}\left(\frac{p_x^4}{\alpha^2} - \frac{12p_x^2}{\alpha} + 12\right)(2\alpha\pi)^{-1/4} e^{-p_x{}^2/4\alpha}$

In the case of Cartesian Gaussians ($\alpha^1 = \alpha^2 = \alpha^3 = \alpha$), Eq. (109) simplifies to

$$\hat{\phi}_{k_1 k_2 k_3}(\alpha; \mathbf{p}) = e^{-i\mathbf{p}\cdot\mathbf{A}} e^{-p^2/4\alpha} \prod_{i=1}^{3} (-i)^{k_i} \bar{K}_{k_i}(p_i/2\sqrt{\alpha}). \tag{117}$$

Furthermore, we note that when $k_1 = k_2 = k_3 = 0$, this expression reduces to the equivalent expression obtained from Eqs. (42), (43) and (96) for a 1s Gaussian-type orbital ($n = 1$ and $l = 0$) as it should:

$$\hat{\phi}(\alpha; \mathbf{p}) = (2\alpha\pi)^{-3/4} e^{-i\mathbf{p}\cdot\mathbf{A}} e^{-p^2/4\alpha}. \tag{118}$$

These are the basis functions used in the so-called Gaussian lobe and floating Gaussian methods for molecules (Preuss, 1956; Whitten, 1963; Frost, 1967). Since the momentum distributions and Compton profiles for such wavefunctions may be calculated as special cases of the formalism for either spherical harmonic basis functions or Cartesian Gaussians, we shall not discuss them specifically in the remainder of this article.

IV. Evaluation of the Momentum Density $\rho(\mathbf{p})$

With the knowledge obtained in Section III of the Dirac–Fourier transforms of the single-particle basis functions $\boldsymbol{\phi}(\mathbf{r})$, we proceed to the calculation of $\rho(\mathbf{p})$ starting from Eqs. (27)–(29) for wavefunctions constructed from spherical harmonic basis orbitals and Cartesian Gaussian functions.

A. Spherical Harmonic Basis Functions

The momentum density contributions $\rho_{ii}(\mathbf{p})$ and $\rho_{ij}(\mathbf{p})$ expressed in terms of the Hankel transforms and spherical harmonics are

$$\rho_{ii}(\mathbf{p}) = D_{ii} p^{-2} h_i^2(p) \tilde{Y}_i^2(\hat{\theta}) \tag{119}$$

and

$$\rho_{ij}(\mathbf{p}) = p^{-2} h_i(p) h_j(p) \tilde{Y}_i(\hat{\theta}) \tilde{Y}_j(\hat{\theta}) \times [D_{ij}(-i)^{l_i - l_j} e^{-i\mathbf{p}\cdot(\mathbf{A}_i - \mathbf{A}_j)} e^{i(m_i - m_j)\hat{\phi}} + \text{c.c.}] \tag{120}$$

respectively, where

$$\tilde{Y}_i(\hat{\theta}) = \tilde{Y}_{l_i m_i}(\hat{\theta}) = e^{-i m_i \hat{\phi}} Y_{l_i m_i}(\hat{\theta}, \hat{\phi}). \tag{121}$$

Under the assumption that the D matrix is real and with the notation

$$l_i - l_j = 2\nu - \mu, \tag{122}$$

where μ is 0 or 1 according to whether $l_i - l_j$ is even or odd, the nondiagonal term reduces to

$$\rho_{ij}^{\mathrm{II}}(\mathbf{p}) = p^{-2}h_i(p)h_j(p)\tilde{Y}_i(\hat{\theta})\tilde{Y}_j(\hat{\theta})2D_{ij}(-1)^{\nu} \times \begin{matrix}\cos\\ \sin\end{matrix}[\mathbf{p}\cdot(\mathbf{A}_i - \mathbf{A}_j) - (m_i - m_j)\hat{\phi}], \quad \begin{matrix}\mu = 0\\ \mu = 1.\end{matrix} \tag{123}$$

In the case of a single center,

$$\rho_{ij}^{\mathrm{I}}(\mathbf{p}) = p^{-2}h_i(p)h_j(p)\tilde{Y}_i(\hat{\theta})\tilde{Y}_j(\hat{\theta})2D_{ij}(-1)^{\nu} \times \begin{matrix}\cos\\ \sin\end{matrix}[(m_j - m_i)\hat{\phi}], \quad \begin{matrix}\mu = 0\\ \mu = 1.\end{matrix} \tag{124}$$

We use the superscripts I and II to indicate the one- and two-center contributions, respectively, to $\rho_{ij}(\mathbf{p})$. Although $\rho_{ij}^{\mathrm{I}}(\mathbf{p})$ is a special case of $\rho_{ij}^{\mathrm{II}}(\mathbf{p})$, it is an advantage in the calculation to treat the two cases separately.

For certain types of analysis of the momentum density, it is convenient to examine the terms in the expansion

$$\rho(\mathbf{p}) = \sum_{L=0}^{\infty} \sum_{M=-L}^{L} \rho_{LM}(p)Y_{LM}(\hat{\theta}, \hat{\phi}). \tag{125}$$

This form is always possible to obtain since the spherical harmonics constitute a complete set of functions on the unit sphere. The product of two spherical harmonics may thus be expanded as

$$Y_{l_1m_1}(\hat{\theta}, \hat{\phi})Y_{l_2m_2}(\hat{\theta}, \hat{\phi}) = \sum_{L=0}^{\infty} \sum_{M=-L}^{L} C_{Ll_1l_2}^{Mm_1m_2} Y_{LM}(\hat{\theta}, \hat{\phi}). \tag{126}$$

The coefficients $C_{Ll_1l_2}^{Mm_1m_2}$ are the so-called Gaunt coefficients defined by

$$C_{Ll_1l_2}^{Mm_1m_2} = \int_0^{\pi} \sin\hat{\theta}\, d\hat{\theta} \int_0^{2\pi} d\hat{\phi}\, Y_{LM}^*(\hat{\theta}, \hat{\phi})Y_{l_1m_1}(\hat{\theta}, \hat{\phi})Y_{l_2m_2}(\hat{\theta}, \hat{\phi}) \tag{127}$$

$$= (-1)^M \left[\frac{(2L+1)(2l_1+1)(2l_2+1)}{4\pi}\right]^{1/2} \begin{pmatrix} L l_1 l_2 \\ -M m_1 m_2 \end{pmatrix} \begin{pmatrix} L l_1 l_2 \\ 0\,0\,0 \end{pmatrix} \tag{128}$$

where in the last line we have written them in terms of $3-j$ symbols (Rotenberg *et al.*, 1959).

By applying this expansion once in the one-center expressions [Eqs. (119) and (124)], we find

$$\rho_{ii}^{\mathrm{I}}(\mathbf{p}) = D_{ii}p^{-2}h_i^2(p)(-1)^{m_i} \sum_{L=0}^{\infty} C_{Ll_il_i}^{0m_i-m_i} Y_{L0}(\hat{\theta}, \hat{\phi}) \tag{129}$$

and

$$\rho^{\mathrm{I}}_{ij}(\mathbf{p}) = p^{-2}h_i(p)h_j(p)2D_{ij}(-1)^{\nu+m_j} \times \sum_{L=0}^{\infty}\sum_{M=-L}^{L} C^{Mm_i-m_j}_{Ll_il_j} \times [i^{\mu}Y_{LM}(\hat{\theta},\hat{\phi}) + (-i)^{\mu}Y^*_{LM}(\hat{\theta},\hat{\phi})] \tag{130}$$

$$= p^{-2}h_i(p)h_j(p)2D_{ij}(-1)^{\nu+m_j}\sum_{L=0}^{\infty}\sum_{M=-L}^{L} C^{Mm_i-m_j}_{Ll_il_j} \times \tilde{Y}_{LM}(\hat{\theta}) \begin{matrix}\cos\\ \sin\end{matrix}(-M\hat{\phi}), \quad \begin{matrix}\mu=0\\ \mu=1.\end{matrix} \tag{131}$$

For the two-center term, we apply relation (41) once and relation (126) twice in order to obtain

$$\rho^{\mathrm{II}}_{ij}(\mathbf{p}) = p^{-2}h_i(p)h_j(p)D_{ij}(-1)^{\nu+m_j}\sum_{L'=0}^{\infty}\sum_{M'=-L'}^{L'} C^{M'm_i-m_j}_{L'l_il_j} \times [(-i)^{-\mu}e^{-i\mathbf{p}\cdot(\mathbf{A}_i-\mathbf{A}_j)}Y_{L'M'}(\hat{\theta},\hat{\phi}) + \text{c.c.}] \tag{132}$$

$$= p^{-2}h_i(p)h_j(p)4\pi D_{ij} \times (-1)^{\nu+m_j}\sum_{L=0}^{\infty}\sum_{M=-L}^{L}\sum_{L'=0}^{\infty}\sum_{M'=-L'}^{L'}\sum_{L''=0}^{\infty}\sum_{M''=-L''}^{L''} \times (-1)^{M''}j_{L''}(pA_{ij}) \times C^{M'm_i-m_j}_{L'l_il_j}C^{MM'-M''}_{LL'L''} \times \{(-i)^{L''-\mu}Y_{L''M''}(\theta_{A_{ij}},\phi_{A_{ij}})Y_{LM}(\hat{\theta},\hat{\phi}) + \text{c.c.}\}, \tag{133}$$

where $\mathbf{A}_i - \mathbf{A}_j = (A_{ij}, \theta_{A_{ij}}, \phi_{A_{ij}})$ in polar coordinates.

B. Cartesian Gaussian-Type Functions

A product of six Hermite polynomials is contained in each contribution to the momentum density for these basis functions

$$\rho_{kl}(\mathbf{p}) = D_{kl}\exp[-i\mathbf{p}\cdot(\mathbf{A}_k-\mathbf{A}_l)]\prod_{i=1}^{3}\prod_{j=1}^{3}\hat{\phi}_{k_i}(p_i)\hat{\phi}^*_{l_j}(p_j) + \text{c.c.} \tag{134}$$

$$= \exp\left[-\left(\frac{1}{\alpha_k}+\frac{1}{\alpha_l}\right)p^2/4\right] \times \left[\prod_{i=1}^{3} B^i_{kl}H_{k_i}\left(\frac{p_i}{2\sqrt{\alpha_k}}\right)H_{l_i}\left(\frac{p_i}{2\sqrt{\alpha_l}}\right)\right] \times 2\,\mathrm{Re}\left\{D_{kl}\exp[-i\mathbf{p}\cdot(\mathbf{A}_k-\mathbf{A}_l)]\prod_{j=1}^{3}(-i)^{k_j-l_j}\right\}, \tag{135}$$

where we have denoted by B_{kl}^i the real factor,

$$B_{kl}^i = \frac{N_{k_i}(\alpha_k)N_{l_i}(\alpha_l)}{2^{k_i+l_i+1}\alpha_k^{(k_i+1)/2}\alpha_l^{(l_i+1)/2}}. \tag{136}$$

In terms of the parameters,

$$k_j - l_j = 2v_j - \mu_j \qquad \begin{cases} \mu_j = 0 & \text{when } k_j - l_j \text{ even} \\ \mu_j = 1 & \text{when } k_j - l_j \text{ odd,} \end{cases} \tag{137}$$

$$\mu = \mu_1 + \mu_2 + \mu_3, \tag{138}$$

$$v = v_1 + v_2 + v_3, \tag{139}$$

the last factor in Eq. (135) may be written

$$\begin{aligned} \operatorname{Re}\left\{D_{kl}e^{-i\mathbf{p}\cdot(\mathbf{A}_k-\mathbf{A}_l)}\prod_{j=1}^{3}(-i)^{2v_j-\mu_j}\right\} &= (-1)^v \operatorname{Re}\{D_{kl}e^{-i\mathbf{p}\cdot(\mathbf{A}_k-\mathbf{A}_l)}i^{\mu}\} \\ &= (-1)^v D_{kl}\begin{Bmatrix} \cos & & 0 \\ \sin & & 1 \\ -\cos[\mathbf{p}\cdot(\mathbf{A}_k-\mathbf{A}_l)]; & \mu = & 2 \\ -\sin & & 3 \end{Bmatrix}, \end{aligned} \tag{140}$$

where the assumption of a real **D** matrix has been made in the final step. In the single-center case ($k \neq l$) this factor is nonzero only for even μ and equal to $(-1)^{v+\mu/2}D_{kl}$. The diagonal contribution [Eq. (28)] reduces to

$$\rho_{kk}^1(\mathbf{p}) = e^{-p^2/2\alpha_k}\left[\prod_{i=1}^{3} B_{kk}^i H_{k_i}^2\left(\frac{p_i}{2\sqrt{\alpha_k}}\right)\right]D_{kk}. \tag{141}$$

The expansion of the momentum density according to Eq. (120) may be obtained by the transformation,

$$\begin{aligned} p_x &= p\left(\frac{2\pi}{3}\right)^{1/2}[Y_{1-1}(\hat{\theta}, \hat{\phi}) - Y_{11}(\hat{\theta}, \hat{\phi})], \\ p_y &= ip\left(\frac{2\pi}{3}\right)^{1/2}[Y_{1-1}(\hat{\theta}, \hat{\phi}) + Y_{11}(\hat{\theta}, \hat{\phi})], \\ p_z &= p\left(\frac{4\pi}{3}\right)^{1/2}Y_{10}(\hat{\theta}, \hat{\phi}), \end{aligned} \tag{142}$$

and (perhaps multiple) use of relation (126). Further, we introduce coefficients F_{LM}^{kl} defined by

$$\prod_{i=1}^{3} B_{kl}^{i} H_{k_i}\left(\frac{p_i}{2\sqrt{\alpha_k}}\right) H_{l_i}\left(\frac{p_i}{2\sqrt{\alpha_l}}\right) = p^m \sum_{L=0}^{\infty} \sum_{M=-L}^{L} F_{LM}^{kl} Y_{LM}(\hat{\theta}, \hat{\phi}), \tag{143}$$

where

$$m = \sum_{i=1}^{3} (k_i + l_i). \tag{144}$$

From their construction it is easy to prove that the nonvanishing F_{LM}^{kl} all have L of the same parity as $\mu = \mu_1 + \mu_2 + \mu_3$. As the left-hand side of Eq. (143) is real, so is the right-hand side and we may replace the expression in square brackets in Eq. (135) by the sum and insert it within the curly brackets. The same procedure as in Section IV,A results in

$$\begin{aligned}\rho_{kl}^{\mathrm{II}}(\mathbf{p}) = (-1)^{\nu} 8\pi p^m \exp\left[-\left(\frac{1}{\alpha_k} + \frac{1}{\alpha_l}\right) p^2/4\right] \\ \times \operatorname{Re}\left\{\sum_{L=0}^{\infty} \sum_{M=-L}^{L} \sum_{L'=0}^{\infty} \sum_{M'=-L'}^{L'} \sum_{L''=0}^{\infty} \sum_{M''=-L''}^{L''} (-1)^{M''} (-i)^{L''-\mu} \right. \\ \left. \times j_{L''}(pA_{ij}) Y_{L''M''}(\theta_{A_{ij}}, \phi_{A_{ij}}) F_{L'M'}^{kl} C_{LL'L''}^{MM'-M''} Y_{LM}(\hat{\theta}, \hat{\phi})\right\}\end{aligned} \tag{145}$$

with the simpler one-center terms of the form ($\mu = 0$ or 2)

$$\rho_{kl}^{\mathrm{I}}(\mathbf{p}) = (-1)^{\nu+\mu/2} 2D_{kl} p^m \exp\left[-\left(\frac{1}{\alpha_k} + \frac{1}{\alpha_l}\right) p^2/4\right] \sum_{L=0}^{\infty} \sum_{M=-L}^{L} F_{LM}^{kl} Y_{LM}(\hat{\theta}, \hat{\phi}) \tag{146}$$

and

$$\rho_{kk}^{\mathrm{I}}(\mathbf{p}) = D_{kk} p^m e^{-p^2/4\alpha_k} \sum_{L=0}^{\infty} \sum_{M=-L}^{L} F_{LM}^{kk} Y_{LM}(\hat{\theta}, \hat{\phi}). \tag{147}$$

V. Evaluation of the Directional Compton Profile $J(\mathbf{q})$

Computational access to directional Compton profiles will be demonstrated in this section for two different schemes. In Subsection A we evaluate the profile from the momentum density itself and in Subsection B from basis overlap integrals.

When one starts from the momentum density $\rho(\mathbf{p})$ there is a further choice between two different methods. One approach is to perform a rotation of the coordinate system once and for all such that one axis is aligned with the scattering vector. The other method considers each density contribution separately. We will illustrate the former for the case of spherical harmonic basis functions and the latter for the Cartesian Gaussian-type functions.

Before describing any of these methods, we note that in analogy with the expansion of the density in the form of Eq. (125) the directional Compton profile can be written as

$$J(\mathbf{q}) = \sum_{L=0}^{\infty} \sum_{M=-L}^{L} J_{LM}(q) Y_{LM}(\theta_q, \phi_q). \tag{148}$$

An interesting observation is that the density functions $\rho_{LM}(p)$ [in Eq. (125)] only contribute to those $J_{LM}(q)$ with the same L and M (Mijnarends, 1967; Seth and Ellis, 1976) via the relation

$$J_{LM}(q) = 2\pi \int_{|q|}^{\infty} p \, dp \rho_{LM}(p) P_L(q/p), \tag{149}$$

where P_L is a Legendre polynomial.

A. From Momentum Densities

The two-dimensional integral over the momentum density [Eq. (1)] may be explicitly written as

$$\begin{aligned} J(\mathbf{q}) &= \int \rho(\mathbf{p}) \, \delta(\mathbf{p} \cdot \hat{\mathbf{q}} - q) \, d\mathbf{p} \\ &= \int_{|q|}^{\infty} p \, dp \int_{-1}^{1} \frac{dt}{(1 - t^2)^{1/2}} \{\rho(p, \cos^{-1} x, \phi_q + \tan^{-1} y) \\ &\quad + \rho(p, \cos^{-1} x, \phi_q + \pi + \tan^{-1} y)\}, \end{aligned} \tag{150}$$

where

$$x = -t(1 - q^2/p^2)^{1/2} \sin \theta_q + (q/p) \cos \theta_q \tag{151}$$

$$y = \frac{[(1 - t^2)(1 - q^2/p^2)]^{1/2}}{t(1 - q^2/p^2)^{1/2} \cos \theta_q + (q/p) \sin \theta_q} \tag{152}$$

and the arguments of ρ are the polar coordinates p, $\hat{\theta}$, and $\hat{\phi}$. The θ_q and ϕ_q stand for the polar angles of the scattering vector. By taking advantage of the inversion symmetry property of the momentum density (Löwdin, 1967; Kaijser and Smith, 1976b), we write Eq. (150) in the alternative form

$$\begin{aligned} J(\mathbf{q}) = \int_{|q|}^{\infty} p \, dp \int_{-1}^{1} \frac{dt}{(1 - t^2)^{1/2}} &\{\rho(p, \cos^{-1} x, \phi_q + \tan^{-1} y) \\ &+ \rho(p, \cos^{-1}(-x), \phi_q + \tan^{-1} y)\}. \end{aligned} \tag{153}$$

Comparison of these expressions indicates that the integrand can be written as a sum of two terms differing only in the polar $\hat{\theta}$ argument or the polar $\hat{\phi}$ argument. It should also be noted that they are well adapted for numerical evaluation, since the inner integral has the appropriate form required for the Chebyshev–Gauss quadrature scheme (Ralston, 1965). However, for certain analytically expressible densities the inner integral may be obtained in closed form, as will be discussed in the following, leaving us with, at most, a single numerical integration.

All molecular densities are not directly suited for analytical integration as they stand in Eqs. (150) and (153) due to the appearance of the interatomic $\mathbf{p} \cdot \mathbf{A}$ factor discussed in Section IV. For molecules with a certain specific symmetry the treatment may be simplified considerably as will be demonstrated later for the case of linear molecules. In the general case, however, it is convenient to perform certain rotational transformations mentioned in the introduction to Section V.

1. *Spherical Harmonic Basis Functions*

The method that we will describe now is to a large extent related to the work on crystals by Mijnarends (1967, 1977), by Aikala (1975b), and by Berggren *et al.* (1976, 1977). The idea is to express the momentum density in a new spherical coordinate system p, $\hat{\theta}'$, $\hat{\phi}'$ chosen so that the polar axis ($\hat{\theta}' = 0$) is aligned with the scattering vector $\hat{\mathbf{q}}$. The integration then reduces to

$$\int \delta(\mathbf{p} \cdot \hat{\mathbf{q}} - q)\, d\mathbf{p} = \int_0^{2\pi} d\hat{\phi}' \int_{|q|}^{\infty} p\, dp \Big|_{\theta' = \cos^{-1}(q/p)} \tag{154}$$

$$= \int_0^{2\pi} d\hat{\phi}' \int_0^{\infty} p_\perp\, dp_\perp \Big|_{\theta' = \cos^{-1}(q/p)}, \tag{155}$$

where

$$p^2 = p_\perp^2 + q^2. \tag{156}$$

In as much as all the density contributions involve spherical harmonics, we make use of the relation

$$Y_{LM}(\hat{\theta}, \hat{\phi}) = \sum_{\sigma=-L}^{L} e^{iM\phi_q}\, d_L^{M\sigma}(\cos\theta_q) Y_{L\sigma}(\hat{\theta}', \hat{\phi}') \tag{157}$$

to express them in the new coordinate system defined above ($\mathbf{q} = q, \theta_q, \phi_q$ in the old system). For positive M and σ, the coefficients are given by (Harris, 1967)

$$d_L^{M\sigma}(\cos\theta_q) = \left[\frac{(L+M)!\,(L-M)!}{(L+\sigma)!\,(L-\sigma)!}\right]^{1/2} \left(\frac{1+\cos\theta_q}{2}\right)^L$$

$$\times \sum_s (-1)^{M+s} \binom{L-\sigma}{M+s} \binom{L+\sigma}{L-s} \left(\frac{1-\cos\theta_q}{1+\cos\theta_q}\right)^{s+[(M+\sigma)/2]} \tag{158}$$

The sum over s is assumed to go over those integers for which the factorials are defined.

We apply this relation to the two-center density contribution written in the form of Eq. (132) and obtain

$$\rho_{ij}^{\mathrm{II}}(\mathbf{p}) = p^{-2}h_i(p)h_j(p)D_{ij}(-1)^{v+m_j} \sum_{L'=0}^{\infty} \sum_{M'=-L'}^{L'} C_{L'l_il_j}^{M'm_i-m_j}$$
$$\times \sum_{\sigma=-L'}^{L'} d_{L'}^{M'\sigma}(\cos\theta_q)$$
$$\times \{(-i)^{-\mu}e^{-i\mathbf{p}\cdot(\mathbf{A}_i-\mathbf{A}_j)}e^{iM\phi_q}Y_{L'\sigma}(\hat{\theta}', \hat{\phi}') + \text{c.c.}\}. \tag{159}$$

The braces contain all the $\hat{\phi}'$ dependence and the integration over this variable results in

$$\int_0^{2\pi} d\hat{\phi}'\{(-i)^{-\mu}e^{-i\mathbf{p}\cdot\mathbf{A}}e^{iM'\phi_q}Y_{L'\sigma}(\hat{\theta}', \hat{\phi}'') + \text{c.c.}\}$$
$$= 4\pi\tilde{Y}_{L'\sigma}(\hat{\theta}') \begin{matrix}\cos\\ \sin\end{matrix}\left(p_qA_q - M'\phi_q - \sigma\phi'_A + \sigma\frac{\pi}{2}\right) \cdot J_\sigma(p_\perp A_\perp), \quad \begin{matrix}\mu = 0\\ \mu = 1,\end{matrix} \tag{160}$$

where A_q is the component of the vector $\mathbf{A} = \mathbf{A}_i - \mathbf{A}_j$ along $\hat{\mathbf{q}}$ and $A_\perp$ is the length of the projection onto a plane orthogonal to $\hat{\mathbf{q}}$. The polar coordinate of $\mathbf{A}$ in the new coordinate system is ϕ'_A. The Compton profile thus becomes

$$J_{ij}^{\mathrm{II}}(\mathbf{q}) = (-1)^{v+m_j}4\pi D_{ij} \sum_{L'=0}^{\infty} \sum_{M'=-L'}^{L'} C_{L'l_il_j}^{M'm_i-m_j} \sum_{\sigma=-L'}^{L'} d_{L'}^{M'\sigma}(\cos\theta_q)$$
$$\times \begin{matrix}\cos\\ \sin\end{matrix}\left[qA_q - M'\phi_q - \sigma\phi'_A + \sigma\frac{\pi}{2}\right]$$
$$\times \int_{|q|}^{\infty} p^{-1}h_i(p)h_j(p)\tilde{Y}_{L'\sigma}(\cos^{-1}(q/p)]J_\sigma(p_\perp A_\perp). \tag{161}$$

If we start from expression (133), we find analogously

$$J_{ij}^{\mathrm{II}}(\mathbf{q}) = (-1)^{v+m_j}8\pi^2 D_{ij} \sum_{L=0}^{\infty} \sum_{M=-L}^{L} \sum_{L'=0}^{\infty} \sum_{M'=-L'}^{L'} \sum_{L''=0}^{\infty} \sum_{M''=-L''}^{L''} (-1)^{M''}$$
$$\times C_{L'l_il_j}^{M'm_i-m_j}C_{LL'L''}^{MM'-M''}$$
$$\times d_L^{M0}(\cos\theta_q)[(-i)^{L''-\mu}e^{iM\phi_q}Y_{L''M''}(\theta_{A_{ij}}, \phi_{A_{ij}}) + \text{c.c.}]$$
$$\times \int_{|q|}^{\infty} p^{-1}h_i(p)h_j(p)\tilde{Y}_{L0}[\cos^{-1}(q/p)]j_{L''}(pA_{ij})\, dp. \tag{162}$$

It should be noted that the arguments of the Bessel functions in these expressions are different. For an off-diagonal single-center term, Eqs. (161) and (162) both reduce to the same expression, namely

$$J^{1}_{ij}(\mathbf{q}) = (-1)^{\nu+m_j} 4\pi D_{ij} \sum_{L=0}^{\infty} \sum_{M=-L}^{L} \times C^{Mm_i-m_j}_{Ll_il_j}\, d^{M0}_{L}(\cos\theta_q) \times \int_{|q|}^{\infty} p^{-1} h_i(p) h_j(p) \tilde{Y}_{L0}[\cos^{-1}(q/p)]\, dp \times \begin{matrix}\cos \\ \sin\end{matrix}(-M\phi_q), \quad \begin{matrix}\mu = 0 \\ \mu = 1.\end{matrix} \tag{163}$$

For the diagonal case, we find

$$J^{1}_{ii}(\mathbf{q}) = (-1)^{m_i} D_{ii} \sum_{L=0}^{\infty} C^{0m_i-m_i}_{Ll_il_i}\, d^{00}_{L}(\cos\theta_q) \times \int_{|q|}^{\infty} p^{-1} h_i^2(p) \tilde{Y}_{L0}[\cos^{-1}(q/p)]\, dp. \tag{164}$$

Since expansion (125) for $\rho(\mathbf{p})$ has the same angular dependence as the one-center terms just discussed, an analogous treatment to the above may be made.

The integrals in the above expressions can sometimes be evaluated analytically. This is easier to see after transformation $p^2 = p_\perp^2 + q^2$ has been performed:

$$\int_{|q|}^{\infty} p^{-1} h_i(p) h_j(p) \tilde{Y}_{L\sigma}[\cos^{-1}(q/p)] J_\sigma(A_\perp p_\perp)\, dp = \int_{0}^{\infty} p_\perp p^{-2} h_i(p) h_j(p) \tilde{Y}_{L\sigma}[\cos^{-1}(q/p)] J_\sigma(A_\perp p_\perp)\, dp_\perp. \tag{165}$$

For Slater-type orbitals, the latter can be expressed in terms of integrals of the form

$$\int_0^{\infty} \frac{p_\perp^{\sigma+1} J_\sigma(A_\perp p_\perp)}{(p^2 + \alpha_i^2)^{n_1+1} (p^2 + \alpha_j^2)^{n_2+1}}\, dp_\perp \tag{166}$$

where factor $p_\perp^\sigma$ arises from $\tilde{Y}_{L\sigma}$ in Eq. (165). By elementary algebra, the even-powered factors of p in the numerator have been included in the factors in the denominator of Eq. (166), so that $n_1 \leq n_i$ and $n_2 \leq n_j$.

When $\alpha_i = \alpha_j$, this Hankel transform yields modified Bessel functions of the third kind (Erdélyi *et al.*, 1954),

$$\int_0^{\infty} \frac{dp_\perp\, p_\perp^{\sigma+1} J_\sigma(A_\perp p_\perp)}{(p_\perp^2 + a_i^2)^{n+1}} = \left(\frac{A_\perp}{2a_i}\right)^{n} \frac{a_i^\sigma}{n!} K_{\sigma-n}(A_\perp a_i) \tag{167}$$

for $\sigma < 2n + \frac{3}{2}$. In this equation, a_i denotes $(\alpha_i^2 + q^2)^{1/2}$. In the case of different orbital exponents α_i and α_j, a partial fraction decomposition of the integrand in Eq. (166) must be performed before applying this formula (Berggren *et al.*, 1976). Thus

$$\int_0^\infty \frac{dp_\perp p_\perp^{\sigma+1} J_\sigma(A_\perp p_\perp)}{(p_\perp^2 + \alpha_i^2)^{n_1+1}(p_\perp^2 + \alpha_j^2)^{n_2+1}}$$

$$= \frac{1}{n_2!} \sum_{k=0}^{n_1} (-1)^k \frac{(n_2 + k)!}{k!(n_1 - k)!} \left(\frac{A_\perp}{2a_i}\right)^{n_1-k}$$

$$\times \frac{a_i^\sigma}{(a_j^2 - a_i^2)^{n_2+k+1}} K_{\sigma+k-n_1}(A_\perp a_i)$$

$$+ \frac{1}{n_1!} \sum_{k=0}^{n_2} (-1)^k \frac{(n_1 + k)!}{k!(n_2 - k)!} \left(\frac{A_\perp}{2a_j}\right)^{n_2-k}$$

$$\times \frac{a_j^\sigma}{(a_i^2 - a_j^2)^{n_1+k+1}} K_{\sigma+k-n_2}(A_\perp a_j). \qquad (168)$$

We note that due to the restriction on σ given above with Eq. (167), formula (168) is only valid when σ is equal to 0 or 1.

When Gaussian-type orbitals are involved, the integrals may also be treated analytically. Since the product of two Gaussians, such as appear in the density, is just another Gaussian, the same kind of Hankel transformation as described in Section III,B is applicable.

We now return to the general Eq. (150) discussed at the beginning of this section and see how the treatment may be simplified in the case of a linear molecule. For such molecules we first recall that the M0's will never couple A0's with different m-quantum numbers (the z axis is assumed to be along the internuclear axis). As a result, the momentum densities are $\hat{\phi}$-independent and ϕ_q may be chosen arbitrarily. The DCP can then be simplified to the form

$$J(\mathbf{q}) = 2\int_{|q|}^\infty p\,dp \int_{-1}^{1} \frac{dt}{(1 - t^2)^{1/2}} \rho(p, \cos^{-1} x, \cdot)$$

$$= 2\int_0^\infty p_\perp\,dp_\perp \int_{-1}^{1} \frac{dt}{(1 - t^2)^{1/2}} \rho(p, \cos^{-1} x, \cdot), \qquad (169)$$

where $p^2 = q^2 + p_\perp^2$ and x was defined in Eq. (151). Because of the form of x, and since the internuclear phase factor is

$$\frac{\sin}{\cos}[\mathbf{p} \cdot \mathbf{A}] = \frac{\sin}{\cos}[qA \cos\theta_q - A \sin\theta_q p_\perp t], \qquad (170)$$

especially simple expressions for the DCP's along and orthogonal to the bond ($\theta_q = 0$ and $\pi/2$, respectively) are obtained.

For the profiles along the bond, $J_{\parallel}(q)$, the only t dependence of the integrand is the factor $(1 - t^2)^{-1/2}$. Integration over t results in the simpler expression

$$J_{\parallel}(q) = 2\pi \int_{|q|}^{\infty} p\, dp \rho(p, \cos^{-1}(q/p), \cdot). \tag{171}$$

The integrand for the directional profile orthogonal to the bond, $J_{\perp}(q)$, can be expressed as a finite sum with terms related to the integral representation of the Bessel function of the first kind, namely

$$\int_{-1}^{1} (1 - t^2)^{\nu-(1/2)} \cos(at)\, dt = \pi^{1/2}\Gamma(\nu + \tfrac{1}{2})(a/2)^{\nu} J_{\nu}(a). \tag{172}$$

Some explicit expressions for σ- and π-type molecular orbitals have been given by Kaijser and Smith (1976a).

2. *Cartesian Gaussian-Type Functions*

The momentum density for wavefunctions constructed with these basis functions may be written in the form of Eq. (125) and can then be treated in the same way as described above for the spherical harmonic-type orbitals. As mentioned in the introduction to this section, we prefer to illustrate an alternative approach here. The basic difference is that while above we made a rotation once and for all, we now perform a different rotation for each two-center product. That is to say we make use of the knowledge of the direction of the internuclear vector $\mathbf{A}_{ij} = \mathbf{A}_i - \mathbf{A}_j$.

Consider a general term of the momentum density [Eqs. (135)–(140)]. Except for a constant, it may be written in the form (we drop the indices on the vector $\mathbf{A}_{ij}$),

$$e^{-ap^2} p_1^{k_1} p_2^{k_2} p_3^{k_3} \begin{matrix}\cos\\ \sin\end{matrix} (\mathbf{p} \cdot \mathbf{A}), \tag{173}$$

and its contribution to the Compton profile with scattering vector $\hat{\mathbf{q}}$ is

$$J(\mathbf{q}) = \int e^{-ap^2} p_1^{k_1} p_2^{k_2} p_3^{k_3} \begin{matrix}\cos\\ \sin\end{matrix} (\mathbf{p} \cdot \mathbf{A})\, \delta(\mathbf{p} \cdot \hat{\mathbf{q}} - q)\, d\mathbf{p}. \tag{174}$$

We introduce new variables by

$$\mathbf{A} = A_q \hat{\mathbf{q}} + \mathbf{A}_{\perp}, \tag{175}$$

$$\mathbf{p} = p_q \hat{\mathbf{q}} + \mathbf{p}_{\perp} = p_q \hat{\mathbf{q}} + p'\hat{\mathbf{e}}' + p''\hat{\mathbf{e}}'', \tag{176}$$

such that $\hat{\mathbf{e}}'$, $\hat{\mathbf{e}}''$and $\hat{\mathbf{q}}$ are orthonormal unit vectors, and

$$\hat{\mathbf{e}}' = \frac{\mathbf{A}_{\perp}}{|\mathbf{A}_{\perp}|}. \tag{177}$$

In terms of these, the trigonometric function in Eq. (174) may be written as

$$\cos(\mathbf{p}\cdot\mathbf{A}) = \cos(p_q A_q)\cos(p' A_\perp) - \sin(p_q A_q)\sin(p' A_\perp) \tag{178}$$

or

$$\sin(\mathbf{p}\cdot\mathbf{A}) = \sin(p_q A_q)\cos(p' A_\perp) + \cos(p_q A_q)\sin(p' A_\perp) \tag{179}$$

and the product $p_1^{k_1} p_2^{k_2} p_3^{k_3}$ becomes

$$\prod_{i=1}^{3} p_i^{k_i} = \prod_{l=1}^{3} (p'c_i' + p''c_i'' + qc_i^q), \tag{180}$$

where

$$c_i' = \hat{\mathbf{e}}' \cdot \hat{\mathbf{e}}_i; \quad c_i'' = \hat{\mathbf{e}}'' \cdot \hat{\mathbf{e}}_i; \quad c_i^q = \hat{\mathbf{q}} \cdot \hat{\mathbf{e}}_i \qquad i = 1, 2, 3. \tag{181}$$

A general term in this expansion of Eq. (174) is

$$\int e^{-ap^2} p_q^l p'^m p''^n \, {\sin \atop \cos}(p_q A_q) \, {\sin \atop \cos}(p' A_\perp)\, \delta(\mathbf{p}\cdot\hat{\mathbf{q}} - q)\, d\mathbf{p}$$

$$= q^l \, {\sin \atop \cos}(qA_q) e^{-aq^2} \int_{-\infty}^{\infty} dp' e^{-ap'^2} p'^m \, {\sin \atop \cos}(p' A_\perp) \int_{-\infty}^{\infty} dp'' e^{-ap''^2} p''^m. \tag{182}$$

Both of the integrals that appear on the right-hand side have analytical solutions, namely

$$\int_{-\infty}^{\infty} dp' e^{-ap'^2} p'^m \, {\sin \atop \cos}(p' A_\perp)$$

$$= {\mathrm{Im} \atop \mathrm{Re}} \int_{-\infty}^{\infty} e^{-ap'^2} e^{ip'A_\perp} p'^m \, dp'$$

$$= \exp(-A_\perp^2/4a) 2^{-[m+(1/2)]} a^{-(m+1)/2} H_m\left(\frac{A_\perp}{2\alpha^{1/2}}\right) {\mathrm{Im} \atop \mathrm{Re}} \{(-i)^m\} \tag{183}$$

in analogy with Eq. (105) and

$$\int_{-\infty}^{\infty} dp'' e^{-ap''^2} p''^n = a^{(n+1)/2}\Gamma((n+1)/2), \qquad \text{if } n \text{ even}$$

$$= 0, \qquad \text{if } n \text{ odd.} \tag{184}$$

With these expressions in hand, it is purely a matter of bookkeeping to evaluate the directional Compton profile $J(\mathbf{q})$ by this method.

B. From Basis Overlap Integrals

When only the directional Compton profiles $J(\mathbf{q})$ are of interest, the intermediate step of the calculation of $\rho(\mathbf{p})$ may be eliminated and the problem reduced (Thulstrup, 1976) to the determination of overlap integrals between the basis functions.

In the definition [Eq. (1)] of $J(\mathbf{q})$, we use the Fourier integral representation of the Dirac δ-function. Then

$$J(\mathbf{q}) = \int \rho(\mathbf{p})\, \delta(\mathbf{p} \cdot \hat{\mathbf{q}} - q)\, d\mathbf{p}$$

$$= (2\pi)^{-1} \iint_{-\infty}^{\infty} \rho(\mathbf{p}) e^{i(\mathbf{p} \cdot \hat{\mathbf{q}} - q)s}\, ds\, d\mathbf{p}$$

$$= (2\pi)^{-1} \int_{-\infty}^{\infty} \tilde{\gamma}(s\hat{\mathbf{q}}) e^{-isq}\, ds, \qquad (185)$$

where (Benesch *et al.*, 1971)

$$\tilde{\gamma}(s\hat{\mathbf{q}}) = \int \rho(\mathbf{p}) e^{i\mathbf{p} \cdot \hat{\mathbf{q}}s}\, d\mathbf{p}. \qquad (186)$$

In turn, we recall relation (11) between $\rho(\mathbf{p})$ and $\gamma(\mathbf{r} \mid \mathbf{r}')$ in order to write

$$\tilde{\gamma}(s\hat{\mathbf{q}}) = \int d\mathbf{r}\, \gamma(\mathbf{r} \mid \mathbf{r} - s\hat{\mathbf{q}}). \qquad (187)$$

Comparison of Eq. (186) with Eq. (12) shows that $\tilde{\gamma}(s\hat{\mathbf{q}})$ is the Fourier transform of $\rho(\mathbf{p})$ as $F(\mathbf{s})$ was of $\rho(\mathbf{r})$. The similarity between Eqs. (187) and (13) should be noted as well (Benesch *et al.*, 1971).

If the expansion [Eq. (16)] of $\gamma(\mathbf{r} \mid \mathbf{r}')$ in terms of the basis functions $\boldsymbol{\phi}(\mathbf{r})$ is substituted in Eq. (187), we have

$$\tilde{\gamma}(s\hat{\mathbf{q}}) = \sum_{i,j} D_{ij} \int \phi_i(\mathbf{r}) \phi_j^*(\mathbf{r} - s\hat{\mathbf{q}})\, d\mathbf{r}$$

$$= \sum_{i,j} D_{ij} \int \phi_i(\mathbf{r}_i + \mathbf{A}_i) \phi_j^*(\mathbf{r}_j + \mathbf{A}_j - s\hat{\mathbf{q}})\, d\mathbf{r}$$

$$= \sum_{i,j} D_{ij}\, \Delta_{ij}(\mathbf{A}_j - \mathbf{A}_i - s\hat{\mathbf{q}}). \qquad (188)$$

Substitution of this expression in Eq. (185), together with the observation that $\tilde{\gamma}(s\hat{\mathbf{q}})$ is an even function of s, leads to the result

$$J(\mathbf{q}) = \pi^{-1} \sum_{i,j} D_{ij} \int_0^{\infty} \Delta_{ij}(\mathbf{A}_j - \mathbf{A}_i - s\hat{\mathbf{q}}) \cos(sq)\, ds. \qquad (189)$$

In the last two equations, Δ_{ij} is a conventional two-center overlap function for which well-known analytical or numerical methods exist (see, e.g., Roothaan, 1951; Löwdin, 1956; Thulstrup, 1975).

VI. The Spherically Averaged Momentum Density $\overline{\rho(p)}$ and the Isotropic Compton Profile $\overline{J(q)}$

A. Spherical Harmonic Basis Functions

The spherically averaging process for the momentum density is most easily performed after the expansion of $\rho(\mathbf{p})$ in a single set of spherical harmonics [Eq. (125)]. The integration over the angular parts of $\mathbf{p}$ [Eq. (5)] results in

$$\overline{\rho(p)} = (4\pi)^{-1} \int_{|\mathbf{p}|=p} \rho(\mathbf{p})\, d\Omega_{\hat{\mathbf{p}}} = (4\pi)^{-1/2} \rho_{00}(p). \tag{190}$$

The one-center contributions are [Eqs. (129) and (131)]

$$\overline{\rho_{ii}^{\mathrm{I}}(p)} = (4\pi)^{-1} D_{ii} p^{-2} h_i^2(p) \tag{191}$$

and

$$\overline{\rho_{ij}^{\mathrm{I}}(p)} = \delta_{l_i, l_j}\, \delta_{m_i, m_j} (4\pi)^{-1} (D_{ij} + D_{ji}) p^{-2} h_i(p) h_j(p), \tag{192}$$

whereas the two-center term [Eq. (133)] reduces to

$$\overline{\rho_{ij}^{\mathrm{II}}(p)} = p^{-2} h_i(p) h_j(p) (-1)^{\nu + m_j} \sum_{L=0}^{\infty} \sum_{M=-L}^{L} C_{Ll_il_j}^{Mm_i-m_j} j_L(pA_{ij})$$
$$\times \{(-i)^{L-\mu} D_{ij}\, Y_{LM}(\theta_{A_{ij}}, \phi_{A_{ij}}) + \text{c.c.}\}. \tag{193}$$

Two kinds of integrals are involved in the calculation of the isotropic Compton profile,

$$\overline{J(q)} = 2\pi \int_{|q|}^{\infty} p\, \overline{\rho(p)}\, dp, \tag{4}$$

namely,

$$\int_{|q|}^{\infty} p^{-1} h_i(p) h_j(p) j_L(pA_{ij})\, dp \tag{194}$$

and

$$\int_{|q|}^{\infty} p^{-1} h_i(p) h_j(p)\, dp \tag{195}$$

which result, respectively, from the two- and one-center contributions to $\rho(\mathbf{p})$. The latter integral may be evaluated analytically for both Slater- and Gaussian-type orbitals. For the former type of orbitals, Eq. (195) involves integrals of the form

$$\int_{|q|}^{\infty} \frac{p^{2m+1}}{(p^2 + \alpha_i^2)^{n_i+1} (p^2 + \alpha_j^2)^{n_j+1}}\, dp = \int_0^{\infty} \frac{(x^2 + q^2)^m x}{(x^2 + a_i^2)^{n_i+1} (x^2 + a_j^2)^{n_j+1}}\, dx \tag{196}$$

where the change of variables, $x^2 = p^2 - q^2$, was made. In the equation, m is an integer and $a_i^2 = \alpha_i^2 + q^2$. The n_i and α_i are the defining parameters for the ith Slater-type orbital [Eq. (51)]. These quantities are such that the integrand $f(x) \to 0$ when $x \to 0$, $x^2 f(x) \to 0$ when $x \to \infty$, and there are no poles of $f(x)$ on the positive axis. Under these conditions, Cox (1976) has shown that

$$\int_0^\infty f(x)\,dx = -(\text{sum of the residues of } f(z) \ln z \text{ in the complex plane}). \tag{197}$$

With the same change of variables, the integral (195) for Gaussian-type orbitals involves contributions of the form

$$\begin{aligned} \int_{|q|}^\infty p^{2m+1} e^{-p^2/2c}\,dp &= e^{-q^2/2c} \int_0^\infty (x^2 + q^2)^m x e^{-x^2/2c}\,dx \\ &= e^{-q^2/2c} \int_0^\infty \sum_{k=0}^m \binom{m}{k} x^{2k+1} q^{2(m-k)} e^{-x^2/2c}\,dx \\ &= e^{-q^2/2c} (2c)^m c \sum_{k=0}^m \frac{m!}{k!} \left(\frac{q^2}{2c}\right)^k \end{aligned} \tag{198}$$

where

$$c = 2\alpha_i \alpha_j / (\alpha_i + \alpha_j). \tag{199}$$

The integral in Eq. (194) is more complicated, but for the special case $q = 0$ we recognize it as a Hankel transform [Eq. (45)].

B. Cartesian Gaussian Basis Functions

The expressions of the momentum density, for which the spherically averaging procedure is most easily performed, are found in Eqs. (145)–(147). They turn out to be

$$\begin{aligned} \overline{\rho_{kl}^{\mathrm{II}}(p)} = {} & (-1)^\nu 2p^m \exp\left[-\left(\frac{1}{\alpha_k} + \frac{1}{\alpha_l}\right)p^2/4\right] \\ & \times \operatorname{Re}\left\{D_{kl} \sum_{L=0}^\infty \sum_{M=-L}^{L} (-i)^{L-\mu} \right. \\ & \left. \times j_L(pA_{ij}) F_{LM}^{kl} Y_{LM}(\theta_{A_{ij}}, \phi_{A_{ij}})\right\} \end{aligned} \tag{200}$$

$$\begin{aligned} \overline{\rho_{kl}^{\mathrm{I}}(p)} = {} & (-1)^{\nu+\mu/2} \pi^{-1/2} D_{kl} F_{00}^{kl} p^m \\ & \times \exp\left[-\left(\frac{1}{\alpha_k} + \frac{1}{\alpha_l}\right)p^2/4\right] \end{aligned} \tag{201}$$

and

$$\overline{\rho_{kk}^{\mathrm{I}}(p)} = (4\pi)^{-1/2} D_{kk} F_{00}^{kk} p^m e^{-p^2/2\alpha_k}, \tag{202}$$

respectively.

As in the previous section, the evaluation of $\overline{J(q)}$ requires the integrals

$$\int_{|q|}^{\infty} \exp\left[-\left(\frac{1}{\alpha_k} + \frac{1}{\alpha_l}\right)p^2/4\right] p^{m+1} j_L(pA_{ij})\, dp \tag{203}$$

and

$$\int_{|q|}^{\infty} \exp\left[-\left(\frac{1}{\alpha_k} + \frac{1}{\alpha_l}\right)p^2/4\right] p^{m+1}\, dp, \tag{204}$$

respectively, which may be compared with Eqs. (194) and (195) and the remarks made there.

VII. Concluding Remarks

Interest in the study of Compton profiles has increased markedly in the last 10 years (Cooper, 1971; Benesch and Smith, 1973; Epstein, 1975; Williams, 1977). Atoms, molecules, and solids have been intensively investigated, both experimentally and theoretically. The experiments have so far been concentrated on the isotropic Compton profile, but there is a growing interest in directional studies with emphasis on the solid state.

Solids, especially ionic crystals, can with success be investigated theoretically within the so-called LCAO method (Löwdin, 1956). The calculation of momentum densities and Compton profiles for these wavefunctions are, as might be expected, very similar to the molecular LCAO wavefunctions treated here. Aikala (1975a,b) has discussed in some detail the case of completely filled shells of spherical harmonic basis orbitals.

Molecular systems still have to wait for experimental directional Compton profiles. The fact that there is much information in the momentum density not revealed by the isotropic profile (Coulson and Duncanson, 1941; Henneker and Cade, 1968; Epstein, 1973; Benesch and Smith, 1973; Kaijser and Lindner, 1975) should provide an exciting challenge for the experimentalist. Calculations on directional Compton profiles have recently been reported (Langhoff and Tawil, 1975; Kaijser and Smith, 1976a).

It is our hope that the methods developed and reviewed here will provide new insight into momentum density and Compton profile evaluation, an area that offers the exciting possibility of experimental and theoretical tests of the quality of calculated wavefunctions.

Acknowledgments

Support of this research by the National Research Council of Canada is gratefully acknowledged. We would like to thank O. Aikala and R. F. Stewart for helpful comments on this manuscript.

References

Abramowitz, M., and Stegun, I. A. (1964). "Handbook of Mathematical Functions." Dover, New York.

Aikala, O. (1975a). *Philos. Mag.* [8] **31**, 935.

Aikala, O. (1975b). *Philos. Mag.* [8] **32**, 333.

Bender, C. F., and Davidson, E. R. (1968). *J. Chem. Phys.* **49**, 4222.

Benesch, R. (1976a). *J. Phys. B.* **9**, 2587.

Benesch, R. (1976b). To be published.

Benesch, R., and Smith, V. H., Jr. (1970). *Chem. Phys. Lett.* **5**, 601.

Benesch, R., and Smith, V. H., Jr. (1971). *Int. J. Quantum Chem.* **4S**, 131.

Benesch, R., and Smith, V. H., Jr. (1973). *In* "Wave Mechanics—The First Fifty Years" (W. C. Price *et al*., eds.), p. 357. Butterworth, London.

Benesch, R., Singh, S. R., and Smith, V. H., Jr. (1971). *Chem. Phys. Lett.* **10**, 151.

Berggren, K.-F., Manninen, S., Paakari, T., Aikala, O., and Mansikka, K. (1977). *In* "Compton Scattering: The Investigation of Electron Momentum Distributions" (B. Williams, ed.), p. 139. McGraw-Hill, New York.

Berggren, K.-F., Martino, F., Eisenberger, P., and Reed, W. A. (1976). *Phys. Rev. B* **13**, 2292.

Boys, S. F. (1950). *Proc. R. Soc. London, Ser. A* **200**, 542.

Browne, J. C., and Poshusta, R. D. (1962). *J. Chem. Phys.* **36**, 1933.

Cade, P. E., and Wahl, A. C. (1974). *At. Data* **13**, 339.

Clementi, E. (1965). *IBM J.* **9**, Suppl., 2.

Cooper, M. (1971). *Adv. Phys.* **20**, 453.

Coulson, C. A., and Duncanson, W. E. (1941). *Proc. Cambridge Philos. Soc.* **37**, 67.

Cox, H. L. (1976). *Phys. Rev. A* **13**, 229.

Dirac, P. A. M. (1958). "The Principles of Quantum Mechanics," p. 97. Oxford Univ. Press, London and New York.

Epstein, I. R. (1973). *Acc. Chem. Res.* **6**, 145.

Epstein, I. R. (1975). *Phys. Chem., Ser. Two* **1**, 105.

Erdélyi, A., Magnus, W., Oberhettinger, F., and Tricomi, F. G. (1953). "Higher Transcendental Functions," Vols. 1, 2, and 3. McGraw-Hill, New York.

Erdélyi, A., Magnus, W., Oberhettinger, F., and Tricomi, F. G. (1954). "Tables of Integral Transforms," Vols. 1 and 2. McGraw-Hill, New York.

Froese-Fischer, C. (1972). *Comput. Phys. Commun.* **4**, 107.

Frost, A. A. (1967). *J. Chem. Phys.* **47**, 3707.

Harris, F. E. (1963). *Rev. Mod. Phys.* **35**, 558.

Harris, F. E. (1967). "Computational Methods of Quantum Chemistry" (unpublished lecture notes).

Hartree, D. R. (1957). "Calculation of Atomic Structures." Wiley, New York.

Henderson, M. G., and Scherr, C. W. (1960). *Phys. Rev.* **120**, 150.

Henneker, W. H., and Cade, P. E. (1968). *Chem. Phys. Lett.* **2**, 575.
Herman, F., and Skillman, S. (1963). "Atomic Structure Calculations." Prentice-Hall, Englewood Cliffs, New Jersey.
Holøien, E. (1956). *Phys. Rev.* **104**, 1301.
Hylleraas, E. A. (1929). *Z. Phys.* **54**, 347.
Kaijser, P., and Lindner, P. (1975). *Philos. Mag.* [8] **31**, 871.
Kaijser, P., and Smith, V. H., Jr. (1976a). *Mol. Phys.* **31**, 1557.
Kaijser, P., and Smith, V. H., Jr. (1976b). *In* "Quantum Science: Methods and Structure" (J.-L. Calais *et al.*, eds.), pp. 417–425. Plenum, New York.
Kato, T. (1957). *Commun. Pure Appl. Math.* **10**, 151.
Kutzelnigg, W., and Smith, V. H., Jr. (1964). *J. Chem. Phys.* **41**, 896.
Langhoff, S. R., and Tawil, R. A. (1975). *J. Chem. Phys.* **63**, 2745.
Larsson, S., and Smith, V. H., Jr. (1969). *Phys. Rev.* **178**, 137.
Linderberg, J., and Bystrand, F. W. (1964). *Ark. Fys.* **26**, 383.
Liu, J. W., and Smith, V. H., Jr. (1977). *Mol. Phys.*, in press.
Löwdin, P.-O. (1953). *Phys. Rev.* **94**, 1600.
Löwdin, P.-O. (1955). *Phys. Rev.* **97**, 1474.
Löwdin, P.-O. (1956). *Adv. Phys.* **5**, 1.
Löwdin, P.-O. (1967). *Adv. Quantum Chem.* **3**, 323.
Löwdin, P.-O., and Appel, K. (1956). *Phys. Rev.* **103**, 1746.
McCarthy, I. E., and Weigold, E. (1976). *Phys. Rep.*, **27c**, 275.
McWeeny, R. (1950). *Nature (London)* **166**, 21.
McWeeny, R., and Coulson, C. A. (1949). *Proc. Phys. Soc. London* **62A**, 509.
Messiah, A. (1967). "Quantum Mechanics," Vol. I. North-Holland Publ., Amsterdam.
Mijnarends, P. E. (1967). *Phys. Rev.* **160**, 512.
Mijnarends, P. E. (1977). *In* "Compton Scattering: The Investigation of Electron Momentum Distributions" (B. Williams, ed.), p. 323. McGraw-Hill, New York.
Mulliken, R. S. (1955). *J. Chem. Phys.* **23**, 1833.
Podolsky, B., and Pauling, L. (1929). *Phys. Rev.* **34**, 109.
Preuss, H. (1956). *Z. Naturforsch., Teil A* **11**, 823.
Ralston, A. (1965). "A First Course in Numerical Analysis," p. 98. McGraw-Hill, New York.
Reed, W. A. (1976). *Acta Crystallogr., Sect. A* **32**, 676.
Richards, W. G., Walker, T. E. H., and Hinkley, R. K. (1971). "A Bibliography of *ab initio* Molecular Wave Functions." Oxford Univ. Press (Clarendon), London and New York.
Richards, W. G., Walker, T. E. H., Farnell, L., and Scott, P. R. (1974). "Bibliography of *ab initio* Molecular Wave Functions: Supplement for 1970–1973." Oxford Univ. Press (Clarendon), London and New York.
Roothaan, C. C. J. (1951). *J. Chem. Phys.* **19**, 1445.
Rotenberg, M., Bivins, R., Metropolis, N., and Wooten, J. K. (1959). "The 3-j and 6-j Symbols." Crosby Lockwood, London.
Schrödinger, E. (1926). *Ann. Phys. (Leipzig)* [8] **79**, 361.
Seth, A., and Ellis, D. E. (1976). *Phys. Rev. A* **13**, 1083.
Shull, H., and Löwdin, P.-O. (1955). *J. Chem. Phys.* **23**, 1362.
Slater, J. C. (1932). *Phys. Rev.* **42**, 33.
Smith, V. H., Jr. (1971). *Chem. Phys. Lett.* **9**, 365.
Thakkar, A. J., and Smith, V. H., Jr. (1975). *Comput. Phys. Commun.* **10**, 73.
Thakkar, A. J., and Smith, V. H., Jr. (1976). To be published.
Thulstrup, P. W. (1975). *Int. J. Quantum Chem.* **9**, 789.
Thulstrup, P. W. (1976). *J. Chem. Phys.* **65**, 3386.
Tsapline, B. (1971). *Chem. Phys. Lett.* **11**, 75.

Watson, G. N. (1944). "A Treatise on the Theory of Bessel Functions," 2nd ed. Cambridge Univ. Press, London and New York.

Whitten, J. L. (1963). *J. Chem. Phys.* **39**, 349.

Williams, B., ed. (1977). "Compton Scattering: The Investigation of Electron Momentum Distributions." McGraw-Hill, New York.

Analysis of Electron–Atom Collisions

HANS KLEINPOPPEN*

Institute of Atomic Physics
University of Stirling
Stirling, Scotland

I. Introduction . . . 77
II. Polarization of Impact Radiation . . . 78
III. Spin Effects in Electron–Atom Collisions . . . 92
A. Scattering and Excitation Amplitudes: One-Electron Atom . . . 92
B. Scattering and Excitation Amplitudes: Two-Electron Atom . . . 101
IV. Electron–Photon Coincidences and Angular Correlations . . . 114
A. Theory . . . 114
B. Experimental Results . . . 118
V. Coherent Electron-Impact Excitation of Atomic Hydrogen . . . 131
VI. Electron Scattering from Laser-Excited Atoms . . . 132
A. Theory . . . 133
B. Experimental Scheme and Results . . . 136
References . . . 139

I. Introduction

Recent developments in the experimental techniques for investigating electron–atom collision processes have enabled physicists to study in detail characteristics such as resonances, spin effects, electron–photon angular correlations, and coherence effects. The analysis of experimental data from such studies has brought us to a much deeper understanding of the details of the electron–atom collision process. By "details" of the electron atom collision process is meant, for example, an analysis of the different types of interactions occurring during the collision, namely the Coulomb or the direct interaction, the exchange and the spin-orbit interaction as investigated in the electron and atomic spin analysis experiments (recent reviews in this field were given by Bederson, 1973; Farago, 1971; Kleinpoppen, 1973). They may also entail an analysis of coherence effects or the atomic alignment and orientation parameters from electron–photon coincidence measurements.

It appears that the new types of analyzing experiments require a complex theoretical and experimental analysis of the observables. The situation may

* This paper was in part written at the Center for Theoretical Studies, University of Miami, Coral Gables, Florida. The author acknowledges the hospitality and the financial support of the Center for Theoretical Studies.

be characterized by the fact that some recently published papers deal with concepts that can be called theories of the measurement of electron–atom collisions (Fano and Macek, 1973; Macek and Jaecks, 1971; Burke and Mitchell, 1974b; Farago, 1974; Kleinpoppen, 1971; Blum and Kleinpoppen, 1975a; Hertel and Stoll, 1974a). In these theories, links have been established between observables, on the one hand, and collision parameters (e.g., scattering amplitudes and their phase differences) and target parameters (e.g., alignment and orientation parameters or state multipoles of excited atoms), on the other hand.

Of course, the relevant theoretical approximations for electron–atom collision processes provide predictions of these amplitudes, their phase differences, and of atomic alignment and orientation parameters. The normal differential cross section, for example, is an averaging quantity consisting of a term such as

$$\sigma = \sum_{i,\,j} |a_i|^2 + |a_i - a_j|^2 \tag{1}$$

with the excitation or scattering amplitudes a_i, a_j describing a given physical process of the electron–atom collisions. Theoreticians normally carry out calculations of these amplitudes based on given theoretical approximation and then "hide" data on interesting features and structure of the amplitudes by summing up the preceding terms and thereby averaging over detailed information. It would certainly be advisable for future comparisons between experimental and theoretical data to have tables of amplitudes and their phase differences available.

In this paper we describe examples for the analysis of the following processes in electron atom collisions: (1) electron-impact polarization studies of line radiation, including high-resolution experiments (Section II); (2) electron and atomic spin effects (Section III); (3) electron–photon angular correlation from electron-impact excitation of atoms (Section IV); (4) coherent electron-impact excitation of atomic hydrogen (Section V); (5) electron scattering from laser-excited atoms (Section VI).

II. Polarization of Impact Radiation

The revived interest in polarization studies of impact radiation since about the beginning of the sixties originates from the following facts. As shown in Fig. 1, the threshold polarization of electron-impact line radiation is governed by the selection rule $\Delta m_l = 0$. Accordingly, threshold polarization can easily be calculated for many simple cases, for example, the threshold polarization of the excitation/deexcitation process ${}^1S \rightarrow {}^1P \rightarrow {}^1S$ should be 100%, because only the $m_l = 0$ sublevel of the P state is excited,

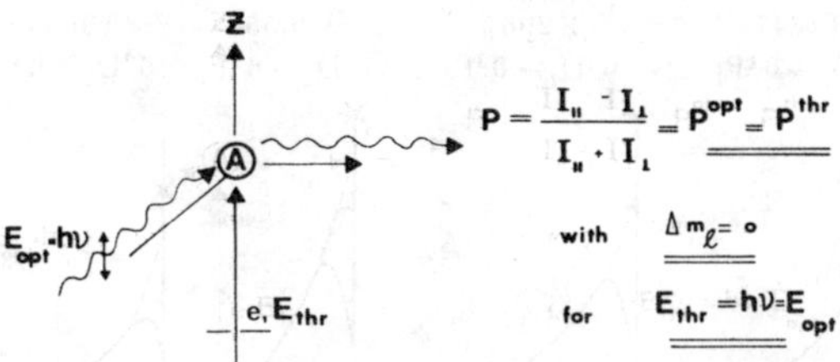

Fig. 1. Equivalence of threshold impact excitation by unidirectional electrons and resonance excitation by linearly polarized light (z axis parallel to electron beam direction and the light vector) with respect to the selection rule $\Delta m_l = 0$. As a consequence the threshold polarization P_{thr} of the electron-impact excitation is equal to the polarization P_{opt} of the resonance fluorescence radiation. Table II provides experimental evidence for $P_{thr} = P_{opt}$ for the first resonance lines of ^{6}Li, ^{7}Li, and ^{23}Na.

and subsequently only the π transition to the S state occurs. There were several experiments in the twenties, particularly that of Skinner and Appleyard (1927), which were in complete disagreement with the threshold polarization as required from angular momentum conservation. Figure 2 illustrates the situation with regard to several mercury lines: the crosses in Fig. 2 represent the threshold polarization of the various mercury lines and the full curves give the experimental results of Skinner and Appleyard showing complete disagreement with what was to be expected theoretically for the polarization close to threshold. This startling discrepancy remained unexplained for several decades. The other important factor that stimulated renewed interest in polarization studies was the new theoretical approach to the problem mainly due to the papers by Percival and Seaton (1958) and by Baranger and Gerjuoy in 1958. The theory developed by Percival and Seaton overcame the deficiencies of the previous Oppenheimer (1927a,b, 1928) and Penney (1932) theory by allowing for radiation damping or what is equivalent to taking the finite level width of the excited fine and hyperfine structure states into account. The particular importance of the paper by Baranger and Gerjuoy (1958) is often referred to the introduction of the atomic compound model for atomic collision processes. However, the most important applications and consequences of this model were related to the angular distribution and polarization of light emitted following excitation through compound states (see the following).

Although there is a large amount of evidence that the existing theory on polarization of impact radiation describes the experimental data correctly, there are still problems left which might either call for extensions or modifications of certain aspects of the theory or, alternatively, for an improvement in the experimental conditions. It therefore appears reasonable to review selected examples that illustrate cases of agreement and disagreement between theory and experiment.

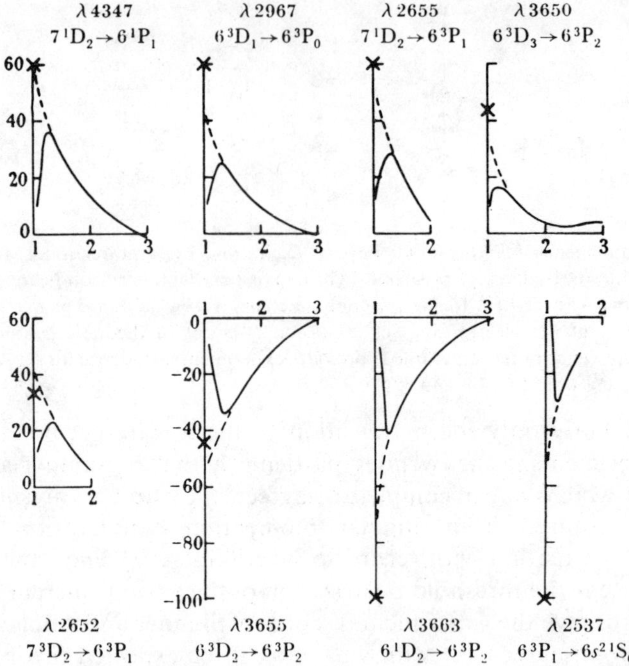

Fig. 2. Polarization of electron-impact radiation of Hg as functions of $(E/E_{\text{thr}})^{1/2}$ (E, electron impact energy; E_{thr}, threshold excitation energy). The full lines represent experimental data of Skinner and Appleyard (1927). Crosses give threshold polarization for mercury isotopes with zero nuclear spin and assuming *LS* coupling. Dashed lines represent extrapolations of full lines, neglecting the experimental drop of polarization near threshold. (From Percival and Seaton, 1958.)

The theoretical basis for impact line polarization lies within the framework of quantum mechanics. The theory is well formulated for cases where spin-orbit interactions between the projectile electron and the atom can be neglected.

The polarization of the line radiation depends on the anisotropic population of the magnetic sublevels excited during the collision, and also on the relative transition probabilities for π and σ transitions of these substates into the magnetic sublevels of the state into which the atom decays. In order to discuss recent results on line polarization, we concentrate on examining applications for atoms in which the ground state has no orbital angular momentum. The intensity components polarized parallel or perpendicular to the direction of the incoming beam (z direction; see Fig. 1) is given by

$$I_{\parallel} \propto \sum_m \frac{A_m^{\pi}}{A} Q_m, \qquad I_{\perp} \propto \sum_m \frac{1}{2} \frac{A_m^{\sigma}}{A} Q_m. \tag{2}$$

Here A is the total transition probability, A_m^π and A_m^σ refer to magnetic substates for π and σ transitions, and Q_m is the excitation cross section for magnetic substate with magnetic quantum m.

It then follows that for the polarization of the line radiation, with $A = A_m^\pi + A_m^\sigma$ (the total transition probability does not depend on m)

$$P = \frac{I_{\parallel} - I_{\perp}}{I_{\parallel} + I_{\perp}} = \frac{3K^\pi - K}{K^\pi + K}, \tag{3}$$

$$K^\pi = \frac{1}{A} \sum_m A_m^\pi Q_m \quad \text{and} \quad K = \sum_m Q_m = Q. \tag{4}$$

Note that in the preceding equations we did not specify the magnetic quantum number that could be taken for m_l, m_j, or m_F. Quantities K^π and K are proportional to the production rates of the total linear polarized and total unpolarized line intensity of the emission, respectively.

The recent emphasis on line polarization measurements according to the scheme as in Fig. 1 is mostly connected with some simple examples and arguments that can best be illustrated as follows: Let us consider the excitation/deexcitation process $^1S \to {}^1P \to {}^1S$ with the total excitation cross sections,[1]

$$Q(S \to P) = Q_{m_l=0} + Q_{m_l=+1} + Q_{m_l=-1} = Q_0 + 2Q_1.$$

Only the Q_0 component results in the excitation of π light, therefore, with $K^\pi \propto Q_0$ and $K \propto Q$, the polarization of the foregoing excitation process is given by

$$P = \frac{Q_0 - Q_1}{Q_0 + Q_1} = \frac{(Q_0/Q_1) - 1}{(Q_0/Q_1) + 1}. \tag{5}$$

At threshold the excitation is governed by the selection rule $\Delta m_l = 0$ or $Q_1 = 0$ and $Q = Q_0$ and $P = 100\%$. A beautiful example proving that threshold polarization of the foregoing process should be 100% is revealed in the study of the polarization of the first resonance line of calcium, investigated as a function of the electron energy by Ehlers and Gallagher (1973) (Fig. 3). The high-energy limit (E_∞) of the polarization should approach $P_\infty = -100\%$ because $Q_0(E_\infty)/Q_1(E_\infty) \to 0$. Including fine and hyperfine structure interactions in the $S \to P \to S$ transitions, the theory must be

[1] Parity invariance requires that the $m_l = \pm 1$ substates are excited with equal probability; note $Q_{m_l=\pm 1} = Q_1$ and $Q_{m_l=0} = Q_0$.

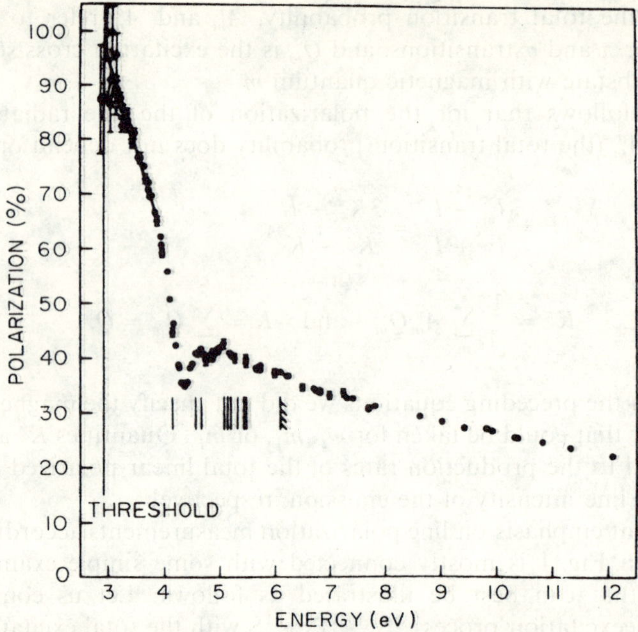

Fig. 3. Polarization of the 4^1P–4^1S, $\lambda = 4227$ Å radiations of calcium for low electron-bombardment energies ($1\,\sigma$ statistical error bars arising from counting statistics). (From Ehlers and Gallagher, 1973.)

modified by taking into account the finite level width and also the jI recoupling of the excited states after the excitation. For the alkali resonance lines the polarization is given by the formula (Flower and Seaton, 1967)

$$P = \frac{3(9\alpha - 2)(Q_0 - Q_1)}{12Q_0 + 24Q_1 + (9\alpha - 2)(Q_0 - Q_1)} \tag{6}$$

with

$$\alpha = \sum_{F,\,F'} \frac{\zeta(I, F, F')}{1 + \varepsilon_{F,\,F'}^2}, \quad \varepsilon_{F,\,F'} = \frac{2\pi\,\Delta\nu_{F,\,F'}}{A}, \tag{7}$$

and $h\,\Delta\nu_{F,\,F'}$ the energy separation of the hyperfine structure states and $\zeta(I, F, F')$ expressed in terms of Racah and vector coupling coefficients. The value of α depends largely on the ratios of the hyperfine structure separations to the natural line width, which can be calculated from the theory of Flower and Seaton (1967) for cases of interest and which are summarized in Table I.

TABLE I

$2^2P_{1/2,\,3/2}$ of atomic hydrogen, with hfs[a]	$\alpha = 0.441$
without hfs	$\alpha = 4/9 = 0.445$
$2^2P_{1/2,\,3/2}$ of ^{6}Li, with hfs	$\alpha = 0.413$
$2^2P_{1/2,\,3/2}$ of ^{7}Li, with hfs	$\alpha = 0.326$
$3^2P_{1/2,\,3/2}$ of ^{23}Na, with hfs	$\alpha = 0.288$

[a] Hyperfine structure.

Table II gives results for theoretical and experimental threshold and optical line radiation of the first resonance line of ^{6}Li, ^{7}Li, and ^{23}Na, and also for the Lyman-α impact polarization of atomic hydrogen. Notice the excellent agreement between theory and experiment of the alkali resonance lines. This indicates that the relation $P_{\text{opt}} = P_{\text{thr}}$ is valid for the alkali resonance lines, at least for light alkalis. Very good agreement also exists for the

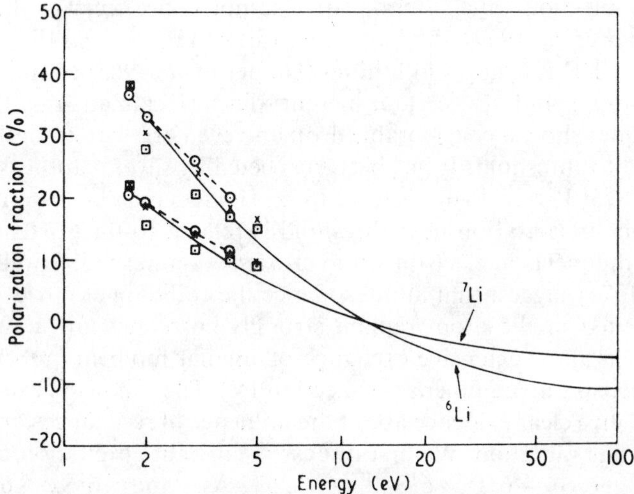

Fig. 4. Polarization of first resonance lines of ^{6}Li and ^{7}Li. (——) Glauber approximation (Tripathi *et al.*, 1973); (×) close-coupling calculations incorporating the dipole polarizability of atomic states (Feautrier, 1970); (⊡) close-coupling calculations (Burke and Taylor, 1969); (⊙---⊙) experimental data (Hafner *et al.*, 1965; Hafner and Kleinpoppen, 1967).

experimental and theoretical polarization data of the first resonance lines of ^{6}Li and ^{7}Li in an energy range just above threshold, as seen in Figs. 4 and 5, whereas in the case of the sodium D line, the spread of experimental data precludes a clear comparison of them with theory.

TABLE II

THRESHOLD AND RESONANCE FLUORESCENCE POLARIZATION (P^{opt}) OF THE FIRST RESONANCE LINE OF ^{6}Li, ^{7}Li, ^{23}Na, AND H[a]

Polarization	^{23}Na	^{6}Li	^{7}Li	H
$P^{thr}_{cal} =$	14.1%	37.5%	21.6%	42.1%
$P^{thr}_{exp} =$	(14.8 ± 1.8)%	(39.7 ± 3.8)%	(20.6 ± 3.0)%	13%(?)[b]
$P^{opt}_{cal} =$	(14.2 ± 0.2)%	(37.7 ± 1.0)%	(21.7 ± 1.5)%	42.1%
$P^{opt}_{exp} =$	(14.0 ± 0.8)%	(37.5 ± 1.7)%	(23.1 ± 1.3)%	—

[a] The P_{cal}, P_{exp}, theoretical, from Flower and Seaton (1967), Hafner *et al.* (1965), Hafner and Kleinpoppen (1967); experimental data from Hafner *et al.* (1965) and Hafner and Kleinpoppen (1967).

[b] From Ott *et al.* (1963).

Contrary to the preceding examples, there are quite a number of atomic lines that appear not to satisfy the threshold polarization required by the $\Delta m_l = 0$ selection rule. Important examples are certain helium lines (Heideman *et al.*, 1969), the Lyman-α radiation (Ott *et al.*, 1963) (Table II), the $3^2D \rightarrow 2^2P$ transition in lithium (Hafner *et al.*, 1965; Hafner and Kleinpoppen, 1967), and also certain mercury lines (Heidman *et al.*, 1969). Most of these lines show a considerable drop and even oscillations in the polarization close to threshold. It has been suspected by several authors (Fano and Macek, 1973; Percival and Seaton, 1958; Heideman *et al.*, 1969) that such a drop in the polarization near threshold is related to the existence of resonances in the inelastic electron–atom cross section near threshold. Fano and Macek (1973) suggest that at a resonance the colliding electron and the one being excited in the atom remain strongly correlated for a time interval sufficient to allow extensive exchange of angular momentum between them with a decrease in alignment (see Section IV). The following examples to be discussed give clear evidence about the influence of resonances on the polarization of line radiation. We first discuss the mercury excitation/deexcitation processes $6^1S_0 \rightarrow 6^3P \rightarrow 6^1S_0$ ($\lambda = 2537$ Å) and $6^1S_0 \rightarrow 6^1P_1 \rightarrow 6^1S_0$ ($\lambda = 1849$ Å). For threshold excitation of the 6^1P state, only the $M_j = 0$ is excited because of the selection rule $\Delta m_l = 0$; for the excitation of the 6^3P_1 state, only $M_j = \pm 1$ states can be excited at threshold because the Clebsch–Gordan coefficient is zero.

$$\mathbf{C}^{S=1,\,L=1,\,J=L}_{M_S=0,\,M_L=0,\,M_J=0} = 0, \tag{8}$$

but all the other angular momentum coupling coefficients are not equal to zero. This has the consequence that the threshold polarization of the aforementioned mercury lines should provide the largest possible difference ($\pm 100\%$) for the line polarization P_{thr} ($\lambda = 2537$ Å) $= -100\%$ and P_{thr}

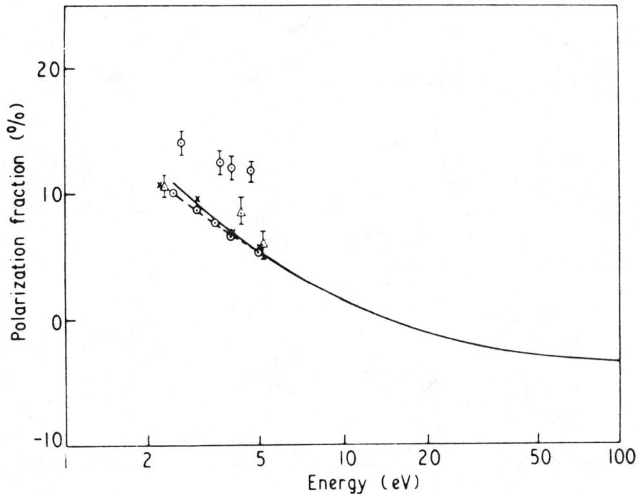

Fig. 5. Polarization of the first resonance of ^{23}Na. (——) Glauber approximation (Tripathi *et al.*, 1973); (×) four-state exchange close-coupling approximation (Moores and Norcross, 1972). Experimental data: (⊙---⊙) Enemark and Gallagher, (1972); (△) Gould (1970); (⊙) Hafner, *et al.* (1965).

($\lambda = 1849$ Å) $= +100\%$. Allowance for departures from *LS* coupling was made for the threshold polarization of the 2537 Å line by Penney (1932). He obtains $P_{\text{thr}} = -92\%$ for zero nuclear spin, -53% for nuclear spin $I = \frac{1}{2}$, and -48% for $I = \frac{3}{2}$; with the normal isotope mixture Penney's calculation for threshold polarization of the 2537-Å line gives $P_{\text{thr}} = -80\%$.

Figures 6 and 7 shows the latest results for the experimental studies with high resolution of the mercury 2537-Å line (Ottley, 1974; Ottley and Kleinpoppen, 1975). The structure of the polarization curve above threshold coincides with resonance structure of an electron-mercury transmission experiment carried out by P. D. Burrow and J. A. Michejda (private communications; similar resonance structure was also observed in the angular dependence in low-energy electron-mercury scattering by Düweke *et al.*, 1973). Baranger and Gerjuoy (1958) first developed a theory in which they associated the failure of experimental data to approach the required threshold polarization with the formation of atomic compound states. They assumed that two negative ion compound states are formed just above the excitation threshold of the 2537-Å line and predicted the polarization of the 2537-Å line in the resonance states and polarization of 60% and 0% in the center of resonances with total angular momenta $j = \frac{3}{2}$ and $j = \frac{1}{2}$, respectively.

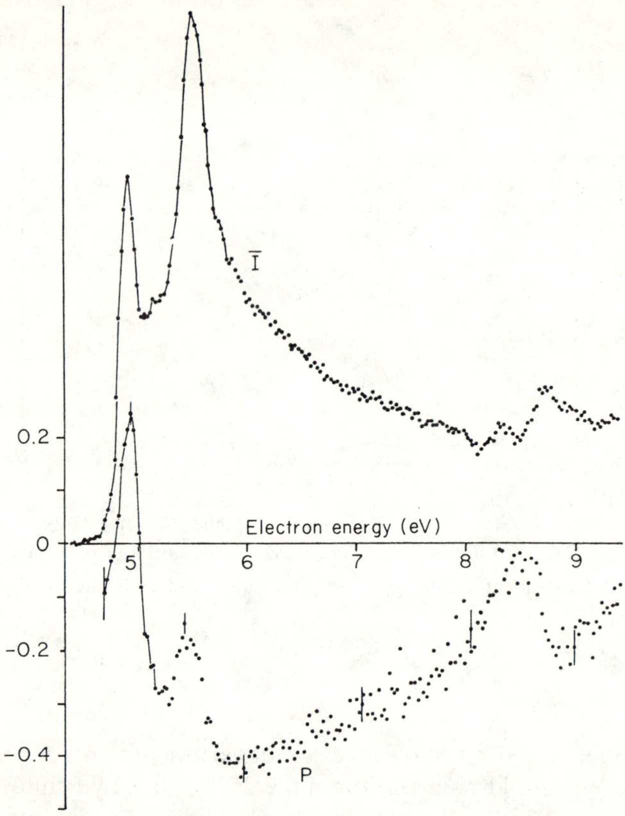

Fig. 6. Excitation function (upper part) and polarization curve (lower part) of the mercury intercombination line $6^3P_1 \rightarrow 6^1S_0$, $\lambda = 2537$ Å versus electron energy (Ottley and Kleinpoppen, 1975). Energy resolution of electron monochromator $\Delta E = 140$ meV. The full line represents the theoretical prediction by McConnell and Moiseiwitsch (1968). Error bars indicate 90% confidence limits plus a small uncertainty in the systematic correction.

The calculations of Baranger and Gerjuoy (1958) assumed zero orbital angular momentum of the outgoing electrons. The experimental data of the polarization curve do not confirm these theoretical predictions at the moment; one must, however, bear in mind that the natural isotope composition and the remaining finite energy resolution of the electron beam still affect the polarization. As seen by comparing Fig. 6 and Fig. 7, the polarization peak at the first resonance close to the threshold rises from about 25% where the energy spread of the electron beam is about 140 meV to almost 40% where the energy spread of the electron is about 100 meV. A rough estimate shows that the predicted polarization of 60% in the $j = \frac{3}{2}$ resonance should be reduced to approximately 45%, taking into account the hyperfine

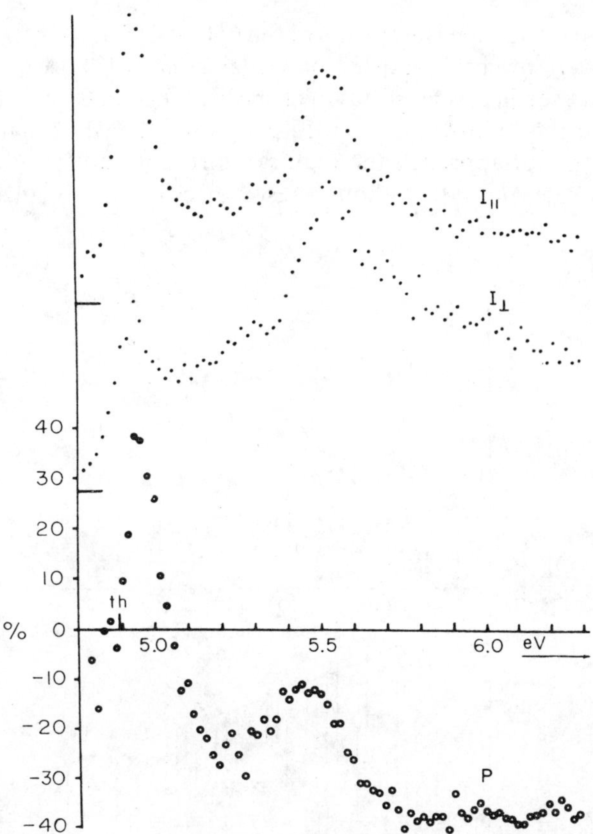

Fig. 7. Measured intensity components polarized parallel and perpendicular (upper part) to the exciting electron beam for the mercury intercombination line $\lambda = 2537$ Å. The lower part of the figure displays the polarization of the intercombination line. Energy resolution of electron monochromator $\Delta E = 100$ meV. The error of the measured polarization at the peak (4.92 eV) is $\pm 7\%$ with 90% confidence limits plus a small uncertainty in the systematic correction. (From Ottley and Kleinpoppen, 1975.)

structure splitting for the normal isotope mixture of mercury. Fano and Cooper (1965) have suggested the identification $^4P_{1/2,\,3/2,\,5/2}$ for the three lowest resonances associated with the compound state configuration 6S $6P^2$. Heddle (1975) offered a new interpretation for the classification of electron-mercury resonances. According to his suggestions the peaks or features in the polarization and excitation function of the λ 2437-Å line are due to Hg compound states with the configurations $^2D_{3/2}$ (at 4.92 eV), $^4P_{5/2}$ (at 5.23 eV, a weakly pronounced feature in the intensity components and the polarization), and $^2D_{5/2}$ (at 5.50 eV).

First polarization measurements of the $6^1P \rightarrow 6^1S$, $\lambda = 1850$ Å transition of mercury were recently reported by Ottley *et al.* (1974). A quartz Rochon polarizer was used in this far ultraviolet spectral region. Clearly the polarization curve for the 1850-Å line, as in Fig. 8, shows that the polarization near threshold does not approach the required threshold polarization based on the selection rule $\Delta m_l = 0$. Although resonances obviously affected the po-

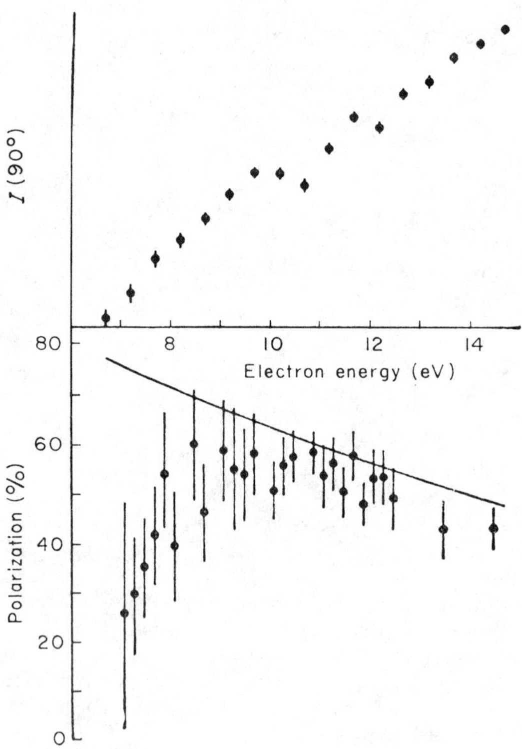

Fig. 8. Excitation function $I(90°)$ (upper part) and polarization curve (lower part) of the mercury $6^1P_1 \rightarrow 6^1S_0$, $\lambda = 1850$-Å line as measured by Ottley *et al.* (1974). The full line represents the theoretical prediction by McConnell and Moiseiwitsch (1968). Error bars indicate 90% confidence limits plus a systematic error not exceeding 4%.

larization of the 2537 Å, it is not so clear if this is the case for the 1850-Å line according to Fig. 8. Future studies with higher energy resolution should enable a more clear-cut conclusion.

High-resolution studies of polarization curves near threshold for line transitions from the 4D states of helium were reported by Heddle *et al.*

(1974). The polarization of the line transition from the D states of helium can be expressed as follows:

$$P(4^1\mathrm{D} \rightarrow 2^1\mathrm{P}) = \frac{3Q_0 + 3Q_1 - 6Q_2}{5Q_0 + 9Q_1 + 6Q_2}$$

$$P_{\mathrm{thr}}(4^1\mathrm{D} \rightarrow 2^1\mathrm{P}) = \frac{3}{5} = 60\% \tag{9}$$

$$P(4^3\mathrm{D} \rightarrow 2^3\mathrm{P}) = \frac{213Q_0 + 213Q_1 - 426Q_2}{671Q_0 + 1271Q_1 + 1058Q_2}$$

$$P_{\mathrm{thr}}(4^3\mathrm{D} \rightarrow 2^3\mathrm{P}) = \frac{213}{671} = 32\%.$$

Figures 9 and 10 show excitation and polarization function lines associated with the preceding transition. Although the total errors of the data in these figures is approximately 30% close to threshold, the measurements agree well with threshold predictions. The aforementioned authors suggest that the sharpness of the fall in the polarization immediately above threshold is associated with a $1s4d^2\ ^2\mathrm{S}$ resonance, which strongly decays into $m_l = |2|$ substates of the D state. Previous studies reveal structure in the

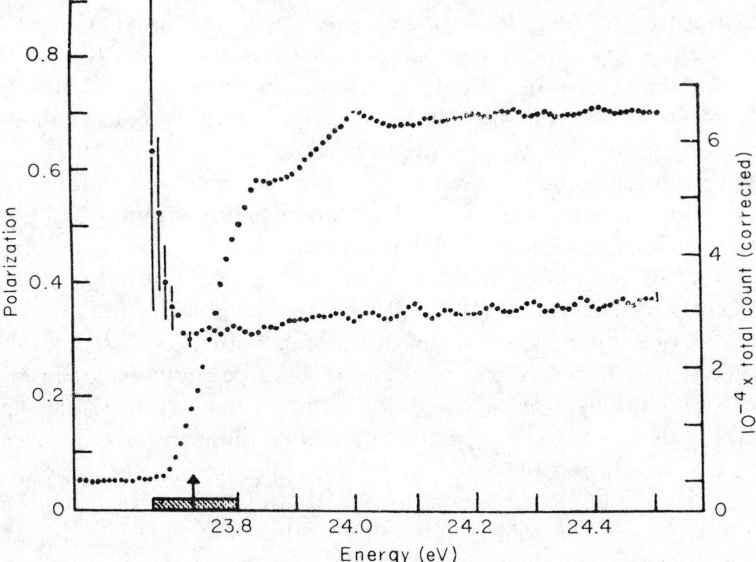

Fig. 9. The excitation function and polarization curve of the $\lambda = 4922$-Å, $4^1\mathrm{D} \rightarrow 2^1\mathrm{P}$ line of helium as measured by Heddle *et al.* (1974) after 120 hours.

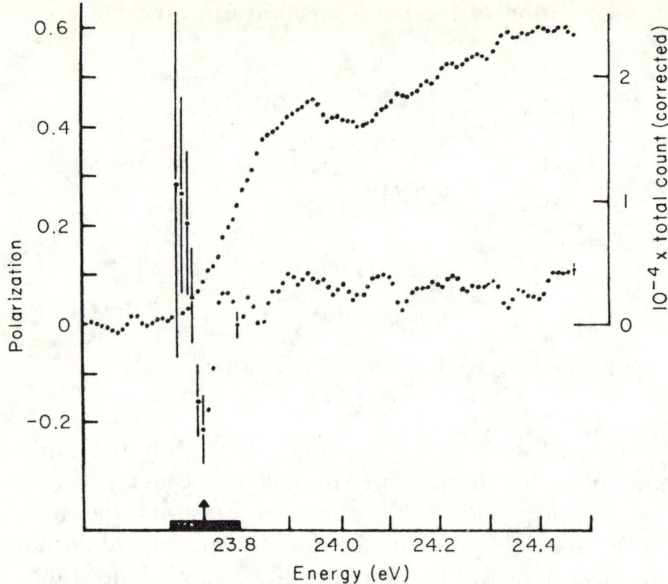

Fig. 10. The excitation function and polarization curve of the $\lambda = 4472$-Å, $4^3D \rightarrow 2^3P$ line of helium as measured by Heddle *et al.* (1974) after 170 hours.

excitation functions of the line transitions [Eqs. (9)] that coincide with the energy position of the polarization curves. The lowest possible polarization values occur if Q_2 is large compared with Q_0 and Q_1, which is -100 and -40% for the singlet and triplet transitions, respectively. The initial fall to -22% in the triplet transition is followed by the return of a positive value. The true minimum polarization might be approached only with even higher energy resolution as demonstrated, for example, in the case of the mercury excitation as described previously. The double feature around 23.8 eV in the polarization of the helium triplet transitions can also be associated with structure in the excitation function.

Recently Mumma *et al.* (1974) determined the polarization of $n^1P \rightarrow 1^1S$ transitions of helium by measuring the angular intensity distribution of the photons. It was, however, not possible to separate the different lines of the $nP \rightarrow 1S$ transitions from each other in the vacuum ultraviolet spectral region. Figure 11 shows the results obtained by these authors. It is clear that the lack of separation of the vacuum ultraviolet lines in the experiment of Mumma *et al.* makes a comparison with theoretical predictions rather difficult. However, the polarization of the collective $\sum_n n^1P \rightarrow 1^1S$ transition might become independent of n at higher energies as indicated in comparing the experimental data with the theoretical calculation of Vriens and Carriere (1970).

Similar oscillatory structure as seen in the polarization curves of helium

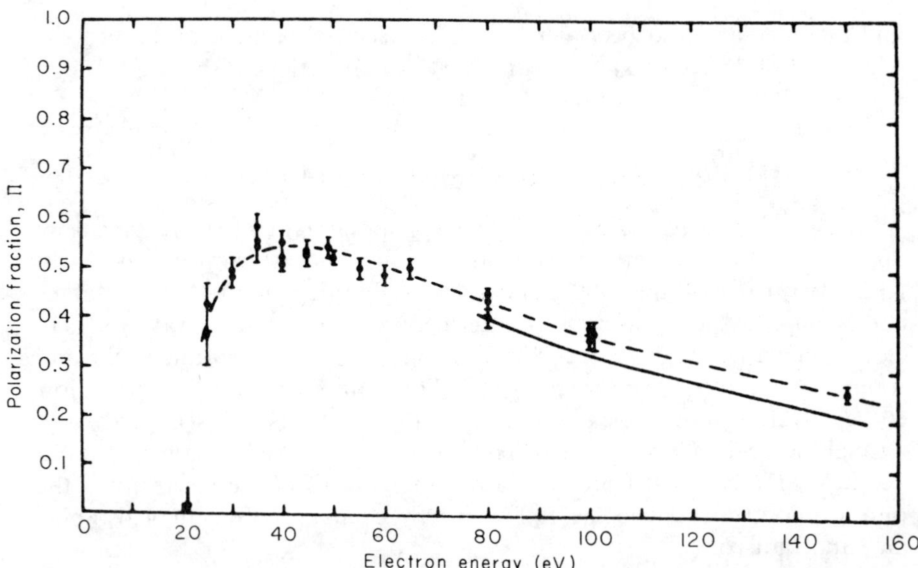

Fig. 11. Polarization curve of the collective $\sum_n n^1P \rightarrow 1^1S$ transition of helium as investigated by Mumma *et al.* (1974) and compared with the calculation of Vriens and Carriere (1970).

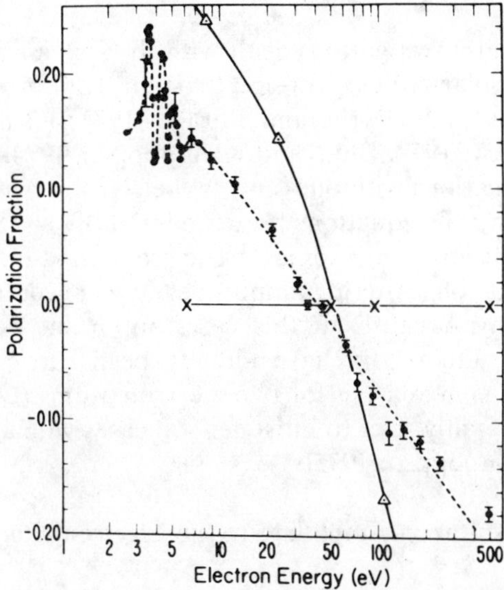

Fig. 12. Polarization of Be^+ ion lines as observed by Crandall *et al.* (1974). (●) Polarization of 4554 Å radiation ($6^2P_{3/2} \rightarrow 6^2S_{1/2}$); (×) polarization of 4934 Å radiation ($6^2P_{1/2} \rightarrow 6^2S_{1/2}$); (△) polarization of the mixture of 4131 Å radiation ($6^2D_{5/2} \rightarrow 6^2P_{3/2}$) with 4166 Å radiation ($6^2D_{3/2} \rightarrow 6^2P_{3/2}$). Bars shown are 1 rms error added to a small systematic error.

and mercury has also been detected in an electron–ion excitation process. Figure 12 shows a striking feature in the polarization of the 4554-Å Ba^+ $2^2P_{3/2} \rightarrow 2^2S_{1/2}$ resonance line (Crandall *et al.*, 1974).

III. Spin Effects in Electron–Atom Collisions

Early theoretical studies on electron spin polarization by scattering from unpolarized targets were carried out for high-energy scattering by Mott (1929, 1932) around the 1930's, and at the beginning of the 1940's by Massey and Mohr (1941) for low-energy scattering. The term "high energy" refers to electron energies high enough for electrons to pass the atomic shells and where the electrons experience a pure Coulomb field; and the term "low energy" refers to energies so low that the electrons are scattered by a strongly screened Coulomb field. The polarization at high energies of about 100 keV mainly result from spin-orbit interactions of the electrons in the pure Coulomb field, whereas, at low energy, exchange effects may also play an important role.

In this section we concentrate our discussion on investigations in which either the incoming electrons or the atomic target system or both of the collision partners are polarized. The extensive study of spin-orbit effects in the electron scattering with heavy atoms has been reviewed by Kessler (1969).

Several theoretical investigations dealt with the spin analysis of the scattering of (partially) polarized electrons on (partially) polarized atoms. These papers (Bederson, 1969a,b; Byrne and Farago, 1971; Kleinpoppen, 1971; Drukarev and Obedov, 1972; Blum and Kleinpoppen, 1974, 1975b) referred particularly to applications with one- and two-electron systems. It shall not be our task to rederive the equations presented in those papers in full detail; we would rather concentrate on some "basic reactions" that are described or quantified in terms of scattering amplitudes. We concentrate our discussions on alkali atoms because, in this case, important experiments that permit comparison with theory have already been carried out. We will review the theory of spin analysis for two-electron atoms (Blum and Kleinpoppen, 1974) but we only refer to most general cases with arbitrary atomic spin (Blum and Kleinpoppen, 1975b).

A. Scattering and Excitation Amplitudes; One-Electron Atom

1. *Elastic Scattering*

We begin by defining scattering amplitudes for the case in which the colliding electrons and atoms are completely polarized. We restrict ourselves to one-electron atoms and we exclude the following interactions: spin-orbit interactions [to be taken into account only for heavy atoms (Burke and Mitchell, 1974a)], scattering interactions producing longitudinal polarization (parity violation), and spin flips without electron exchange between the projectile and the atomic electron, for example, the collision process with polarized electrons e(↑) and polarized atoms A(↓), e(↑) + A(↓) → e(↓) + A(↑). The remaining types of scattering processes to be included are

$$\mathrm{e}(\downarrow) + \mathrm{A}(\uparrow) \rightarrow \mathrm{A}(\uparrow) + \mathrm{e}(\downarrow), \qquad \text{scattering amplitude } f, \tag{10}$$

$$\mathrm{e}(\downarrow) + \mathrm{A}(\uparrow) \rightarrow \mathrm{A}(\downarrow) + \mathrm{e}(\uparrow), \qquad \text{scattering amplitude } g, \tag{11}$$

$$\mathrm{e}(\uparrow) + \mathrm{A}(\uparrow) \rightarrow \mathrm{A}(\uparrow) + \mathrm{e}(\uparrow), \qquad \text{scattering amplitude } f - g. \tag{12}$$

The cross sections for these scattering processes are then equal to the square of the magnitude of the scattering amplitudes. We call f the direct, g the exchange, and $f - g$ the interference or triplet amplitude ($f - g = {}^3F$).

TABLE III

SPIN ANALYSIS FOR THE SCATTERING OF ELECTRONS ON ONE-ELECTRON ATOMS

Scattering of unpolarized electrons on partially polarized atoms	$\dfrac{\lvert f\rvert^2}{\sigma} = 1 - \dfrac{P'_\mathrm{e}}{P_\mathrm{A}}, \quad \dfrac{\lvert g\rvert^2}{\sigma} = 1 - \dfrac{P'_\mathrm{A}}{P_\mathrm{A}}$
Scattering of partially polarized electrons on unpolarized atoms	$\dfrac{\lvert g\rvert^2}{\sigma} = 1 - \dfrac{P'_\mathrm{e}}{P_\mathrm{e}}, \quad \dfrac{\lvert f\rvert^2}{\sigma} = 1 - \dfrac{P'_\mathrm{A}}{P_\mathrm{e}}$
Scattering of partially polarized electrons on partially polarized atoms ($S = \sigma_\mathrm{e}^\uparrow + \sigma_\mathrm{e}^\downarrow = \sigma_\mathrm{A}^\uparrow + \sigma_\mathrm{A}^\downarrow$, partial cross sections specified by the spin directions of the scattered electrons and atoms)	$\lvert f\rvert^2 = \dfrac{1}{P_\mathrm{A} - P_\mathrm{e}}\left[\dfrac{S}{P_\mathrm{A}} - M_\mathrm{e} - \sigma\left(\dfrac{1}{P_\mathrm{A}} - P_\mathrm{A}\right)\right],$ $\dfrac{\lvert f\rvert^2}{S} = \dfrac{1}{P_\mathrm{A} - P_\mathrm{e}}\left[\dfrac{1}{P_\mathrm{A}} - P'_\mathrm{e} - \dfrac{\sigma}{S}\left(\dfrac{1}{P_\mathrm{A}} - P_\mathrm{A}\right)\right],$ $\lvert g\rvert^2 = \dfrac{1}{P_\mathrm{e} - P_\mathrm{A}}\left[\dfrac{S}{P_\mathrm{e}} - M_\mathrm{e} - \sigma\left(\dfrac{1}{P_\mathrm{e}} - P_\mathrm{e}\right)\right],$ $\dfrac{\lvert g\rvert^2}{S} = \dfrac{1}{P_\mathrm{e} - P_\mathrm{A}}\left[\dfrac{1}{P_\mathrm{e}} - P'_\mathrm{e} - \dfrac{\sigma}{S}\left(\dfrac{1}{P_\mathrm{e}} - P_\mathrm{e}\right)\right].$ $\lvert f\rvert^2 = \dfrac{1}{P_\mathrm{e} - P_\mathrm{A}}\left[\dfrac{S}{P_\mathrm{e}} - M_\mathrm{A} - \sigma\left(\dfrac{1}{P_\mathrm{e}} - P_\mathrm{e}\right)\right],$ $\dfrac{\lvert f\rvert^2}{S} = \dfrac{1}{P_\mathrm{e} - P_\mathrm{A}}\left[\dfrac{1}{P_\mathrm{e}} - P'_\mathrm{A} - \dfrac{\sigma}{S}\left(\dfrac{1}{P_\mathrm{e}} - P_\mathrm{e}\right)\right],$ $\lvert g\rvert^2 = \dfrac{1}{P_\mathrm{A} - P_\mathrm{e}}\left[\dfrac{S}{P_\mathrm{A}} - M_\mathrm{A} - \sigma\left(\dfrac{1}{P_\mathrm{A}} - P_\mathrm{A}\right)\right],$ $\dfrac{\lvert g\rvert^2}{S} = \dfrac{1}{P_\mathrm{A} - P_\mathrm{e}}\left[\dfrac{1}{P_\mathrm{A}} - P'_\mathrm{A} - \sigma\left(\dfrac{1}{P_\mathrm{A}} - P_\mathrm{A}\right)\right].$ Phase difference δ between f and g: $\cos\delta = (\sigma - S)/P_\mathrm{e}P_\mathrm{A}\lvert f\rvert\,\lvert g\rvert.$

The relation of these amplitudes to the differential cross section $\sigma(\theta, \phi)$ for the scattering with unpolarized colliding particles $\{e(\uparrow\downarrow) + A(\uparrow\downarrow)\}$ is given by

$$\begin{aligned}\sigma(\theta, \phi) &= \tfrac{1}{2}|f|^2 + \tfrac{1}{2}|g|^2 + \tfrac{1}{2}|f-g|^2 \\ &= \tfrac{1}{4}|f+g|^2 + \tfrac{3}{4}|f-g|^2 = \tfrac{1}{4}|{}^1F|^2 + \tfrac{3}{4}|{}^3F|^2,\end{aligned} \tag{13}$$

where ${}^1F$ is the singlet scattering amplitude.

For simplicity we may write Eq. (13) as

$$\sigma(\theta, \phi) = \sigma_d(\theta, \phi) + \sigma_{ex}(\theta, \sigma) + \sigma_{int}(\theta, \phi) \tag{14}$$

and call $\sigma_d(\theta, \phi) = \frac{1}{2}|f|^2 = \sigma_d$ the direct cross section, $\sigma_{ex}(\theta, \phi) = \frac{1}{2}|g|^2 = \sigma_{ex}$ the exchange cross section, and $\sigma_{int}(\theta, \phi) = \frac{1}{2}|f-g|^2 = \sigma_{int}$ the interference cross section. It follows from Eqs. (10)–(12) and from a more detailed discussion in Section III,A,3 that the polarized electron–atom techniques provide a tool for determining these partial cross sections. Table III summarizes the results of spin analysis for different examples of scattering partially polarized and unpolarized collision partners and spin polarization of the electrons and atoms before and after the scattering.

2. *Inelastic Scattering*

The details of the theory for the analysis of scattering amplitudes related to inelastic scattering processes are described extensively in the literature (Moiseiwitsch and Smith, 1968; Burke, 1969; Massey, 1969; Mott and Massey, 1965; Bates, 1968; Geltman, 1969).

From a practical point of view one is primarily interested in the excitation and deexcitation process $S \rightarrow P \rightarrow S$ (excitation to a P state and subsequent transition from the excited P state to an S state). We, therefore, restrict our discussion in detail to the particular excitation process.

It follows from Percival and Seaton (1958) that the total cross section $Q(S \rightarrow P)$ of the excitation $S \rightarrow P$ can be expressed in terms of cross sections for the components of the initial angular momenta of the excited state:

$$Q(S \rightarrow P) = Q(S \rightarrow P)_{m_l=0} + 2Q(S \rightarrow P)_{m_l=\pm 1}, \tag{15}$$

with

$$Q(S \rightarrow P)_{m_l=+1} = Q(S \rightarrow P)_{m_l=-1}.$$

As in the case of the elastic scattering, the excitation process can also be described in terms of scattering amplitudes. General expressions of such scattering amplitudes, including the inelastic process, are, for example, discussed in Burke's paper [1969, see in particular Burke's Eq. (I.12)]. It follows that the excitation process $S \rightarrow P$ of a one-electron atom can also be described in terms of direct (F) and exchange (G) amplitudes. The partial cross sections for the orbital angular momentum components are to be expressed by the amplitudes F_0 and G_0 for $Q(S \to P)_{m_l=0}$, and F_1 and G_1 for $Q(S \to P)_{m_l=\pm 1}$, respectively. In the same way as for elastic scattering, it can be shown that

$$
\begin{aligned}
Q(S \to P)_{m_l=0} &= \tfrac{1}{2}|F_0|^2 + \tfrac{1}{2}|G_0|^2 + \tfrac{1}{2}|F_0 - G_0|^2 \\
&= \tfrac{3}{4}|F_0 - G_0|^2 + \tfrac{1}{4}|F_0 + G_0|^2, \qquad (16)\\
Q(S \to P)_{m_l=\pm 1} &= \tfrac{1}{2}|F_1|^2 + \tfrac{1}{2}|G_1|^2 + \tfrac{1}{2}|F_1 - G_1|^2 \\
&= \tfrac{3}{4}|F_1 - G_1|^2 + \tfrac{1}{4}|F_1 + G_1|^2. \qquad (17)
\end{aligned}
$$

The total cross section is then

$$
Q(S \to P) = \tfrac{1}{2}|F_0|^2 + |F_1|^2 + \tfrac{1}{2}|G_0|^2 + |G_1|^2 + \tfrac{1}{2}|F_0 - G_0|^2 + |F_1 - G_1|^2. \qquad (18)
$$

Analogous to the elastic scattering (Section III,A,1), we call $\frac{1}{2}|F_0|^2 + |F_1|^2 = Q_d(S \to P)$ the direct excitation cross section, $\frac{1}{2}|G_0|^2 + |G_1| = Q_{ex}(S \to P)$ the exchange excitation cross section, and $\frac{1}{2}|F_0 - G_0|^2 + |F_1 - G_1|^2 = Q_{int}(S \to P)$ the interference excitation cross section. Relations (15)–(17) remain valid even if the atoms show fine- and hyperfine-structure splitting. As will be seen in Section IV, the analysis of the excitation process carried out by polarized electrons and atoms will be affected by the fine- and hyperfine-structure interaction. In our description of the excitation process $S \to P$ we will take into account the fine-structure separation (it can even be used to simplify the analysis of the excitation process), whereas the complication caused by the hyperfine-structure can be overcome by some special experimental arrangements (e.g., by decoupling the hyperfine structure by an external magnetic field or preferably by observing the inelastically scattered electrons).

We can also express the total excitation cross section in terms of the cross sections for the fine-structure states:

$$
Q(S \to P) = Q(^2S_{1/2} \to {}^2P_{1/2}) + Q(^2S_{1/2} \to {}^2P_{3/2}). \qquad (19)
$$

The fine-structure cross sections are proportional to their statistical weights (Mott and Massey, 1965; Bates, 1968; Geltman, 1969):

$$
\begin{aligned}
Q(^2S_{1/2} \to {}^2P_{1/2}) &= \tfrac{1}{3}Q(S \to P) \\
&= \tfrac{1}{3}Q(S \to P)_{m_l=0} + \tfrac{2}{3}Q(S \to P)_{m_l=\pm 1}, \\
Q(^2S_{1/2} \to {}^2P_{1/2}) &= \tfrac{1}{6}|F_0|^2 + \tfrac{1}{3}|F_1|^2 + \tfrac{1}{6}|G_0|^2 + \tfrac{1}{3}|G_1|^2 \\
&\quad + \tfrac{1}{6}|F_0 - G_0|^2 + \tfrac{1}{3}|F_1 - G_1|^2 \\
&= \tfrac{1}{3}(Q_d + Q_{ex} + Q_{int}), \qquad (20)
\end{aligned}
$$

$$
\begin{aligned}
Q(^2S_{1/2} \to {}^2P_{3/2}) &= \tfrac{2}{3}Q(S \to P) \\
&= \tfrac{2}{3}Q(S \to P)_{m_l=0} + \tfrac{4}{3}Q(S \to P)_{m_l=\pm 1}, \\
Q(^2S_{1/2} \to {}^2P_{3/2}) &= \tfrac{1}{3}|F_0|^2 + \tfrac{2}{3}|F_1|^2 + \tfrac{1}{3}|G_0|^2 + \tfrac{2}{3}|G_1|^2 \\
&\quad + \tfrac{1}{3}|F_0 - G_0|^2 + \tfrac{2}{3}|F_1 - G_1|^2 \\
&= \tfrac{2}{3}(Q_d + Q_{ex} + Q_{int}). \qquad (21)
\end{aligned}
$$

By describing the excitation process in terms of differential cross sections, we can write the following relations between the differential excitation amplitudes (f_0, f_1) and (g_0, g_1) and the differential cross sections (including the partial differential cross section σ_{m_l}):

$$
\begin{aligned}
\sigma(S \to P) &= \sigma(^2S_{1/2} \to {}^2P_{1/2}) + \sigma(^2S_{1/2} \to {}^2P_{3/2}) \\
&= \sigma(S \to P)_{m_l=0} + 2\sigma(S \to P)_{m_l=1} \\
&= \tfrac{1}{2}|f_0|^2 + |f_1|^2 \qquad (= \sigma_d) \\
&\quad + \tfrac{1}{2}|g_0|^2 + |g_1|^2 \qquad (= \sigma_{ex}) \\
&\quad + \tfrac{1}{2}|f_0 - g_0|^2 + |f_1 - g_1|^2 \qquad (= \sigma_{int}), \qquad (22)
\end{aligned}
$$

$$\sigma(^2S_{1/2} \to {}^2P_{3/2}) = \tfrac{2}{3}(\sigma_d + \sigma_{ex} + \sigma_{int}), \qquad (23)$$

$$\sigma(^2S_{1/2} \to {}^2P_{1/2}) = \tfrac{1}{3}(\sigma_d + \sigma_{ex} + \sigma_{int}). \qquad (24)$$

Transformation of Eqs. (22)–(24) to those of Eqs. (20) and (21) is to be carried out by means of the relations between the integral and the differential amplitudes ($i = 0$ or 1, θ is the scattering angle of the inelastically scattered electron):

$$|F_i|^2 = 2\pi \int_0^\pi |f_i(\theta)|^2 \sin\theta \, d\theta, \qquad |G_i|^2 = 2\pi \int_0^\pi |g_i(\theta)|^2 \sin\theta \, d\theta.$$

By using completely polarized electrons and atoms, we can then describe the excitation processes $^2S_{1/2} \to {}^2P_{1/2,\,3/2}$ or $^2S_{1/2} \to {}^2P_{1/2}$ or $^2S_{1/2} \to {}^2P_{3/2}$ by means of the amplitudes just defined:

$$
\begin{aligned}
&e(\uparrow) + A(^2S_{1/2}\uparrow) \to A(^2P_{1/2,\,3/2}, \uparrow) + e(\uparrow), \\
&\qquad |f_0 - g_0|^2 + 2|f_1 - g_1|^2, \qquad (25) \\
&e(\uparrow) + A(^2S_{1/2}, \downarrow) \to A(^2P_{1/2,\,3/2}, \downarrow) + e(\uparrow), \\
&\qquad |f_0|^2 + 2|f_1|^2, \qquad (26) \\
&e(\uparrow) + A(^2S_{1/2}, \downarrow) \to A(^2P_{1/2,\,3/2}, \uparrow) + e(\downarrow), \\
&\qquad |g_0|^2 + 2|g_1|^2. \qquad (27)
\end{aligned}
$$

The corresponding excitation cross sections for the separated excitation of the fine-structure levels $^2P_{3/2}$ and $^2P_{1/2}$ are the same as in Eqs. (25)–(27) multiplied, however, by factors $\frac{2}{3}$ and $\frac{1}{3}$, respectively.

Figure 13 shows, for example, how the excitation amplitudes are related to the excitation processes for transitions from a given magnetic sublevel of the ground state to a magnetic sublevel m_l of the excited states $^2P_{1/2}$. It should be noted that the coefficients of these amplitudes are correct for excitation of unpolarized atoms by unpolarized electrons. These coefficients can be derived from the expression relating the cross section for fine-structure states to those for the excitation of the components of the orbital angular momentum (Mott and Massey, 1965; Bates, 1968; Geltman, 1969).

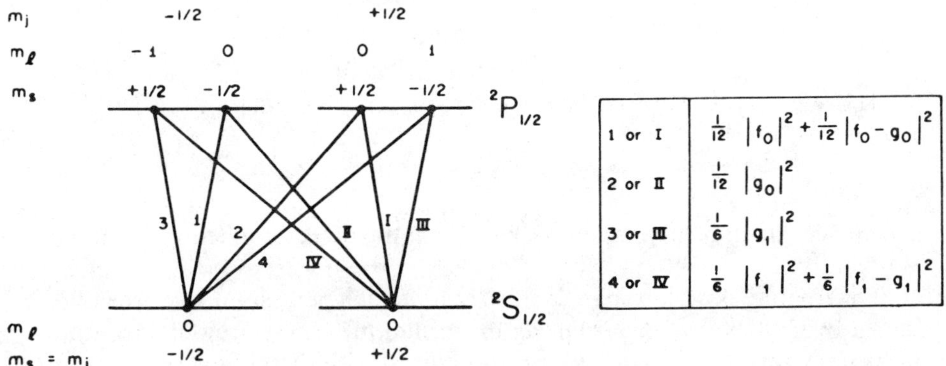

1 or I	$\frac{1}{12}\|f_0\|^2 + \frac{1}{12}\|f_0 - g_0\|^2$
2 or II	$\frac{1}{12}\|g_0\|^2$
3 or III	$\frac{1}{6}\|g_1\|^2$
4 or IV	$\frac{1}{6}\|f_1\|^2 + \frac{1}{6}\|f_1 - g_1\|^2$

Fig. 13. Two magnetic sublevels of the $^2S_{1/2}$ ground state and the excited fine structure state $^2P_{1/2}$ labeled by their quantum numbers m_j, m_l, and m_s. The table on the right gives the excitation amplitudes squared times the intensity coefficients of the different transitions when the excitation is carried out with unpolarized electrons and atoms.

3. *Experimental Methods and Applications*

The first successful experiment in which unpolarized electrons were scattered on polarized electrons was described by the N.Y.U. group (Rubin *et al.*, 1969). Figure 14 is a schematic diagram of their apparatus. An alkali beam is velocity- and spin-state selected by a Stern-Gerlach magnet (two-wire type magnet), before being cross-fired by an unpolarized beam of electrons. After scattering the atom is spin analyzed by an E-H gradient balance magnet (Rubin *et al.*, 1969). The entire analyzer–detector assembly rotates about the scattering center in the X-Y plane. The electron gun is rotatable in X-Y plane about the Y axis. The detector is also translatable in $\pm Z$ direction. By using simple kinematic relations, we may obtain the electron polar scattering angle from the atomic recoil scattering angle. Normally the polarizer and analyzer are set to transmit opposite spin states so that only spin-

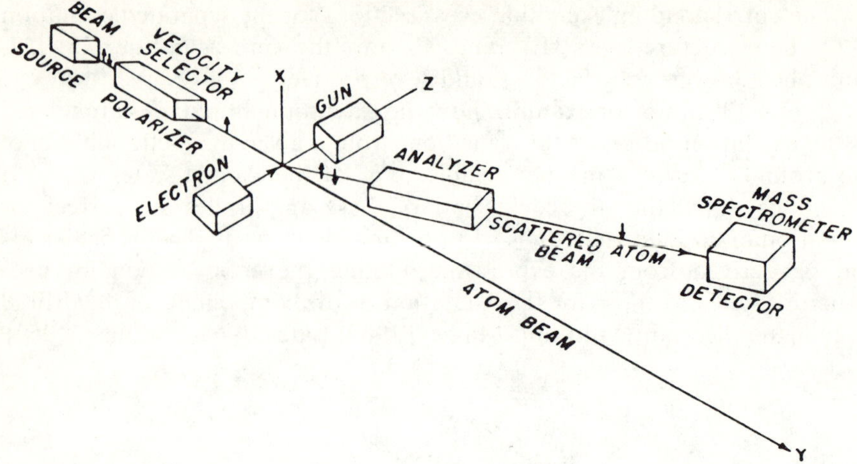

Fig. 14. Schematic diagram of the N.Y.U. recoil experiment (Rubin *et al.*, 1969) with polarized alkali atoms.

exchanged (or spin-flipped) atoms can reach the detector when the analyzer is operative.

Differential elastic exchange cross sections for potassium have been obtained by Collins *et al.* (1971) using this technique. These are obtained from measurements of $R(\theta) = \sigma_{ex}(\theta)/\sigma(\theta)$, combined with direct measurements of $\sigma(\theta)$ normalized to the total cross section at 1 eV. The $R(\theta)$ is simply the ratio of the atom beam current at a given detector position with and without the spin analyzer operative, corrected for transmission and residual depolarization. Partial depolarization caused by hyperfine coupling is avoided in this work by the use of a high ($>$ 1000 gauss) magnetic field in the interaction region. Figure 15 shows these results, compared to Karule's (1970) calculation. The bump in the 1.2-eV experimental curve in the vicinity of 90° is attributable to spin-flip caused by inelastic scattering from the high-energy tail of the electron energy distribution.

Campbell *et al.* (1971) at Edinburgh University used an electron trap in which the polarization of a polarized beam of alkali atoms is transferred by exchange processes to initially unpolarized trapped electrons; the exchange collisions gradually polarize the trapped electrons, which are subsequently extracted to form a polarized beam. From the degree of polarization, as measured by a Mott analyzer, the total exchange cross section can be obtained. Figure 16 includes typical data from the Edinburgh group compared to experimental data of the N.Y.U. group (Collins *et al.*, 1971) (obtained by

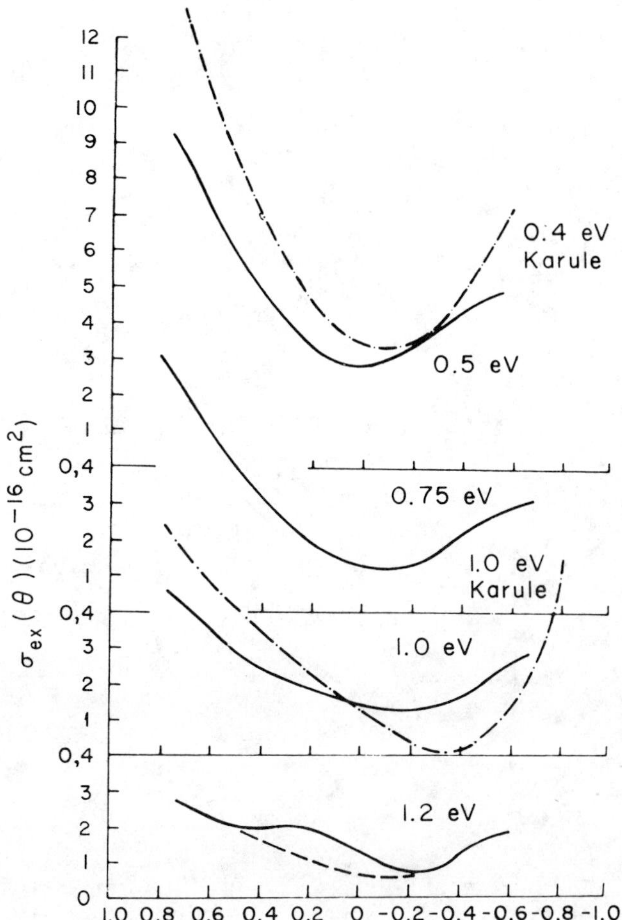

Fig. 15. Differential spin-exchange cross sections (solid curves for given electron energies) studied by means of the N.Y.U. recoil techniques with polarized atoms (Collins *et al.*, 1971). Dot-dashed curves are Karule's (1970) theoretical predictions at 0.4 and 1.0 eV.

integrating results such as those of Fig. 15 over all angles) and to the theory of Karule and Peterkop (1965).

Figure 17 shows the scheme of the JILA–Stirling type experiment (Hils *et al.*, 1972) for the measurement of the elastic direct differential cross section of alkali atoms. A beam of potassium atoms is polarized by the field of a hexapole magnet and the polarization is aligned parallel to the atomic beam direction by a weak magnetic field (a few milligauss). An unpolarized beam of electrons is cross-fired and the polarization of the scattered beam is

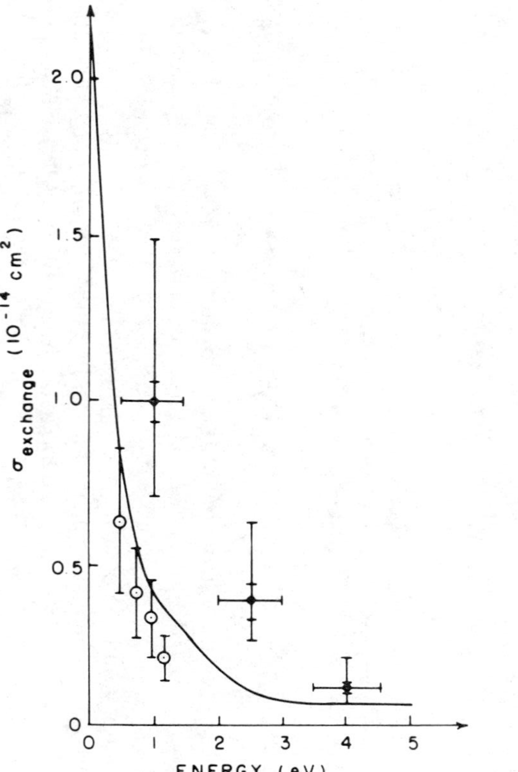

Fig. 16. Total elastic exchange cross section for electron scattering on potassium. (⊢$\bar{\text{I}}$⊣ Campbell *et al.*, 1971; —theory of Karule and Peterkop, 1965; $\bar{\odot}$ Collins *et al.*, 1971).

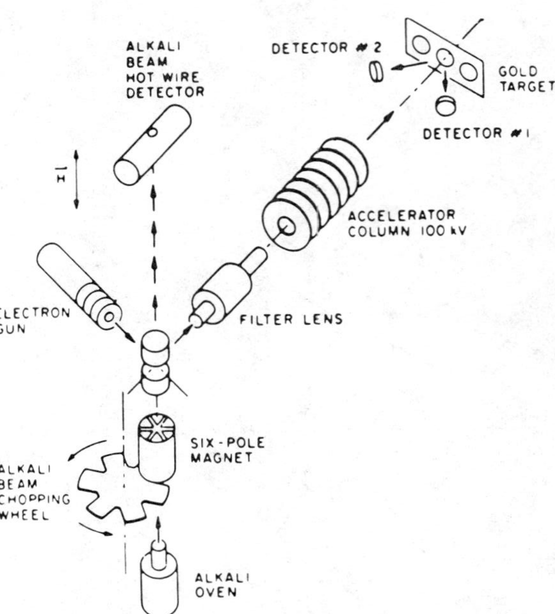

Fig. 17. Experimental scheme of the JILA-Stirling experiment (Hils *et al.*, 1972) to measure $|f|^2/\sigma$.

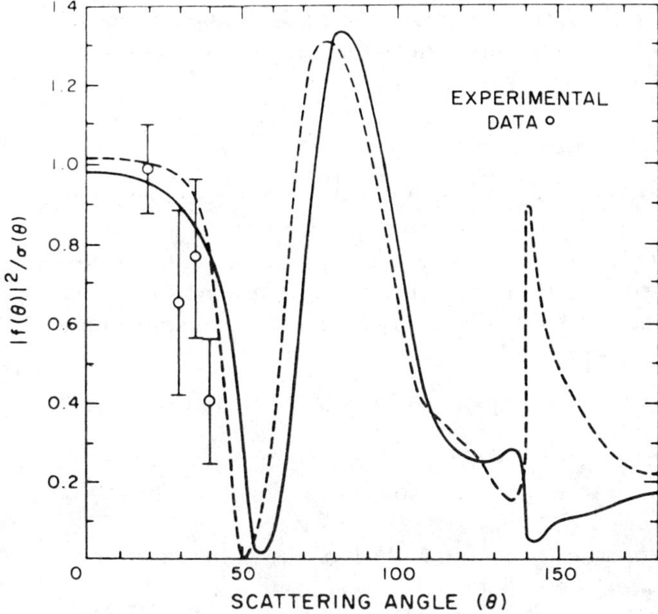

Fig. 18. Experimental data of $|f|^2/\sigma$ versus scattering angle θ for elastic electron-potassium scattering at 3.3 eV (Hils *et al.*, 1972). Continuous and dashed curves—theoretical values from Karule and Peterkop (1965) for 3 and 4 eV, respectively. Error bars 1 rms error.

measured by a Mott detector. From the measured polarization were derived the relative elastic direct differential cross sections for a given energy. Figure 18 presents the first pioneering data on the measurement of $|f|^2/\sigma$ for potassium atoms.

B. Scattering and Excitation Amplitudes: Two-Electron Atoms

A general analysis of spin-polarization effects in collisions between electrons and two-electron atoms has been given by Blum and Kleinpoppen (1974) using the density matrix techniques. A special example of spin-flip processes in the inelastic scattering of polarized electrons on mercury has been studied theoretically and by experiment by Hanne and Kessler (1974, 1976a,b).

1. *Characterization of the Polarization State*

Spin-orbit coupling, hyperfine interaction, and nuclear spin effects are neglected. Under these conditions the spin state of an ensemble of spin-S particles in general can be described as a statistical mixture of the pure spin states in which the particle may be found. By representing any pure state of the beam in the form of a column vector $a_i^{(n)}$ (where i runs from 1 to 2S + 1 and n denotes the nth pure state of the mixture), we may express the elements of the (2S + 1)-dimensional density matrix as

$$\rho_{ij} = \sum_n p_n a_i^{(n)} a_j^{(n)*}, \tag{28}$$

where p_n is the probability of finding the nth pure state in the given ensemble and the asterisk denotes the complex conjugate.

The density matrix characterizing an ensemble of spin-1 particles is a 3 × 3 matrix; thus nine basis matrices are needed to expand ρ. Besides the three-dimensional identity matrix E and the three spin-1 matrices S_i, we use the following set S_{ij}:

$$S_{ij} = \tfrac{3}{2}(S_i S_j + S_j S_i) - 2\delta_{ij} E \qquad (i, j = x, y, z). \tag{29}$$

We are using the standard representation of these matrices. Definition (5) is in accordance with the "Madison convention" in nuclear physics (Barschall and Haeberli, 1971).

The properties of S_i and S_{ij}, which are required for the calculations in this section, may be condensed to the form

$$S_i S_j S_k = \tfrac{1}{2}(\delta_{jk} S_i + \delta_{ij} S_k + i\varepsilon_{ijm} S_k S_m + i\varepsilon_{ikm} S_m S_j + i\varepsilon_{jkm} S_i S_m). \tag{30}$$

This equation governs the algebra of these matrices.

For spin-$\frac{1}{2}$ particles, a quadratic combination of spin operators reduces to a linear combination of the Pauli matrices. In the case of spin-1 particles, however, the expectation values $\langle S_{ij} \rangle$ give new information. Thus, in addition to the polarization vector P_i, knowledge of the components P_{ij} of the polarization tensor, defined by $P_{ij} = \langle S_{ij} \rangle$, are needed for a complete description of a beam of spin-1 particles.

In order to expand ρ in terms of the S_i and S_{ij}, we make the *ansatz*:

$$\rho = aE + \sum_i b_i S_i + \sum_{ij} c_{ij} S_{ij}.$$

Although the S_{ij} are symmetric, we sum over all i, j. The overcompleteness of the basis matrices may be taken into account by requiring $\sum_i C_{ii} = 0$. This constraint enables us to proceed as though we were dealing with an orthogonal set. Using the normalization condition and Eq. (30), we calculate the expectation values and relate the coefficients a, b_i, and c_{ij} to P_i and P_{ij}. We obtain

$$\rho = \frac{1}{3}\left(E + \frac{3}{2}\sum_i P_i S_i + \frac{1}{3}\sum_{ij} P_{ij} S_{ij}\right). \tag{31}$$

It follows that $\sum_i P_{ii} = 0$ and remembering the normalization condition, the spin state of a spin-1 particles in general is characterized by eight real parameters, namely three components of the polarization vector and the five independent components of the symmetric polarization tensor.

As an illustration we discuss the special case where the beam is in one of the pure spin states,

$$\begin{pmatrix}1\\0\\0\end{pmatrix}, \quad \begin{pmatrix}0\\1\\0\end{pmatrix}, \quad \begin{pmatrix}0\\0\\1\end{pmatrix},$$

which we denote in the following as $|+1\rangle$, $|0\rangle$, $|-1\rangle$. In states $|+1\rangle$ and $|-1\rangle$ the polarization vector points along the positive or negative directions of the quantization axis (z axis). In state $|0\rangle$ we can think of the spin vector as perpendicular to the quantization axis but precessing around it. This requires that we consider quantities more complicated than the polarization vector.

In the simple case where the beam is an incoherent superposition of the three pure states $|m\rangle$ ($m = \pm 1, 0$), three numbers are sufficient to describe the ensemble, for example N_+, N_0, N_-, which are the numbers of particles in the three states. An equivalent description is obtained by giving the values of intensity I and of the components of the polarization vector and tensor:

$$I = N_+ + N_0 + N_-,$$

$$P_z = \langle S_z \rangle = \sum_m P_m m|S_z|m = (N_+ - N_-)/I$$

$$P_x = P_y = 0$$

($P_m = Nm/I$ is the probability of finding a particle in the state $|m\rangle$),

$$P_{zz} = \langle S_{zz} \rangle = \sum_m m|S_{zz}|mP_m = (N_+ - 2N_0 + N_-)/I,$$

$$P_{xx} = P_{yy} = -\tfrac{1}{2}P_{zz},$$

$$P_{ij} = 0 \qquad \text{for } i \neq j.$$

In describing systems of electrons and spin-1 atoms, the joint state of the two particles is represented by a 6×6 density matrix, acting in the composite spin space of electrons and atoms. Any operator in this space can be written in the form $\Omega_A \times \Omega_e$, where Ω_A(e) acts on the atomic (electronic) spinor and the $\times$ denotes the direct product. (This type of product is often called a tensor or Kronecker product.)

The density matrix may be written as a linear combination of the 6×6 identity matrix 1 and 35 Hermitian traceless matrices. We choose the following set:

$$1, \quad \beta_i = E \times \sigma_i, \quad \alpha_i = S_i \times \varepsilon,$$

$$\alpha_{ij} = S_{ij} \times \varepsilon, \alpha_i\beta_j, \alpha_{ij}\beta_k \qquad (i, j, k = x, y, z) \tag{32a}$$

As in the case of spin-1 particles, this set is overcomplete and it follows that

$$\sum_i \alpha_{ii} = 0. \tag{32b}$$

In terms of these operators, the density matrix of the ingoing beam is given by the expression,

$$\rho_{\text{in}} = \frac{1}{6}\Big(1 + \sum_i P_i^{(e)}\beta_i + \frac{3}{2}\sum_i P_i^{(A)}\alpha_i + \frac{1}{3}\sum_{ij} P_{ij}\alpha_{ij} + \frac{3}{2}\sum_{ij} Q_{ij}\alpha_i\beta_j + \frac{1}{3}\sum_{ijk} R_{ijk}\alpha_{ij}\beta_k\Big), \tag{33}$$

which may be derived in a similar way to Eq. (31). The $P_i^{(e)}$ and $P_i^{(A)}$ are the i components of the electronic and atomic polarization vector:

$$P_i^{(e)} = \text{tr}\ \rho_{\text{in}}\beta_i\,, \qquad P_i^{(A)} = \text{tr}\ \rho_{\text{in}}\alpha_i\,. \tag{34a}$$

The P_{ij} is a component of the atomic polarization tensor,

$$P_{ij} = \text{tr}\ \rho_{\text{in}}\alpha_{ij}\,, \tag{34b}$$

and Q_{ij} and R_{ijk} are correlation terms,

$$Q_{ij} = \text{tr}\ \rho_{\text{in}}\alpha_i\beta_j\,, \qquad R_{ijk} = \text{tr}\ \rho_{\text{in}}\alpha_{ij}\beta_k\,. \tag{34c}$$

From (32b) it follows that

$$\sum_i P_{ii} = 0, \qquad \sum_i R_{iij} = 0. \tag{35}$$

We note that the correlation tensors are zero if at least one of the particle beams (electrons or atoms) is unpolarized. If both are completely polarized, these terms are given by

$$Q_{ij} = P_i^{(A)}P_i^{(e)}, \qquad R_{ijk} = P_{ij}P_k^{(e)}.$$

These relations may be proved with methods similar to those of Burke and Schey (1962).

2. *Polarization State after the Scattering*

Expansion of the M Matrix. After having characterized the initial beam, we start now with an analysis of the scattering process between electrons and spin-1 atoms. We consider elastic scattering. If the ingoing electrons are in one of the pure states $|m^{(e)}\rangle (m^{(e)} = \pm\frac{1}{2})$ and the atoms in one of the pure states $|m^{(A)}\rangle (m^{(A)} = \pm 1, 0)$, there will be ten amplitudes for each energy and scattering angle, because there are ten possible processes:

$$|+1\rangle|+\tfrac{1}{2}\rangle \rightarrow |+1\rangle|+\tfrac{1}{2}\rangle,$$

$$|0\rangle|+\tfrac{1}{2}\rangle \rightarrow \begin{matrix} |0\rangle|+\tfrac{1}{2}\rangle \\ |+1\rangle|-\tfrac{1}{2}\rangle \end{matrix},$$

$$|-1\rangle|+\tfrac{1}{2}\rangle \rightarrow \begin{matrix} |-1\rangle|+\tfrac{1}{2}\rangle \\ |0\rangle|-\tfrac{1}{2}\rangle \end{matrix}, \tag{36}$$

and five further reactions with $|-\frac{1}{2}\rangle$ instead of $|+\frac{1}{2}\rangle$. These processes are the analogs to the six "basic reactions" listed in Section III,A.

In order to catalog reactions (36), we define an operator M in spin space so that its matrix elements $\langle m^{(A)\prime} m^{(e)\prime} | M | m^{(A)} m^{(e)} \rangle$ are the amplitudes for a transition from an initial state $|m^{(A)}\rangle |m^{(e)}\rangle$ to the final state $|m^{(A)\prime}\rangle |m^{(e)\prime}\rangle$ for a given energy and scattering angle and that

$$\sigma(m^{(A)} m^{(e)} \rightarrow m^{(A)\prime} m^{(e)\prime}) = |\langle m^{(A)\prime} m^{(e)\prime} | M | m^{(A)} m^{(e)} \rangle|^2 \tag{37a}$$

is the differential cross section for this elastic transition. (We suppress here the dependence of M on energy and angle.)

If the wavefunction of the incident beam is given by $e^{i\mathbf{k}\cdot\mathbf{r}}\phi|m^{(A)} m^{(e)}\rangle$, the asymptotic wavefunction after scattering is

$$\psi_{\text{as}} \rightarrow [e^{i\mathbf{k}\cdot\mathbf{r}} + (1/r)e^{ikr}M]\phi|m^{(A)}\rangle|m^{(e)}\rangle, \tag{37b}$$

where ϕ is the wavefunction of the atom except the spin part. Terms $M|m^{(A)}\rangle|m^{(e)}\rangle$ can be regarded as the spin state of the particles after the reaction. Thus operator M transforms the initial spinor into the final one.

Any scattering process for any given ingoing beam in any arbitrary polarization state may be analyzed in terms of the M-matrix elements of the simple reactions (36). This has been discussed in Section III, and by Kleinpoppen (1971) for the case of the one-electron atom.

By using Clebsch–Gordan techniques, we expand the element $\langle m^{(A)\prime} m^{(e)\prime} | M | m^{(A)} m^{(e)} \rangle$ in terms of the amplitudes $f^{(3/2)}$ and $f^{(1/2)}$, where $f^{(S_t)}$ denotes the scattering amplitudes in the channel with total spin S_t, $f^{(S_t)} = \langle S_t, S_{tz} | M | S_t, S_{tz} \rangle$ (independent of S_{tz}). We get

$$M = \begin{bmatrix} f^{(3/2)} & 0 & 0 & 0 & 0 & 0 \\ 0 & \frac{1}{3}(f^{(3/2)} + 2f^{(1/2)}) & (\frac{2}{3})^{1/2}(f^{(3/2)} - f^{(1/2)}) & 0 & 0 & 0 \\ 0 & (\frac{2}{3})^{1/2}(f^{(3/2)} - f^{(1/2)}) & \frac{1}{3}(2f^{(3/2)} + f^{(1/2)}) & 0 & 0 & 0 \\ 0 & 0 & 0 & \frac{1}{3}(2f^{(3/2)} + f^{(1/2)}) & (\frac{2}{3})^{1/2}(f^{(3/2)} - f^{(1/2)}) & 0 \\ 0 & 0 & 0 & (\frac{2}{3})^{1/2}(f^{(3/2)} - f^{(1/2)}) & \frac{1}{3}(f^{(3/2)} + 2f^{(1/2)}) & 0 \\ 0 & 0 & 0 & 0 & 0 & f^{(3/2)} \end{bmatrix}. \tag{37c}$$

Instead of using the amplitudes $f^{(S_t)}$, we may express the elements of M in terms of the direct (f) and exchange (g) amplitudes, defined by

$$\langle +\tfrac{1}{2}, -1 | M | +\tfrac{1}{2}, -1 \rangle = f,$$
$$\langle -\tfrac{1}{2}, +1 | M | +\tfrac{1}{2}, 0 \rangle = -\sqrt{2}\, g.$$

Physically f and g refer to scattering events in which the ingoing electron can definitely be stated to have, or not to have, suffered an exchange collision with one of the atomic electrons.

In elastic scattering processes between spin-$\frac{1}{2}$ and spin-1 particles involving explicit spin-dependent interactions, the expression for the M matrix is more complicated. However, by observing electrons scattered in forward direction, we have the same structure [Eq. (37)] for M. (This follows from the fact that there is no orbital angular momentum component in the direction of motion, which may be chosen as the z axis.)

One may go further in the analysis (without going into the details of the dynamics) by expressing M in terms of the set (32a). Because all explicit spin-dependent forces are neglected, M must have the simple form,

$$M = f^{(3/2)}\eta_{3/2} + f^{(1/2)}\eta_{1/2}.$$

The η_{S_t} are the projection operators upon spin functions with total spin S_t:

$$\eta_{3/2} = \frac{1}{3}\left(2 + \sum_i \alpha_i \beta_i\right),$$

$$\eta_{1/2} = \frac{1}{3}\left(1 - \sum_i \alpha_i \beta_i\right).$$

By reformulating the terms, we get

$$M = \tfrac{1}{3}(2f^{(3/2)} + f^{(1/2)})l + \tfrac{1}{3}(f^{(3/2)} - f^{(1/2)}) \sum_i \alpha_i \beta_i \qquad (38a)$$

$$= (f - g)l - g \sum_i \alpha_i \beta_i. \qquad (38b)$$

Instead of using the projection operators, Eqs. (38) may be derived from invariance principles. The operator M is required to be invariant under rotation. Thus M must be a combination of all scalars that can be built from the set (32a). Apart from the identity matrix, there is only one scalar combination $\boldsymbol{\alpha} \cdot \boldsymbol{\beta} = \sum_i \alpha_i \beta_i$. Thus the most general form of M in terms of the set (32a) is

$$M = a \cdot 1 + b \sum_i \alpha_i \cdot \beta_i.$$

Coefficients a and b follow from

$$\operatorname{tr} M = 6a,$$
$$\operatorname{tr} M\alpha_z\beta_z = b \sum_i \operatorname{tr}(S_i \cdot S_z) \operatorname{tr}(\sigma_i \cdot \sigma_z) = 4b.$$

By using Eq. (15) and the explicit forms of S_z and σ_z, we get

$$a = \tfrac{1}{3}(2f^{(3/2)} + f^{(1/2)}),$$
$$b = \tfrac{1}{3}(f^{(3/2)} - f^{(1/2)})$$

which gives us Eqs. (38).

The main results of the previous discussion are both Eq. (33), which specifies the spin state of the initial beam, and Eq. (38), which extracts from the scattering process the properties independent of the dynamics. From both we obtain expressions for the quantities characterizing the final beam of spin-$\frac{1}{2}$ and spin-1 particles.

Because M transforms incident spinors into final spinors, the density matrix ρ_{out} of the outgoing beam is given by the relation

$$\rho_{\text{out}} = M\rho_{\text{in}} M. \tag{39}$$

We define polarization vectors and tensors for the final particles as

$$\begin{aligned}
\sigma(\theta) &= \operatorname{tr} \rho_{\text{out}}, & \sigma(\theta)P_{ij} &= \operatorname{tr} \rho_{\text{out}}\alpha_{ij}, \\
\sigma(\theta)P^{(A)} &= \operatorname{tr} \rho_{\text{out}}\alpha_i, & \sigma(\theta)Q_{ij} &= \operatorname{tr} \rho_{\text{out}}\alpha_i\beta_j, \\
\sigma(\theta)P_i^{(e)} &= \operatorname{tr} \rho_{\text{out}}\beta_i, & \sigma(\theta)R_{ijk} &= \operatorname{tr} \rho_{\text{out}}\alpha_{ij}\beta_k.
\end{aligned} \tag{40}$$

The ρ_{out} is not normalized; thus, the differential cross section $\sigma(\theta)$ occurs in Eq. (40).

We note that all the calculations are carried through without using the explicit forms of the matrices—all that is needed are the algebraic properties. We can thus see the usefulness of expansions (33) and (38).

In terms of f and g we get the following results for the quantities (40), where ε_{ijk} is the Kronecker tensor:

$$\sigma(\theta) = |f - g|^2 + 2|g|^2 + (|g|^2 - fg^* - f^*g) \sum_n Q_{nn}, \tag{41a}$$

$$\begin{aligned}
\sigma(\theta)P_i^{(e)\prime} = {} & (|f - g|^2 - \tfrac{2}{3}|g|^2)P_i^{(e)} \\
& + (3|g|^2 - fg^* - f^*g)P_i^{(A)} + i(fg^* - f^*g) \\
& \times \sum_{mn} Q_{mn}\varepsilon_{imn} + \tfrac{2}{3}|g|^2 \sum_j R_{ijj},
\end{aligned} \tag{41b}$$

$$
\begin{aligned}
\sigma(\theta)P_i^{(A)\prime} = {} & \tfrac{2}{3}(3|g|^2 - fg^* - f^*g)P_i^{(e)} \\
& + (|f-g|^2 + |g|^2)P_i^{(A)} + \tfrac{1}{2}i(f^*g - fg^*) \\
& \times \sum_{mn} Q_{mn}\varepsilon_{imn} - \tfrac{1}{3}(fg^* + f^*g)\sum_j R_{ijj},
\end{aligned} \tag{41c}
$$

$$
\begin{aligned}
\sigma(\theta)P'_{ij} = {} & (|f-g|^2 - |g|^2)P_{ij} \\
& + \tfrac{1}{2}(4|g|^2 - fg^* - f^*g)\left(-\delta_{ij}\sum_m Q_{mm} + \tfrac{3}{2}(Q_{ij} + Q_{ji})\right) \\
& + \tfrac{1}{2}i(f^*g - fg^*)\sum_{km}(R_{ikm}\varepsilon_{jkm} + R_{jkm}\varepsilon_{ikm}),
\end{aligned} \tag{41d}
$$

$$
\begin{aligned}
\sigma(\theta)Q'_{ij} = {} & \tfrac{2}{3}\delta_{ij}(|g|^2 - fg^* - f^*g) \\
& + \tfrac{2}{3}i(f^*g - fg^*)\sum_k P_k^{(e)}\varepsilon_{ijk} - \tfrac{1}{2}i(f^*g - fg^*) \\
& \times \sum_k P_k^{(A)}\varepsilon_{ijk} + \tfrac{1}{3}(4|g|^2 - f^*g - fg^*)P_{ij} \\
& + (|f-g|^2 - \tfrac{1}{2}(fg^* + f^*g) + |g|^2)Q_{ij} \\
& + \tfrac{1}{2}(fg^* + f^*g)\sum_m Q_{mm}\delta_{ij} - \tfrac{1}{3}i(f^*g - fg^*) \\
& \times \sum_{kl} R_{ikl}\varepsilon_{jkl},
\end{aligned} \tag{41e}
$$

$$
\begin{aligned}
\sigma(\theta)R'_{ijk} = {} & \tfrac{1}{3}|g|^2(-2\delta_{ij}P_k^{(e)} + 3\delta_{jk}P_i^{(e)} + 3\delta_{ik}P_j^{(e)}) \\
& + \tfrac{1}{4}(fg^* + f^*g)(2\delta_{ij}P_k^{(A)} - 3\delta_{jk}P_i^{(A)} - 3\delta_{ik}P_j^{(A)}) \\
& + \tfrac{1}{2}i(fg^* - f^*g)\cdot\sum_i (P_{il}\varepsilon_{jkl} + P_{jl}\varepsilon_{ikl}) \\
& + \tfrac{1}{2}i(fg^* - f^*g) \\
& \times \sum_m\left(\tfrac{3}{2}Q_{im}\varepsilon_{jkm} + \tfrac{3}{2}Q_{jm}\varepsilon_{ikm} + \delta_{ij}\sum_i Q_{lm}\varepsilon_{lmk}\right) \\
& + (|f-g|^2 - \tfrac{1}{2}|g|^2)R_{ijk} \\
& - \tfrac{2}{3}|g|^2\delta_{ij}\sum_l R_{kll} + \tfrac{1}{2}(3|g|^2 - fg^* - f^*g) \\
& \times (R_{ikj} + R_{jki}) + \tfrac{1}{2}(-|g|^2 + fg^* + f^*g) \\
& \times \left(\delta_{jk}\sum_l R_{ill} + \delta_{ik}\sum_l R_{jll}\right).
\end{aligned} \tag{41f}
$$

Equations (41) are our main results. They are the analogs of the expressions given by Burke and Schey (1962) for one-electron atoms.

The main feature of Eqs. (41) is that the dynamical elements and the quantities that characterize the spin state of the ingoing beam are separated. Thus we easily obtain those properties of the scattering process under consideration that are independent of the special dynamics. We note the following:

a. For an unpolarized ingoing beam, the polarization vectors and tensors of the outgoing particles vanish, and we get the well-known result that the polarization of the final beam is zero for unpolarized initial beams.

b. In scattering processes between unpolarized electrons and polarized atoms, only the vector part of the polarization can be transferred; there will be no electron polarization if the atom has only tensor polarization. The direction of the atomic polarization vector remains unaltered, but the beam is depolarized. The polarization vector of the final electron is parallel to that of the target.

c. If polarized electrons are scattered from unpolarized atoms the outgoing atoms can have no tensor polarization. The polarization vectors of both final particles are parallel to that of the incident electron.

A similar discussion for the case of spin-$\frac{1}{2}$ particles in terms of Stokes parameters has been given by Bederson (1969a,b).

Furthermore, from Eqs. (41) we can read off the information about the M-matrix elements, which can be extracted from experiments with polarized beams. Specifying the polarization state of the ingoing particles, and measuring polarization vectors and tensors of the outgoing ones, for each energy and scattering angle, the scattering amplitudes can be calculated from these experimental values with the help of Eqs. (41) and compared with theoretical predictions. In addition the expressions (41) show how the results of special experiments can be combined in order to separate the contributions of direct and exchange scattering.

In order to determine f, g and the relative phase between them, three independent measurements are required. As an example we mention the following set of experiments.

a. Measurement of the unpolarized cross section. Because unpolarized particles are characterized by vanishing values of all the polarization vectors and tensors and correlation terms, we get from Eq. (41a)

$$\sigma_{\text{un}} = |f - g|^2 + 2|g|^2. \tag{42a}$$

b. Scattering of unpolarized electrons from polarized atoms and spin analysis of the outgoing atom give the depolarization ratio:

$$d(\theta) = P_i^{(\text{A})\prime}/P_i^{(\text{A})} = 1 - |g|^2/\sigma(\theta) \tag{42b}$$

which follows from the expression (41c) with $P_i^{(\text{e})} = 0$, $Q_{ij} = 0$, $R_{ijk} = 0$.

c. Scattering of completely polarized electrons and atoms. The differential cross section for the process $|+\frac{1}{2}\rangle|+1\rangle \to |+\frac{1}{2}\rangle|+1\rangle$ follows from (41a):

$$\sigma(\theta) = |f - 2g|^2. \tag{42c}$$

Although this experiment requires completely polarized particles, no spin analysis of the final beam is necessary.

So far we discussed only elastic scattering events. However, the formalism is more general. Thus in the case of an atomic triplet–triplet transition from $\Gamma = n_1\, l_1\, m_{l_1}$ to $\Gamma' = n_1'\, l_1'\, m_{l_1}'$, we only have to replace the elastic amplitudes in Eqs. (41) by the expression

$$\begin{aligned} f_{\Gamma'\Gamma}^{(S_t)} = \frac{i}{(k_n k_n')^{1/2}} \sum_{l_2 l_2' L} i^{(l_2 - l_2')}[4\pi(2l_2 + 1)]^{1/2} \\ \times (l_1 m_{l_1}, l_2 0 | LM_2)(l_1' m_{l_1}', l_2 m_{l_2}' | LM_2) \\ \times (\delta_{n_1' k_n' l_1' l_2', n k_n l_1 l_2} - S^{LS_t}_{n_1' k_1' l_1' l_2', n k_n l_1 l_2}) \gamma_{l_2'}^{m_{l2'}}(\theta, \phi). \end{aligned} \tag{43}$$

Functions γ_l^m are standard spherical harmonics, $(l_1 m_{l_1}, l_2 0 | LM_2)$ is a Clebsch–Gordan coefficient, and k_n' and k_n denote the wave numbers.

Because all spin-dependent forces in the scattering matrix have been neglected, we get the same polarization pattern in both elastic and inelastic processes.

In singlet–triplet and triplet–singlet transitions, there is only one possible spin channel with total spin $S_t = \frac{1}{2}$. Thus measurements of the differential cross section at all energies and angles is sufficient in order to determine the amplitude $f_{\Gamma'\Gamma}^{(1/2)}$ (which is proportional to the exchange amplitude). We get

$$\sigma_{\Gamma'\Gamma} = (k'/k)|f_{\Gamma'\Gamma}^{(1/2)}|^2$$

for singlet–triplet and

$$\sigma_{\Gamma'\Gamma} = \tfrac{1}{3}(k'/k)|f_{\Gamma'\Gamma}^{(1/2)}|^2$$

for triplet–singlet transitions.

In singlet–singlet transitions, $f_{\Gamma'\Gamma}^{(1/2)}$ is a coherent superposition of the direct and exchange amplitude. The polarization state of the electron cannot change in collisions on spin-0 particles, in so far as spin-dependent interactions are neglected. Thus in this case it is not possible to separate direct and exchange contributions by using polarized electrons.

3. *Exchange Scattering in the Excitation of Mercury*

The influence of exchange processes in the excitation of mercury has been studied by Hanne and Kessler (1974, 1976a,b). They based their analysis on experiments that are similar to those described in Section III,A.

Polarized electrons scattered inelastically on mercury atoms may be depolarized and the depolarization measured reveals information on the amplitudes of the excitation process. Neglecting all interactions (e.g., spin-orbit or spin-spin interactions) which could cause the spin of the scattered electron to flip except for the exchange interaction, we can derive the depolarization for simple examples as follows.

a. $^1S \rightarrow {}^3S$ *Excitation.* Under the assumption that *LS* coupling is valid, the excitation amplitude for this excitation process can be separated into the part describing the excitation of states with given orbital angular momentum multiplied by the Clebsch–Gordan coefficients. By using completely polarized electrons, the following magnetic substates can be excited:

$$e(\uparrow) + \mathrm{Hg}(^1\mathrm{S}) \rightarrow \mathrm{Hg}(^3\mathrm{S},\, m = 0) + e(\uparrow)$$

$$\rightarrow \mathrm{Hg}(^3\mathrm{S},\, m = 1) + e(\downarrow).$$

The second reaction channel has a probability twice that of the first channel due to the ratio $1 : \sqrt{2}$ for the relevant Clebsch–Gordan coefficients of the two magnetic substates involved. Having excited the ^3S state, a beam of partially polarized electrons with polarization P should change its polarization to $P' = -\frac{1}{3}P$.

b. $^1S \rightarrow {}^3P$ *Excitation.* Hanne and Kessler (1974, 1975) dealt with this excitation process in detail. For the case in which the spin-orbit interaction in the atom can be neglected, the initially polarized beam of electrons exciting the ^3P states will be depolarized in the same way as in the previous example.

If spin-orbit interaction is taken into consideration in the atomic structure, the depolarization is modified and can be calculated by using a density matrix description of the polarization phenomena in electron–atom scattering. Furthermore, for heavy atoms such as mercury, Russel–Saunders coupling is no longer valid. A singlet ^1P contribution is mixed to the $^3P_1$ state such that the wavefunction of the state is

$$\psi(^3\mathrm{P}_1) = \alpha\psi^0(^3\mathrm{P}_1) + \beta\psi^0(^1\mathrm{P}_1),$$

where ψ^0 are the special functions for pure Russel–Saunders states, and $\alpha = -0.985$ and $\beta = 0.171$ are the coupling coefficients. With the additional assumption that the spin-orbit relaxation time in the atomic structure of the ^3P state is long compared with the excitation time, Hanne and Kessler (1974, 1975) calculated the depolarization ratio of electrons exciting the $^3P_1$ state:

$$\frac{P'}{P} = \frac{\beta^2(|f_0^s - g_0^s|^2 + 2|f_1^s - g_1^s|^2) - \alpha^2|g_1|^2}{\beta^2(|f_0^s - g_0^s|^2 + 2|f_1^s - g_1^s|^2) + \alpha^2(|g_0|^2 + 2|g_1|^2)}.$$

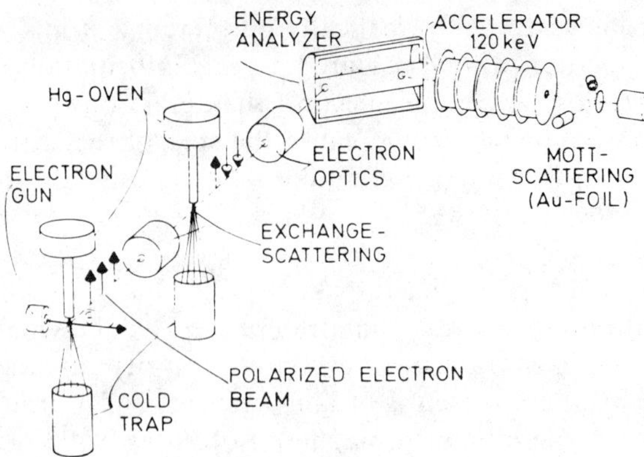

Fig. 19. Schematic diagram of the apparatus of Hanne and Kessler (1974, 1976a,b).

Figure 19 shows a schematic diagram of the apparatus used for the measurement of the depolarization ratio of the 6^3P or 6^1P excitation. Transversally polarized electrons were produced by elastic scattering on mercury (Jost and Kessler, 1966) energy of initial electrons $E = 80$ eV, scattering angle $\theta = 80°$, $P = 0.22$, current $\simeq 10^{-10}$ amp). Inelastically scattered electrons were

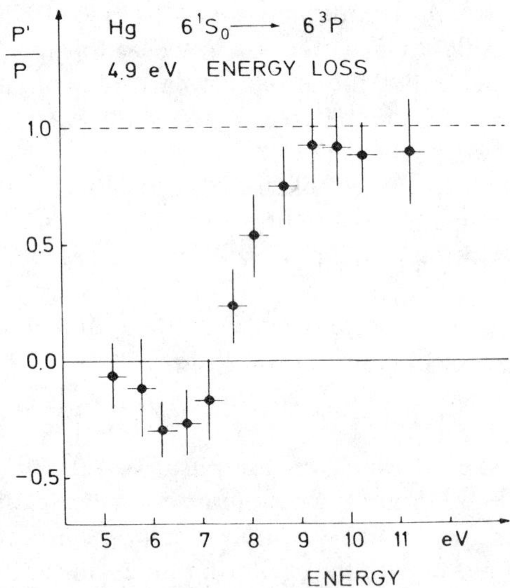

Fig. 20. Depolarization ratio for the $6^1S_0 \rightarrow 6^3P$ excitation of mercury versus electron energy. Error bars include the statistical error and the reproducibility of the measured polarizations as well as the reproducibility of the energy.

removed by a filter lens. The polarized beam of electrons was then decelerated from 5 to 15 eV and focused on a second mercury atomic vapor target (target density 10^{-2}–10^{-3} torr). Electrons scattered in the forward direction on the second target passed a cylindrical mirror energy analyzer. This analyzer was tuned to the energy loss of those electrons having excited the ^3P states (overall energy resolution 0.8 eV). The polarization of the electrons scattered at the second target was measured by a Mott detector. Spin-orbit interaction in the scattering process at the second target could be neglected because it should not occur in forward scattering processes. Figure 20 shows the experimental results for the depolarization ratio P'/P for the $6^1S_0 \rightarrow {}^3P$ excitation processes.

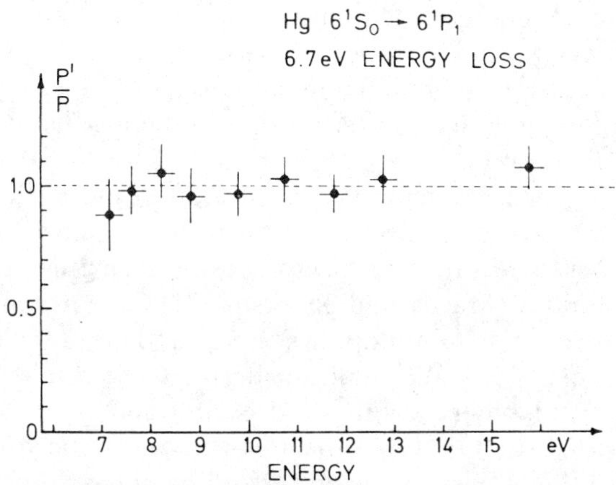

Fig. 21. Depolarization ratio of the mercury $6^1S_0 \rightarrow 6^1P_1$ excitation process as a function of the electron energy.

As can be seen from Fig. 20, at electron energies below 8 eV strong depolarization or even reversal of the sign of the polarization has been observed. In this connection it is worth mentioning the fact that the excitation process $6^1S_0 \rightarrow 6^3P_1$ is highly affected by the occurrence of narrow resonances in the energy range close to threshold and possibly also in the energy range between 8 and 9 eV (see also Section II). A high-resolution study of the depolarization ratio should reflect the structure as seen in light polarization measurements (see Figs. 6 and 7).

Contrary to the foregoing case, the $6^1S_0 \rightarrow 6^1P_1$ excitation process did not show any depolarization within the error bars, as seen in Fig. 21.

IV. Electron–Photon Coincidences and Angular Correlations

A. Theory

Although coincidence and angular correlation measurements have been applied extensively in nuclear and particle physics, the use of these methods is rather new to atomic physics.

The method of electron–photon coincidence techniques will be discussed for a special example of an electron–atom excitation process. It will be demonstrated how it can be used as a tool for investigating the following parameters: (1) the collision parameters of the electron–atom excitation process (i.e., differential inelastic cross sections, excitation amplitudes and their phase differences); (2) the "source parameters" of the excited atomic state (i.e., orbital angular momentum transfer to the atom, orientation and alignment parameters, and multipole states of the excited atom); and (3) the characteristics of electron–photon angular correlation (i.e., symmetry direction of the photon emission and its relation to the momentum transfer and to photon polarization data).

The theory of measurements in which photons are detected in delayed coincidences with scattered electrons has been developed in a form that relates the coincidence rates for electron–photon angular correlations to scattering amplitudes (Macek and Jaecks, 1971) or to the orientation and alignment parameters or to multipole states (Fano and Macek, 1973; Blum and Kleinpoppen, 1975a). We shall not derive these results but will apply them to the following excitation/deexcitation process of helium: $1^1S_0 \rightarrow n^1P_1 \rightarrow {}^1S_0$. In this case, the links between the observable coincidence rate and the excitation parameters follow from simple and plausible arguments. According to theory, the excitation into the excited states can be described as a coherent superposition of excitation onto degenerate magnetic sublevels (neglecting spin-orbit and spin-spin interactions). The atom is prepared into the P state of helium such that the amplitudes a_{m_l} for magnetic sublevel excitation govern the initial distribution of the sublevel excitation of the eigenstate $\psi({}^1P_1)$ of the excited atom:

$$\psi({}^1P_1) = \sum_{m_l} a_{m_l} \psi_{m_l}({}^1P_1) = a_0\,|10\rangle + a_1\,|11\rangle + a_{-1}\,|1-1\rangle. \qquad (44)$$

For the three excitation amplitudes a_0, a_1, and a_{-1} of the P state, mirror symmetry of the electron–atom scattering process requires the restriction $a_1 = a_{-1}$. Eigenstate $\psi({}^1P_1)$ can be normalized such that the amplitudes are related to the inelastic differential excitation cross section σ of the ${}^1P$ state:

$$|a_0|^2 = \sigma_0, \qquad |a_1|^2 = \sigma_1, \qquad (45)$$

$$\sigma = \sigma_0 + 2\sigma_1. \qquad (46)$$

σ_0, σ_1 are partial differential cross sections for exciting the magnetic sublevels $m_l = 0$ and $m_l = \pm 1$, respectively. Amplitudes a_{m_l} are, in general, complex numbers depending on the entire problem of the collision process. For any given excitation process, determined by the excitation energy and the scattering angle of the inelastically scattered electron, amplitudes a_1 and a_0 are expected to have a fixed phase relationship to each other:

$$a_1 = |a_1| e^{i\alpha_1} \quad \text{and} \quad a_0 = |a_0| e^{i\alpha_0} \tag{47}$$

with

$$\chi = \alpha_1 - \alpha_0 \, . \tag{48}$$

By taking $\lambda = \sigma_1/\sigma$ or $1 - \lambda = 2(\sigma_1/\sigma)$, the following coincidence rate N_c for the simultaneous detection of the electron and the photon in the scattering plane (see Fig. 22) can be derived from the general theory of Macek and Jaecks (1971):

$$N_c \propto \lambda \sin^2 \theta_\gamma + (1 - \lambda) \cos^2 \theta_\gamma - 2[\lambda(1 - \lambda)]^{1/2} \cos \theta_\gamma \sin \theta_\gamma \cos \chi, \tag{49}$$

where θ_γ is the angle between photon propagation and initial electron beam.

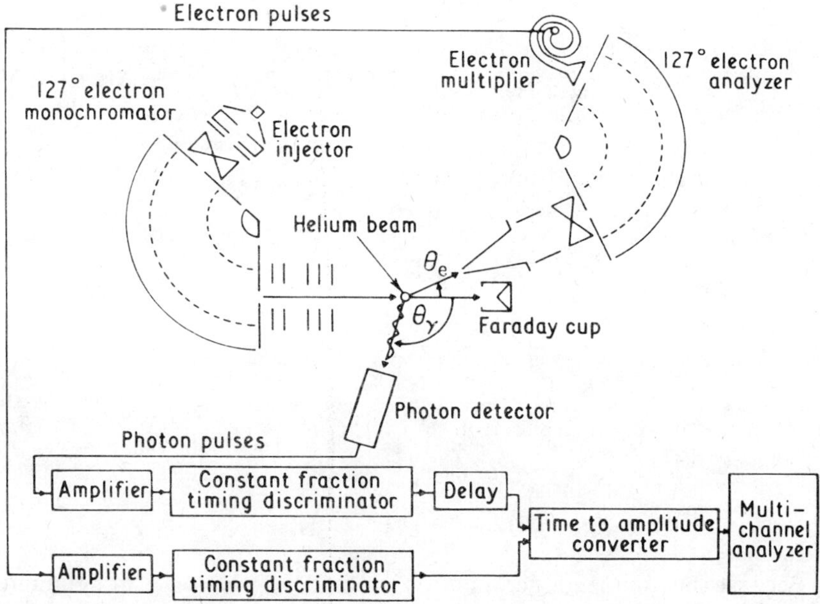

Fig. 22. Schematic diagram for electron–photon coincidence measurements (Eminyan *et al.*, 1972, 1974): the atomic beam (helium) is crossed by the monoenergetic electron beam from the 127° electron monochromator. The electron and photon pulses from the excitation process are fed into the delayed coincidence electronics.

Thus, photons are counted without regard to polarization by a detector placed in any direction in the scattering plane and on the opposite side of the electrons scattered under a fixed angle. It can easily be verified that Eq. (49) is identical to

$$N_c \propto (1/\sigma)|a_0 \sin\theta_\gamma - \sqrt{2}\,a_1 \cos\theta_\gamma|^2. \tag{50}$$

Of course, this last equation assumes that two classical oscillators with relative phase difference χ and amplitudes $|a_0|$ and $\sqrt{2}|a_1|$ are coherently excited in the z and x directions, respectively.

The collision parameters λ and χ which, as shown below, can be determined from the coincidence experiment, can easily be linked with the expectation values of orbital angular momentum quantities of the atom in the excited P state (in units of $\hbar$ and with the z axis parallel to the initial electron beam, the x axis in the scattering plane, and the y axis perpendicular to scattering plane of the electrons):

$$\begin{aligned} &\langle L_x\rangle = 0, && \langle L_x^2\rangle = \lambda,\\ &\langle L_y\rangle = -2[\lambda(1-\lambda)]^{1/2}\sin\chi, && \langle L_y^2\rangle = 1,\\ &\langle L_z\rangle = 0, && \langle L_z^2\rangle = 1-\lambda, \end{aligned} \tag{51}$$

with $L^2 = L_x^2 + L_y^2 + L_z^2 = L(L+1) = 2$ for $L = 1$. Fano and Macek (1973) connected the anisotropic population of the magnetic substates with alignment and orientation parameters of the excited atom. These can also be linked with the above collision parameters λ and χ. The Fano–Macek alignment tensors for our example are given by

$$\begin{aligned} A_0^{\rm col} &= \langle 3L_z^2 - L^2\rangle\{L(L+1)\}^{-1} = \tfrac{1}{2}(1-3\lambda),\\ A_{1+}^{\rm col} &= \langle L_xL_z + L_zL_x\rangle\{L(L+1)\}^{-1} = \{\lambda(1-\lambda)\}^{1/2}\cos\chi,\\ A_{2+}^{\rm col} &= \langle L_x^2 - L_y^2\rangle\{L(L+1)\}^{-1} = \tfrac{1}{2}(\lambda-1). \end{aligned} \tag{52}$$

The orientation vector of the atom in the excited P state is defined by (Fano and Macek, 1973)

$$O_{1-}^{\rm col} = \langle L_y\rangle\{L(L+1)\}^{-1} = -\{\lambda(1-\lambda)\}^{1/2}\sin\chi. \tag{53}$$

Because the excitation/deexcitation process $^1S \rightarrow {}^1P \rightarrow {}^1S$ for helium represents a rather straightforward case, very simple relations for the polarization phenomenon of the angular correlation can directly be obtained from the properties of the wavefunction involved and from symmetry and invariance principles. This can be seen as follows. By taking the initial direction of the electron beam along the z axis and the scattered electrons trajectory in the xz plane, we denote the angular parts of the wavefunctions in Eq. (44) as follows:

$$\psi_z = |10\rangle,$$
$$\pm(1/\sqrt{2})(\psi_x \pm i\psi_y) = |1 \pm 1\rangle,$$
$$-\sqrt{2}\psi_x = |11\rangle - |1-1\rangle. \tag{54}$$

Thus Eq. (44) can be rewritten in the following representations:

$$\begin{aligned}
|\psi\rangle &= a_0\psi_z - \frac{1}{\sqrt{2}}a_1(\psi_x + i\psi_y) + \frac{1}{\sqrt{2}}a_{-1}(\psi_x - i\psi_y) \\
&= a_0\psi_z + \frac{1}{\sqrt{2}}\{a_{-1} - a_1\}\psi_x - \frac{i}{\sqrt{2}}\{a_1 + a_{-1}\}\psi_y \\
&= a_0\psi_z - \sqrt{2}a_1\psi_x \\
&= \frac{1}{2}\left\{a_0 + \frac{\sqrt{2}}{i}a_1\right\}(\psi_z + i\psi_x) + \frac{1}{2}\left\{a_0 - \frac{\sqrt{2}}{i}\right\}(\psi_z + i\psi_x).
\end{aligned} \tag{55}$$

The deexcitation process $^1\mathrm{P} \to {}^1\mathrm{S}$ has polarized components that are proportional to the moduli squared of the excitation amplitudes of the wavefunction components ψ_x, ψ_y, and ψ_z (components linearly polarized along the x, y, or z direction) or the combinations of wavefunctions $\psi_z \pm i\psi_y$, $\psi_x \pm i\psi_z$, and $\psi_x \pm i\psi_y$ (for the right- and left-handed circularly polarized components, $I^{\sigma\pm}$, observed in the x, y, or z directions, respectively):

$$I_x^\pi \propto 2|a_1|^2, \qquad I_y^\pi = 0, \qquad I_z^\pi \propto |a_0|^2,$$
$$I_{zx}^{\sigma+} \propto \left|a_0 + \frac{\sqrt{2}}{i}a_1\right|^2, \qquad I_{zx}^{\sigma-} \propto \left|a_0 - \frac{\sqrt{2}}{i}a_1\right|^2,$$
$$I_{xy}^{\sigma+} = I_{xy}^{\sigma-} \propto 2|a_1|^2, \qquad I_{zy}^{\sigma\pm} = 0. \tag{56}$$

As seen from the preceding equations, the photon radiation observed in any direction in the scattering plane is 100% linearly polarized parallel to the scattering plane, but the circular polarization vanishes. The linear and circular polarization of the photon radiation observed perpendicular to the scattering plane is given by

$$P_{zx}^{\mathrm{lin}} = \frac{I_z^\pi - I_x^\pi}{I_z^\pi + I_x^\pi} = \frac{|a_0|^2 - 2|a_1|^2}{|a_0|^2 + 2|a_1|^2} = 2\lambda - 1, \tag{57}$$

$$P_{zx}^{\mathrm{circ}} = \frac{I_{zx}^{\sigma+} - I_{zx}^{\sigma-}}{I_{zx}^{\sigma+} + I_{zx}^{\sigma-}} = -2\{\lambda(1-\lambda)\}^{1/2}\sin\chi. \tag{58}$$

Equation (58) is identical to Eq. (53). Note that right circularly polarized light has negative helicity in conventional optics.

No circular polarization occurs if the direction of observation of the photons is within the scattering plane; this is equivalent to the fact that no "net" orbital angular momentum transfer takes place with reference to the scattering place:

$$P_{xy}^{\text{circ}} = P_{yz}^{\text{circ}} = 0. \tag{59}$$

Wykes (1972) calculated the photon polarization of alkali resonance radiation detected in coincidence with inelastically scattered electrons. The general equations derived by Wykes describe the polarization for any arbitrary observation angle. Because fine and hyperfine structure and also the finite level width of the excited state have to be taken into account, the polarization of the alkali resonance radiation leads to modified expressions compared to Eqs. (56)–(59). The spin-orbit and nuclear magnetic interactions depolarize the photon radiation with the effect that, for example, P_{xy}^{lin} and P_{zy}^{lin} change and are no longer 100% as in the case of the helium $1^1\text{S} \rightarrow n^1\text{P}$ excitation process previously discussed.

B. Experimental Results

The scheme of the experimental method is determined by the following requirements: (1) a monoenergetic beam of electrons excites a well-defined target of atoms, e.g., a collimated beam of helium atoms; (2) both the inelastically scattered electrons and the photons from the deexcitation process have to be "filtered," the unwanted elastic and inelastic electrons and the photons arising from those processes not being studied. The initial monoenergetic beam of electrons exciting the atoms can be taken from the exit slit of an electron monochromator, and the inelastically scattered electrons having lost the threshold excitation energy can be filtered from unwanted electrons by means of a second electron monochromator tuned to the appropriate electron energy (Fig. 22). Photons with the frequency of the transition to be studied and the energy-selected electrons having excited the atom can be detected by observing them as delayed coincidences in a fast coincidence circuit (Bell, 1968; Imhof and Read, 1971a,b; Eminyan *et al.*, 1973, 1974). A typical energy-loss spectrum obtained by scanning the voltage on the entrance slit of the analyzer is displayed in Fig. 23. Photons from the transition to be studied and the energy-selected electrons are detected by appropriate electron and photon detectors and the corresponding pulses are fed into a time-to-amplitude converter (TAC). Correlated electron and photon pulses from true coincidence events arrive at the TAC with a definite correlation in time. These "real" coincidences fall into a restricted interval Δt of channels of the multichannel analyzer (MCA), whereas chance coincidences are randomly distributed and introduce only a uniform background on the delay-time spectrum of the MCA. The number of real coincidences collected in a given time can be found by subtracting the baseline measured outside Δt from the total number of coincidences, real and chance, within the coincidence peak confined to Δt. A delayed coincidence spectrum between pho-

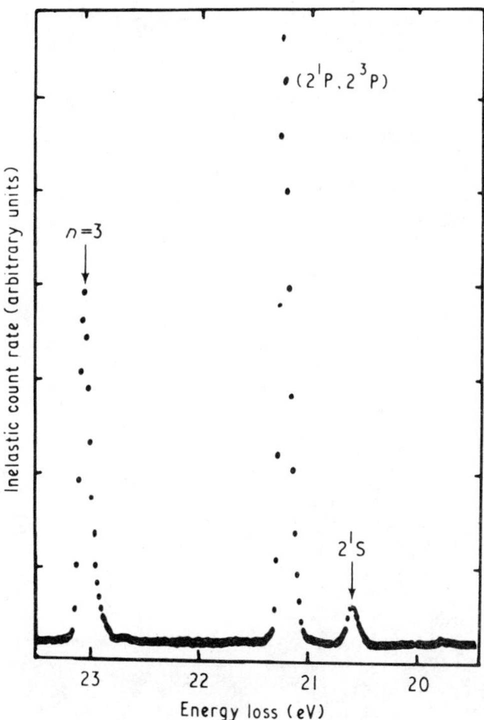

Fig. 23. Typical helium energy-loss spectrum as used for electron–photon coincidences from $2^1P \rightarrow 1^1S$ transitions (see Fig. 22). Initial electron energy, 80 eV; electron scattering angle, $\theta_e = 16°$; energy resolution, 0.25 eV (FWHM).

tons from the $2^1P \rightarrow 1^1S$ transition ($\lambda = 584$ Å) and the electrons from the (2^1P, 2^3P) energy-loss peak is shown in Fig. 24. Note that the insufficient energy resolution between the 2^1P and the 2^3P energy-loss peak does not affect the number of real coincidences since the 2^3P state does not decay with emission of vacuum ultraviolet light.

As discussed in the previous section, the electron–photon angular correlation function N_c [see Eqs. (49) and (50)] can be expressed in terms of the collision parameters λ and χ. The measurement of electron–photon angular correlations can then be used to determine parameters λ and χ from a fit of the angular correlation function N_c to the experimental correlation data. Figure 25 presents examples for electron–photon angular correlations from the helium $2^1P \rightarrow 1^1S$ transition. As seen from the examples of Fig. 25, the deviation between the experimentally observed correlation and the predicted one by first Born approximation is becoming larger with increasing electron scattering angle, thus demonstrating the well-known fact that the Born approximation (dashed curves in Fig. 25) approaches the exper-

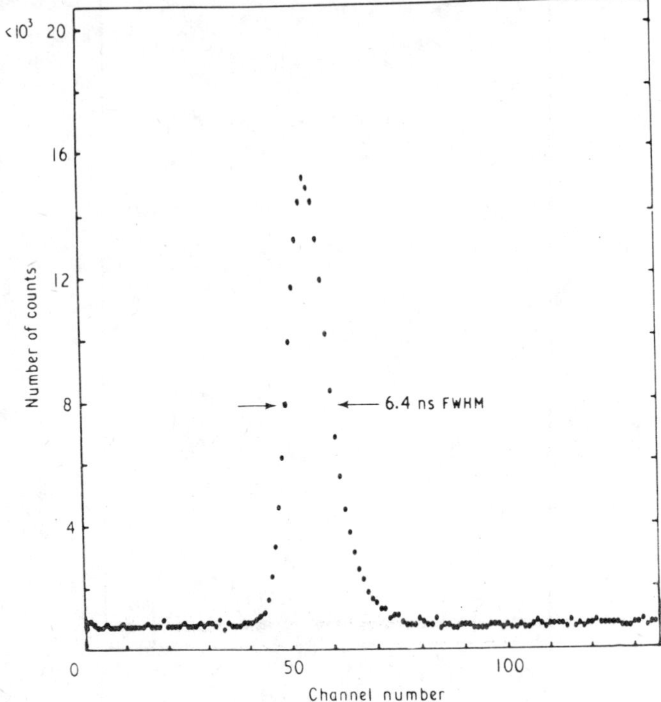

Fig. 24. Delayed coincidence spectrum from the excitation/deexcitation process $1^1\mathrm{S} \rightarrow 2^1\mathrm{P} \rightarrow 1^1\mathrm{S}$ of helium. Initial electron energy, 80 eV; electron scattering angle, $\theta_e = 16°$; photon angle, $\theta_\gamma = 126°$ (opposite to scattered electron beam); accumulation time ~ 12 hours; channel width 0.58 nsec; inelastic electron rate, ~ 6 kHz; photon rate, 21 kHz. (From Eminyan *et al.*, 1973, 1974.)

imental data for small momentum transfer and sufficiently high energy—at the lowest scattering angle in Fig. 25 ($\theta_e = 16°$) the first Born approximation is close to the experimental data, whereas it deviates substantially from reality at larger angles ($\theta_0 = 35°$ and $40°$) at the same energy of 60 eV. It might be useful to recall the fact that the Born approximation requires the selection rule $\Delta m_l = 0$ with respect to the momentum transfer direction. Consequently, the electron–photon angular correlation of our excitation/deexcitation process under consideration may be illustrated by the picture as shown in Fig. 26. For a fixed electron scattering angle the probability of finding the photon is determined by the dipole characteristics of a classical oscillator parallel to the momentum transfer ΔP. Zero intensity of the photon emission occurs in the direction of the oscillating dipole or in the direction of the momentum transfer ΔP, which, together with the direction of the incoming electron beam, define angle $\theta_{\min}$. In the case observed, which is illustrated in the lower part of Fig. 26, the minimum photon inten-

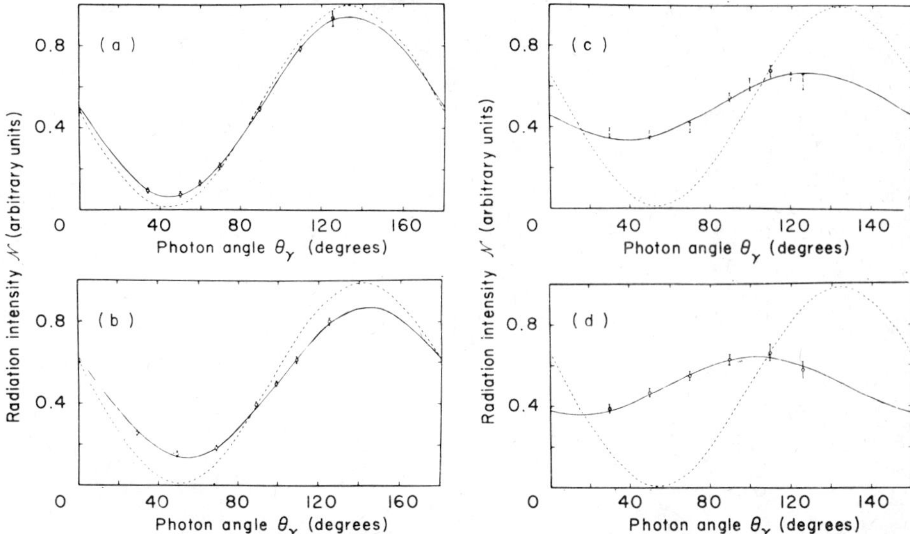

Fig. 25. Electron–photon angular correlation for helium $2^1P \rightarrow 1^1S$ transitions for different electron scattering angles at 60 eV. Error bars, 1 rms error. Full curve, least squares fit of Eq. (49) to the data; dashed curve, predictions of first Born approximation.

sity does not decrease to zero and does not occur in the direction of the momentum transfer. The tilt between the minimum angle, as predicted by the Born approximation, and the actual observed minimum angle can be as large as 40° according to the measurements (see Fig. 25).

In order to extract the collision parameters λ and χ and the minimum angle θ_{min} from the electron–photon angular correlation, a least squares fit of the theoretical angular correlation function [Eq. (49)] to the data points, such as those of Fig. 25, was carried out. Figures 27–31 show experimental data obtained by this procedure for λ, $|\chi|$, and θ_{min} of the He, $1^1S \rightarrow 2^1P \rightarrow 1^1S$ excitation/deexcitation processes.

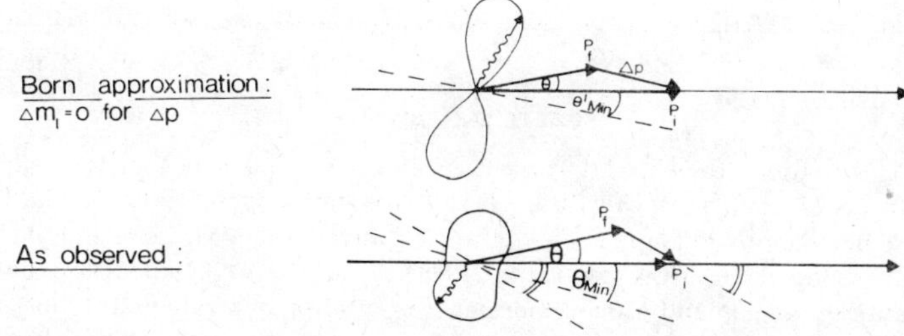

Fig. 26. Momentum analysis (P_i, initial momentum of electron traveling in z direction; P_f, final momentum of electron after collisional excitation; ΔP, momentum transfer to atom) and photon angular distribution curve (arrow with zig-zag stem) of $^1S \rightarrow {}^1P \rightarrow {}^1S$ excitation/de-excitation process of helium.

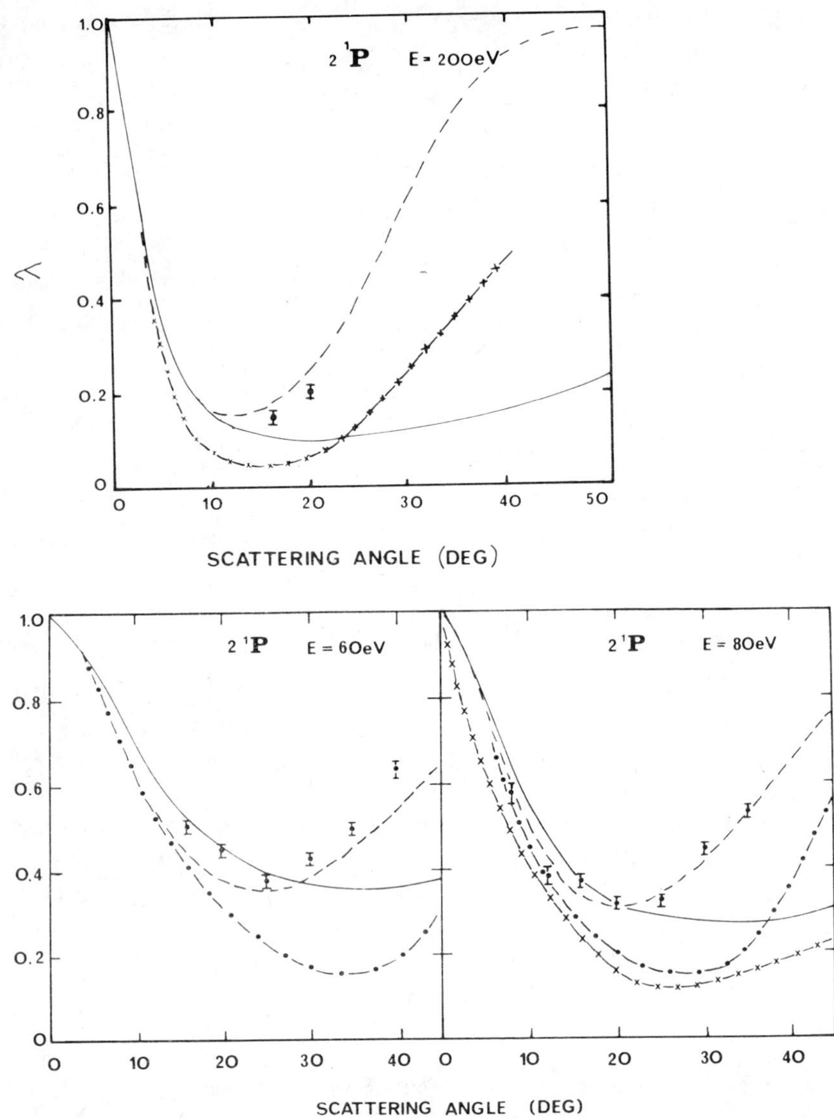

Fig. 27. Variation of λ with electron energy and scattering angle for 2^1P excitations from the experiments of Eminyan *et al.* (1973, 1974). Error bars represent one standard deviation. (——) First Born approximation, (- - - -) distorted wave theory (Madison and Shelton, 1973), (-·-·) many-body theory (Csanak *et al.*, 1973), (×-×-×) eikonal distorted-wave Born approximation (Joachain and Vanderpoorten, 1974, also private communication).

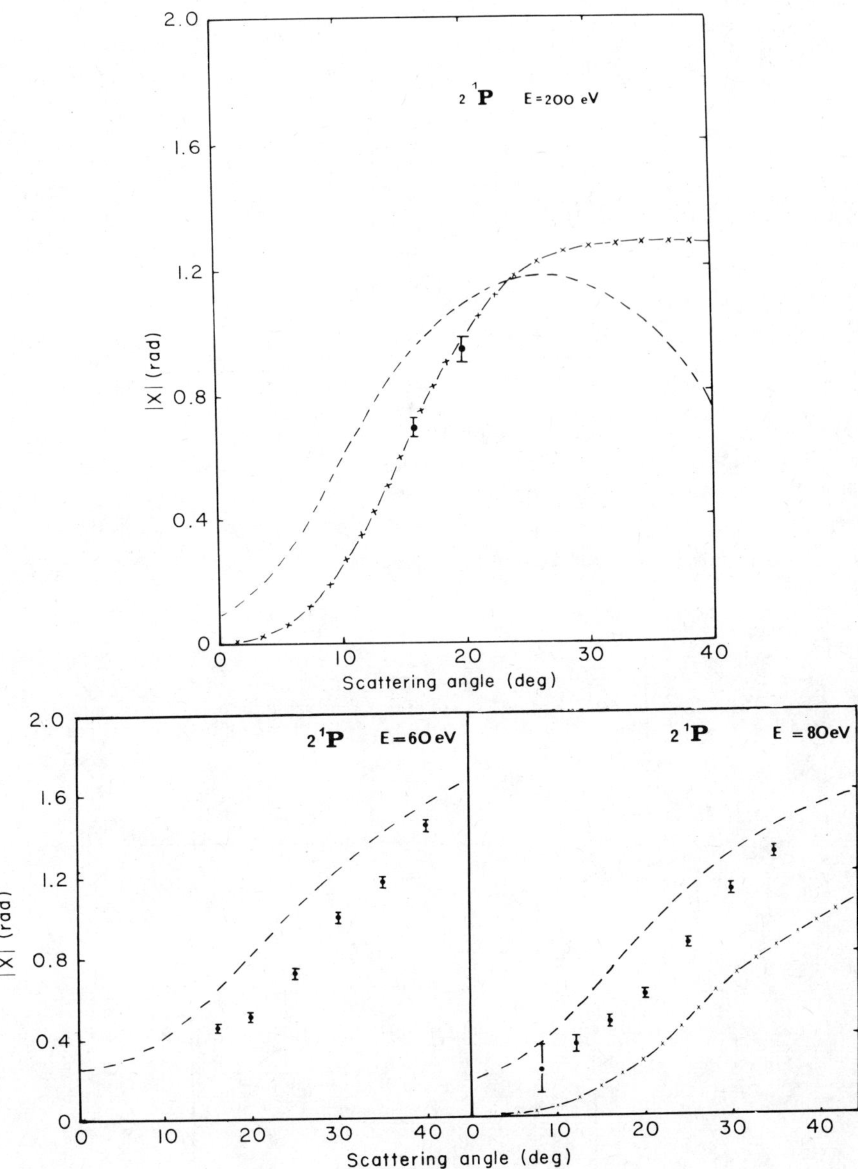

Fig. 28. Variation of $|\chi|$ with electron scattering angle for the 2^1P excitation of helium. (From Eminyan *et al.*, 1973, 1974.) Error bars are 1 SD. Legend as in Fig. 27.

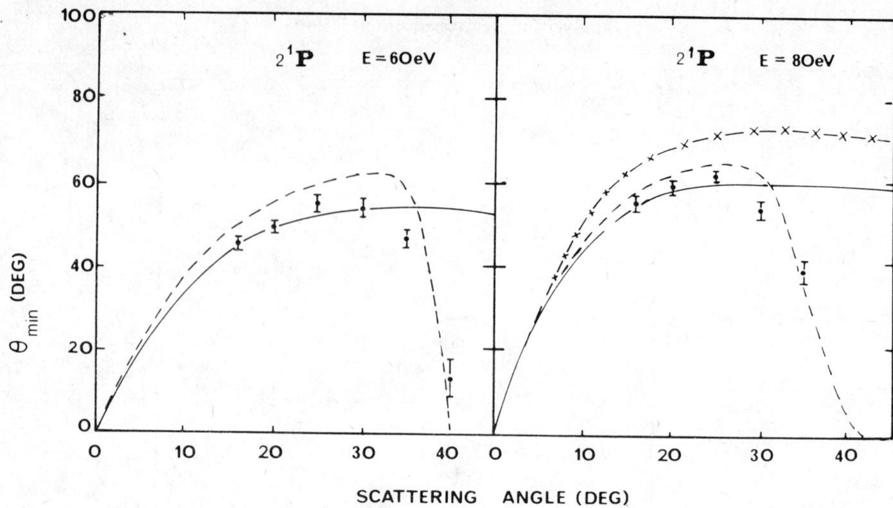

Fig. 29. Variation of θ_{min} with electron energy and scattering angle for 2^1P excitation of helium. Legend as in Fig. 27. (From Eminyan *et al.*, 1973, 1974).

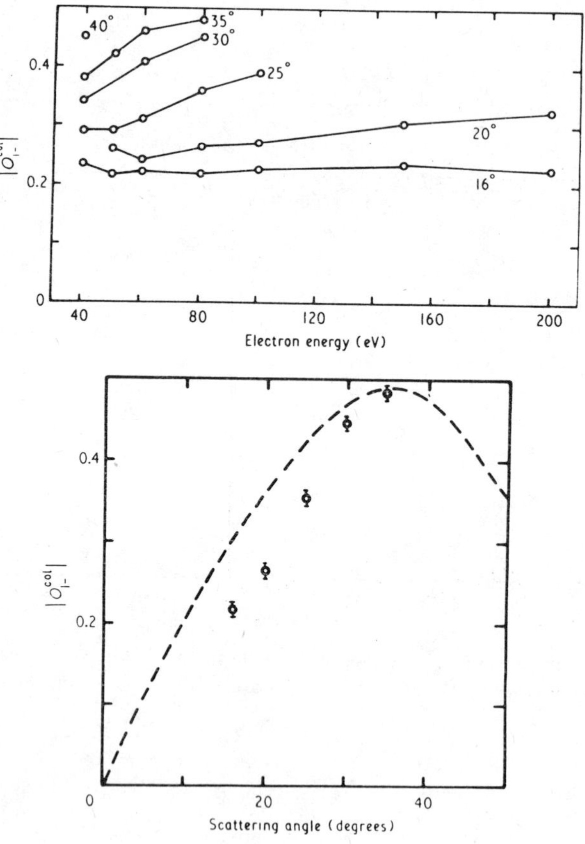

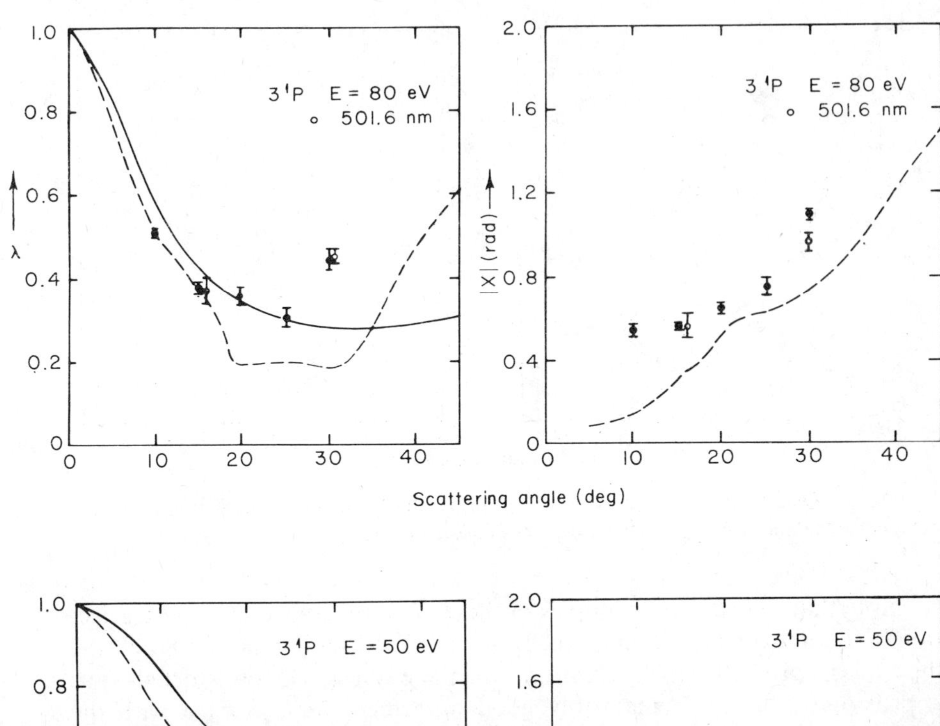

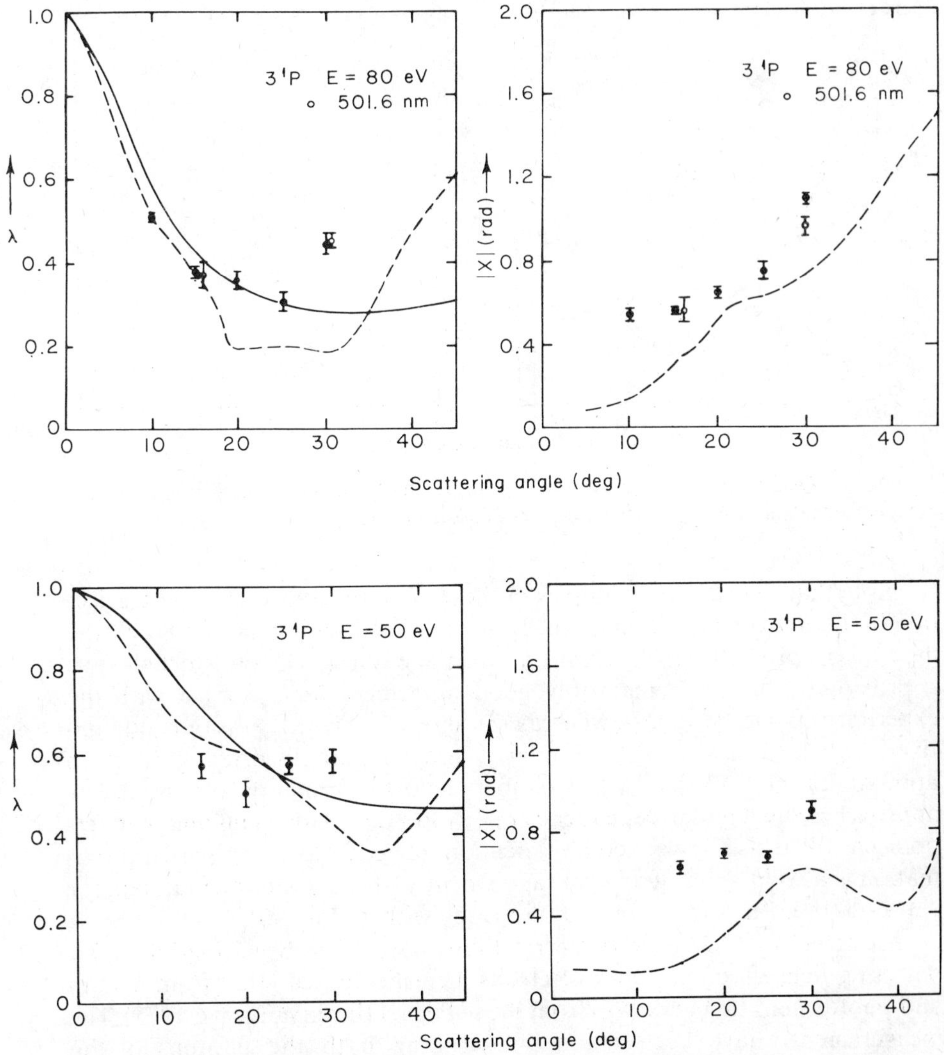

Fig. 31. Values of λ and $|\chi|$ versus electron scattering angle for the 3^1P excitation of helium from the observation of the 537 Å photons ($3^1P \rightarrow 1^1S$ transition) data or 5016 Å photons ($3^1P \rightarrow 2^1S$ transition) data, 50 and 80 eV, by Eminyan *et al.* (1973, 1974). Error bars are 1 SD. Solid curves represent first Born approximation. Dashed curves are the multichannel eikonal predictions of Flannery and McCann (1975).

Fig. 30. Variation of the Fano–Macek orientation parameter with electron energy and different scattering angles (upper part) and with the scattering angle with fixed energy (80 eV) for the helium $1^1S \rightarrow 2^2P$ excitation process. The upper part includes only the experimental data, whereas the lower part also displays theoretical predictions (dashed curve) for 78 eV by Madison and Shelton (1973).

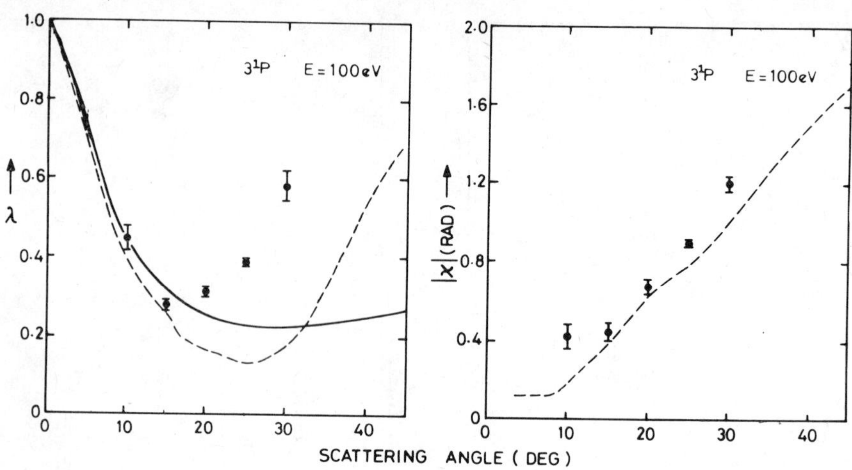

Fig. 32. Quantities λ and $|\chi|$ versus electron scattering angle for the 3^1P excitation of helium. Same data as in Fig. 31 but with 100 eV electron energy.

Note that the various examples of these data are compared with a selection of theoretical approximations. It is worth mentioning that for 80 eV the theoretical prediction of Madison and Shelton (1973) (distorted-wave approximation) is in remarkably good agreement for $\lambda = \sigma_0/\sigma$ with the experimental data (Fig. 27), whereas the agreement with $|\chi|$ is only qualitatively satisfactory (Fig. 29). Similarly, the eikonal distorted-wave Born approximation (DWBA) with Glauber distorting potentials qualitatively approaches the angular dependence of $|\chi|$ at 80 eV (Joachain and Vanderpoorten, 1974, also private communication) (Figs. 28 and 29). This approximation is also in good qualitative agreement with the experimental data for $|\chi|$ at 200 eV but shows some discrepancy with λ (Fig. 30).

Electron–photon angular correlations have also been measured by detecting delayed coincidence electrons inelastically scattered from helium and photons emitted in decays from the 3^1P level (Eminyan *et al.*, 1975). The measurements have been made by detecting both the photons of the 3^1P–1^1S (537 Å) or the 3^1P–2^1S (5016 Å) transition. Of course, the theoretical interpretation of these angular correlations is identical to that described above for the 2^1P–1^1S transition. Figures 32–39 present data for the quantities λ, $|\chi|$, $\theta_{\min}$, and $O_{1-}^{\rm col}$ from the helium 3^1P angular correlation measurements. These data also allow a detailed comparison for the 3^1P excitations for electron scattering in the angular range 10°–30° between various intermediate energy theoretical approximations. At present only the multichannel eikonal calculations of Flannery and McCann (1975) are available for comparison with the 3^1P data.

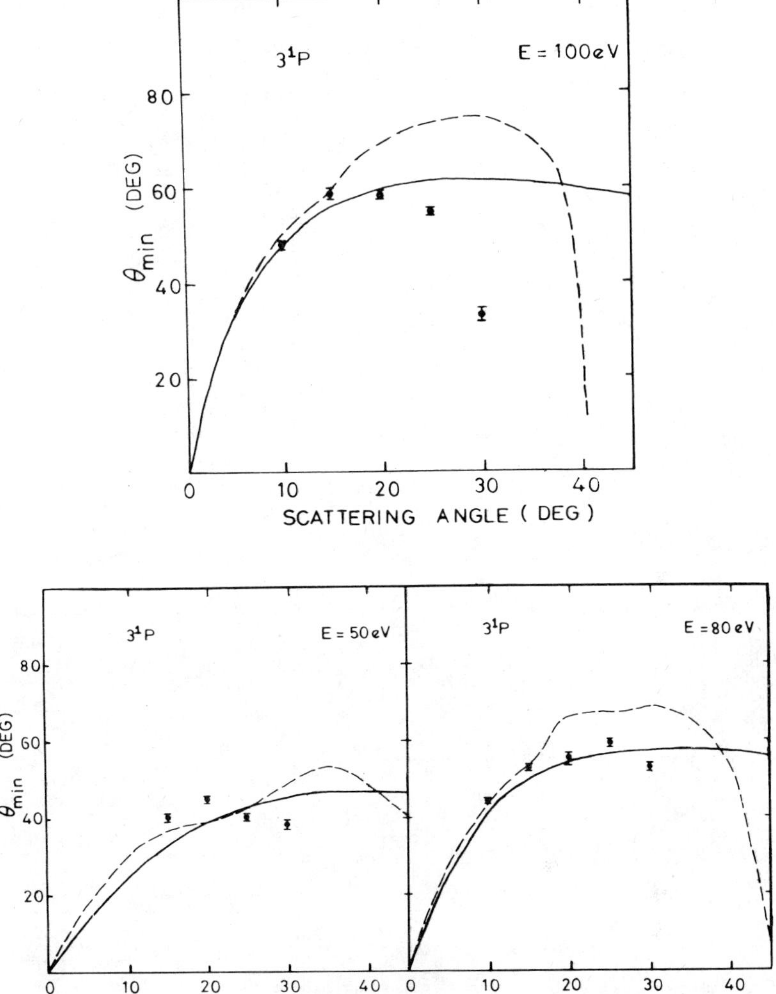

Fig. 33. Values of θ_{min} for the 3^1P excitation of helium versus electron scattering angle at incident electron energy of 50, 80, and 100 eV. Error bars are 1 SD. Solid curves are first Born approximations; dashed curves are multichannel eikonal predictions calculated from the λ and χ values of Flannery and McCann (1975).

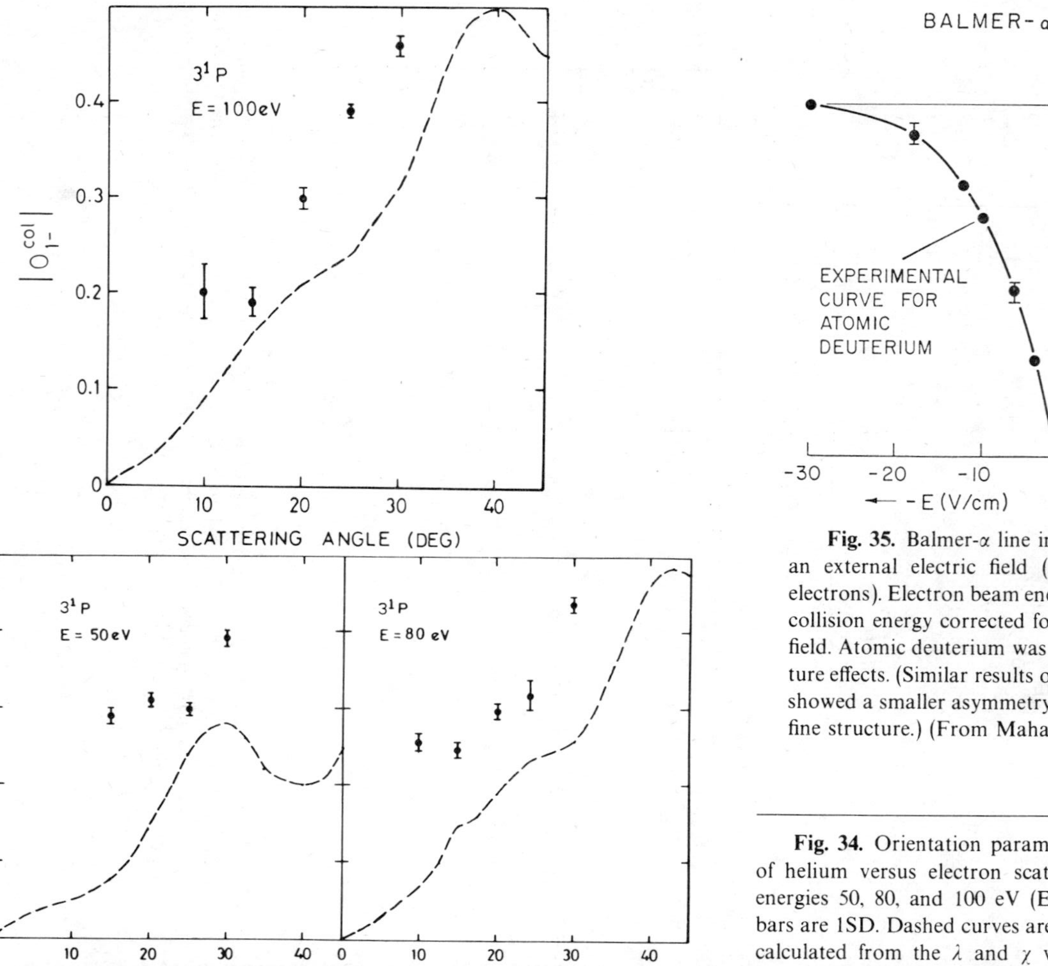

Fig. 35. Balmer-α line intensity measured as a function of an external electric field (axial field in beam direction of electrons). Electron beam energy is 500 eV which is the average collision energy corrected for gain or loss in the applied axial field. Atomic deuterium was used to minimize hyperfine structure effects. (Similar results obtained by using atomic hydrogen showed a smaller asymmetry, due to the more complex hyperfine structure.) (From Mahan *et al.*, 1973, and Smith, 1975.)

Fig. 34. Orientation parameter $|O_{1-}^{col}|$ for the 3^1P excitation of helium versus electron scattering angle at incident electron energies 50, 80, and 100 eV (Eminyan *et al.*, 1972, 1974). Error bars are 1SD. Dashed curves are multichannel eikonal predictions calculated from the λ and χ values of Flannery and McCann (1975).

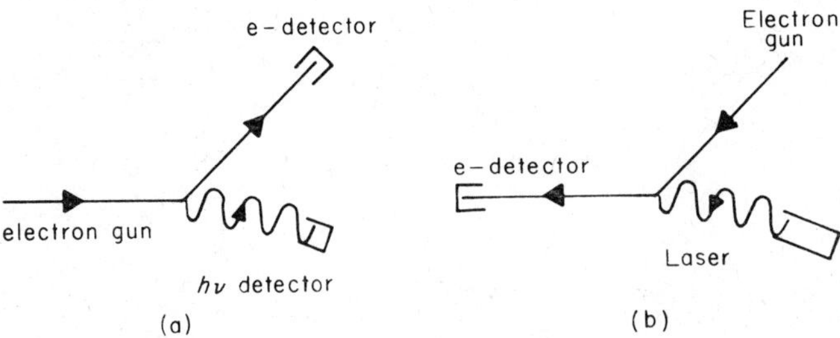

Fig. 36. Comparison of (a) the photon-coincidence experiment with (b) the electron-deexcitation collision experiment for a laser-excited target. As pointed out by Hertel and Stoll (1974b), the latter can be seen to result from (a) by a time reversal operation.

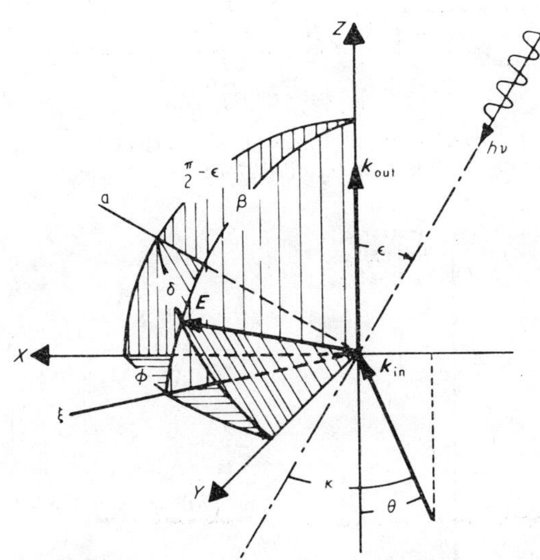

Fig. 37. Geometry of the electron scattering by laser-excited atoms: k_{in}, k_{out} are vectors of in- and outgoing electrons; δ is the angle of the polarization vector of the light wave with respect to the electron scattering plane as defined by the wave vectors.

The 3^1P angular correlation data exhibit the same general trends as the 2^1P data. In comparing these data, it is worthwhile to note that the marked departure of θ_{min} from the momentum transfer axis observed at large scattering angles in the 2^1P 60 and 80 eV data was also observed for the 3^1P 100 eV data and that this behavior seems to be associated with a very high degree of orientation of the excited atom.

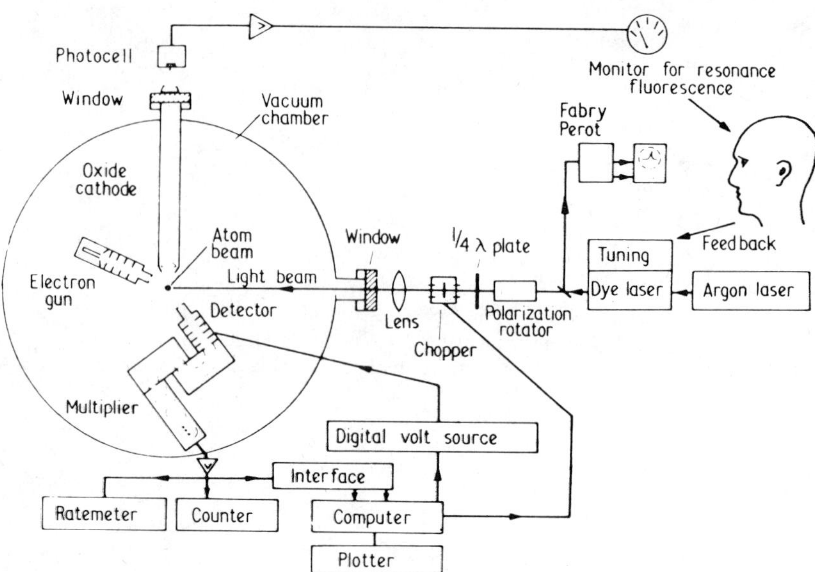

Fig. 38. Schematic diagram of the apparatus for inelastic scattering of slow electrons by laser-excited sodium atoms (Hertel and Stoll, 1974b).

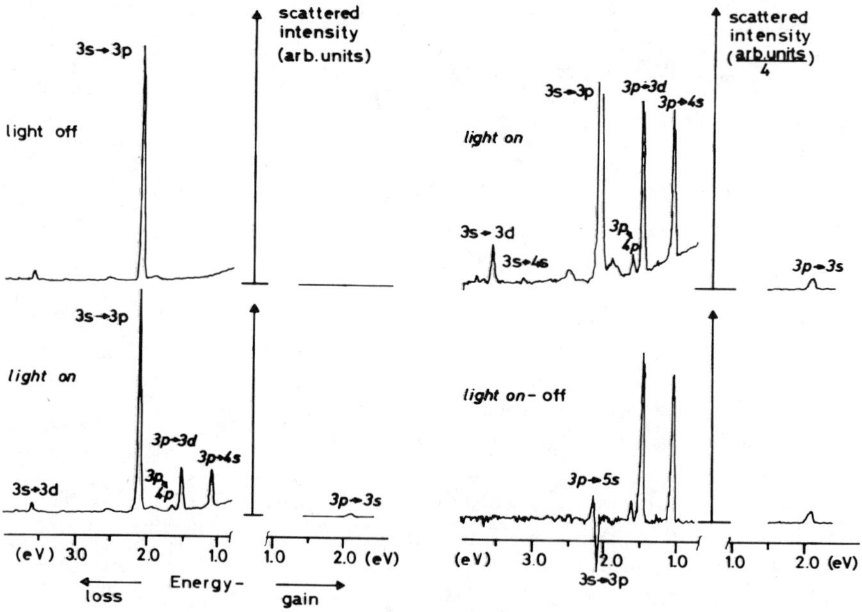

Fig. 39. Energy loss/gain spectra for electrons scattered by ground- and excited-state sodium atoms (Hertel, 1975).

V. Coherent Electron-Impact Excitation of Atomic Hydrogen

Mahan *et al.* (1973) showed how coherent electron-impact excitation of the $n = 3$ shell of atomic hydrogen can be detected by applying a static electric field in the collision region. The characteristic result is the observed asymmetry (Fig. 35) of the Balmer-α radiation with respect to the direction of an electric field applied in the excitation region. It follows from a calculation by Mahan *et al.* (1973) that the asymmetry observed in the Balmer-α emission excited by electron impact is the result of coherent excitation of states of opposite parity and the same axial projection of orbital momentum (m_l), predominantly $P_{3/2,\,3/2}$–$D_{3/2,\,3/2}$, $P_{3/2,\,1/2}$–$D_{3/2,\,1/2}$, and

$$S_{1/2,\,1/2}\text{–}P_{1/2,\,1/2}$$

pairs. A coherent superposition of states of opposite parity implies nonzero electric dipole moment (Eck, 1973). The charge density of the atom in the coherently excited states is proportional to $|\psi_{\text{even}} + \psi_{\text{odd}}|^2$ which leads to cross-terms that change sign across the plane transverse to the electron beam at the nucleus. In a classical picture, the role of the static electric external field is to superpose an additional displacement of the charge cloud which increases or decreases the charge polarization of the upper state during the radiation period. In order to provide some verification of the observed asymmetry in Balmer-α emission, Mahan (1974) has expressed the emission intensity as a function of applied electric field, following from the formulation of Percival and Seaton (1958) using Born excitation cross sections by inserting terms to represent Stark mixing between excitation and spontaneous emission. Thus the total Balmer-α intensity has been expressed by

$$I_{\text{tot}} = \int_0^\infty I(t)\,dt \sim \frac{1}{k_i^2}\int \sum_{\alpha\delta\gamma}$$

$$\times \left[\sum_{\beta\beta'} \int_0^\infty F_{\beta\beta'}(\alpha, \hat{K}) A_{\beta\delta,\,\beta'\delta}(\mathbf{E}, t) G_{\delta\gamma}(\hat{e})\,dt\right] K\,dK\,d\phi.$$

Here the terms are defined as follows:

$f_\beta(\alpha, \hat{K})$ is the scattering amplitude for excitation of an atom from an initial state α to an excited state β by electron impact with a momentum change vector in the direction K;

$g_{\beta\gamma}(\hat{e})$ is the radiation matrix element giving the transition probability per second for emission of an $\hat{e}$ photon by an atom in an excited state β radiating into the final state γ;

$F_{\beta\beta'}(\alpha, K) \sim f_{\beta}(\alpha, \hat{K})f^*_{\beta'}(\alpha, \hat{K})$, where the JM_J scattering amplitudes have been calculated from the appropriate vector-coupled Born LM_L excitation amplitudes in which relative phases of excitation are carefully preserved;

$A_{\beta\delta,\,\beta'\delta}(\mathbf{E}, t) \sim a_{\beta\delta}(\mathbf{E}, t)a^*_{\beta\delta}(\mathbf{E}, t)$, where the appropriate $a_{\beta\gamma}(\mathbf{E}, t) \sim a_i(t)$ are the amplitudes of states coupled by an external electric field;

$G_{\gamma}(\hat{e}) \sim g_{\delta\gamma}(\hat{e})g^*_{\delta\gamma}(\hat{e})$ are the radiation matrix elements connecting these Stark mixed δ levels and final $n = 2$ levels resulting in H_{α} radiation.

This formula predicts an asymmetry as observed in the Balmer-α radiation with respect to the direction of the electric field. It fails at quantitative prediction and this may be attributed to the inadequacy of the Born approximation which does not take into account the substantial charge polarization effect and which does not make reliable predictions for phase differences of excitation amplitudes.

Morgan and McDowell (1975) dealt with the theory of the coherent excitation of nP states of atomic hydrogen by which photons from deexcitations $n\text{P} \rightarrow n\text{S}$ are detected in coincidence with the inelastically scattered electrons. The authors showed that electron–photon coincidence measurements (e.g., Lyman-α photons in coincidence with the inelastically scattered electrons) of such processes do not allow a measurement of the relative phase of the m_l components of the nP state unless the singlet and triplet scattering can be separated from each other by using spin-polarized electrons and atoms.

Morgan and McDowell (1975) also carried out exploratory calculations on electron L_{α} photon coincidences in the distorted-wave-polarized orbital approximation. Morgan and Stauffer (1975) have made similar calculations in the Born, Born-Oppenheimer, Coulomb-projected Born, Coulomb-projected Born exchange (CPBE), and the generalized CPBE approximation.

VI. Electron Scattering from Laser-Excited Atoms

Recently Hertel and Stoll (1974b) reported a successful experiment for the scattering of electrons on laser-excited sodium atoms. As shown by these authors, the scattering process includes superelastically scattered electrons from atoms in the laser-excited states. In fact, this type of experiment was interpreted as a time-reversed electron–photon coincidence experiment (Section IV) as illustrated in Fig. 36.

The angular correlation in the electron–photon coincidence experiment of Eminyan *et al.* (1973, 1974) corresponds to the laser experiment in that the superelastic scattering occurs only when photons excite the atoms from the ground to the excited state (see Fig. 36). We will first explain how the electron scattering from laser-excited atoms can theoretically be interpreted on similar lines as described in Section IV and then we will discuss the initial experimental data of Hertel and Stoll (1974b).

A. Theory

1. *Optical Pumping*

From the knowledge of optical pumping of the $^2S_{1/2} \rightarrow {}^2P_{1/2,\,3/2}$ transitions in alkali atoms we can derive the following conclusions:

a. pumping with right-handed circularly polarized light (σ^+) prepares a preferential population of states with positive magnetic quantum number m_F of both the excited and ground state. In the limiting case of infinite number of absorption and reemission processes, the atoms are in the $m_F = 3$ substate of the excited state $P_{3/2}$ and the $m_F = 2$ substate of the ground or excited $P_{1/2}$ state. In this case the atoms are completely oriented either in the ground or excited states.

b. Pumping with linearly polarized light leads in the limiting case of a large number of absorption and reemission processes to no population of the $m_F = \pm 3$ states but there is overpopulation in the magnetic substates with $m_F = 0$ and $m_F = \pm 1$. Consequently, the atom has been aligned.

2. *Scattering Intensity*

Hertel and Stoll (1974b) developed a theory for the superelastic scattering from an optically pumped excited level n_i to another level n_j in terms of the time-inverse scattering process $n_j \rightarrow n_i$. In the actual experimental situation, level n_i is prepared in a particular state by the optical pumping process, for example, with right or left circularly polarized light this state is an eigenstate with $M_F = \pm F$. The scattered electron intensity is then proportional to the component of the eigenstate $|FM_F)$ contained in the scattering wavefunction Ψ which follows from the asymptotic wavefunction:

$$\exp(ik_j \cdot r)\psi_{m_j} + \frac{1}{r}\exp(ik_i r)\sum_{m_i} \psi_{m_i} f_{im_i,\,jm_j} = \exp(ik_j \cdot r)\psi_{m_j} + \frac{1}{r}\exp(ik_i r)\Psi_i \tag{60}$$

with

$$\Psi_i = \sum_{m_i} \psi_{im_i} f_{im_i,\,jm_j},$$

where ψ_{im_i} are the eigenfunctions of the magnetic substates of n_i. Thus we have for the electron scattering intensity from the state-selected atoms:

$$I = C\langle |FM_F|\Psi_i|^2\rangle, \tag{61}$$

where C is a constant consisting of multiplicative factors such as solid angle, detection efficiencies, the number of atoms in the optically pumped level n_i, the flux of the incident electrons, and the total (at fixed scattering angle) differential superelastic scattering cross section $\sigma_{n_j n_i}$ for the transition $n_i \to n_j$. The last factor $\sigma_{n_j n_i}$ incorporates the sum over final states and averages over initial states appropriate to excitation by scattering from an initial isotropic distribution. All of the effects of the anisotropy of the pumped level are contained in $\langle |(FM_F|\Psi_i)|^2 \rangle$.

Based upon this formula, the differential deexcitation cross section for a polarized initial hyperfine state can be calculated as follows:

$$\sigma(\theta, \delta, \varepsilon; 3\text{p}[FM_F] \to 3\text{s}) = \frac{k_{\text{out}}}{k_{\text{in}}} \sum_{M\bar{M}} K_{M\bar{M}}(\phi, \beta) q_{M\bar{M}}(\theta), \tag{62}$$

where

$$q_{M\bar{M}}(\theta) = \sum_{S=0} \frac{2S+1}{4} f_m(\theta, 0) f_M(\theta, 0) \tag{63}$$

and S is the total electron spin.

We may express $q_{M\bar{M}}$ in terms of direct and exchange amplitudes f_M and g_M, respectively ($f_M^S = f_M + (-)^S g_M$) and obtain

$$q_{M\bar{M}}(\theta) = f_M f_{\bar{M}}^* - 0.5^*(f_M g_{\bar{M}}^* + f_{\bar{M}}^* g_M) + g_M g_{\bar{M}}^* . \tag{64}$$

Or, if we put $f_M = |f_M| \exp(i\phi_M)$ and

$$g_M = |g_M| \exp(i\chi_M) \tag{65}$$

with g_M, ϕ_M, χ_M real, we can describe any $q_{M\bar{M}}$ in our case by the seven values:

$$|f_1| = |f_{-1}|, \ |f_0|, \qquad g_1 = g_{-1}, \qquad \phi_1 = -\phi_{-1} + \pi, \qquad \chi_1 = -\chi_{-1}, \chi_0, g_0 .$$

$$\begin{aligned} q_{MM}(\theta) = {} & |f_M| \, |f_{\bar{M}}| \exp[i(\phi_M - \phi_{\bar{M}})] \\ & \times \{1 - 0.5 g_M \exp(i\chi_{\bar{M}}) + g_{\bar{M}} \exp(-i\chi_{\bar{M}}) \\ & + g_M g_{\bar{M}} \exp i(\chi_M - \chi_{\bar{M}})\}. \end{aligned} \tag{66}$$

The special features of the laser-excited experiment are incorporated in $K_{M\bar{M}}$:

$$\begin{aligned} K_{M\bar{M}}(\phi, \beta) = {} & \\ & \sum_{M_s M_I M_F} C_{M_s\, M\, M_s+M}^{1/2\, L\, J} C_{M_s\, M\, M_s+M}^{1/2\, L\, J} C_{M_1\, M_s+M\, M_s+M+M_1}^{1\ \ J\ \ F} C_{M_1\, M_s+M\, M_s+M+M_1}^{1\ \ J} \\ & \times (M_F)_{M_s+M+M_1,\, M_F}^{F}(\phi, \beta, -\phi)_{M_s+M+M_1,\, M_F}^{F}(\phi, \beta, -\phi). \end{aligned} \tag{67}$$

It can be shown that

$$K_{M\overline{M}} = K^*_{\overline{M}M} \quad \text{and} \quad q_{M\overline{M}} = q^*_{\overline{M}M} .$$

The differential deexcitation cross section of Eq. (62) can be compared with the usual inelastic cross section for the excitation processes:

$$\sigma(\theta, 3\text{S} \to 3\text{P}) = \frac{k_{\text{out}}}{k_{\text{in}}} \; {}_M q_{MM}(\theta). \tag{68}$$

The important difference between this "normal" inelastic cross section and the deexcitation cross section in Eq. (62) is based upon the fact that in the inelastic case only diagonal terms $q_{MM}(\theta)$ occur, whereas in the deexcitation process of the scattering, the cross section can contain nondiagonal terms, which can be related to phase differences between the scattering amplitudes.

Macek and Hertel (1974) showed that the scattering intensity [Eq. (61)] can be related to state multipoles of the excited atoms defined as expectation values of irreducible tensors that are constructed from components of angular momentum vectors. To this end the authors write Eq. (61) as the expectation value of the projection operator $\tau = |FM_F)(FM_F|$, τ as a sum of its irreducible components $\tau_q^{[k]}$, and apply the Wigner–Eckart theorem to relate $\tau_q^{[k]}$ to operators $T_q^{[k]}$ constructed from components of angular momentum operators, and transform these operators from the reference frame defined by the pumping light to the frame defined by the collision. The scattering intensity is then given by

$$I = C \sum_k W(k) V^{(k)}(F) T_{0+}^{[k]}(\text{ph}) \tag{69}$$

where

$$\begin{aligned} V^{(k)}(F) &= \frac{(F||\tau^{[k]}||F)}{(F||T^{[k]}||F)} \\ &= \frac{2^k(2k+1)^{1/2}}{k!} \left[\frac{(2F-k)!}{(2F+k+1)!}\right]^{1/2} \end{aligned} \tag{70}$$

and

$$W^{(k)} = \sum_{M_F} (-)^{k-F-M_F} W(M_F)(F - M_F F M_F | kq). \tag{71}$$

The final step consists of transforming to the collision frame. Since only $q = 0$ terms occur in Eq. (69), transformation to the collision frame introduces spherical harmonics rather than the more general rotation functions. Then

$$\langle T_{0+}^{[k]}(\text{ph})\rangle = \sqrt{(4\pi)}(2k+1)^{-1/2} \sum_{pq=0}^{k} T_{qp}^{[k]}(\text{col}) Y_{qp}^{[k]}(\theta, \phi), \tag{72}$$

where θ, ϕ are the spherical coordinates of the laser light axis for circularly polarized light and of the electric vector for linearly polarized light.

Equation (69) together with the transformation relation [Eq. (72)] relates the scattered electron intensity to the $2(F+1)^2 - 1$, F = integer or $2(F+\frac{3}{2})(F+\frac{1}{2}) - 1$, F = half-integer, state multipoles $\langle T_{qp}^{[k]} \rangle$. For $k = 0$, 1, and 2 parameters in Eq. (72) the state multipoles can be connected with the Fano–Macek orientation and alignment parameters (see Section IV). They are

$$\begin{aligned} O_{1-}^{\text{col}} &= \langle T_{1-}^{[1]}(\text{col}) \rangle / F(F+1), \\ A_0^{\text{col}} &= 2\langle T_0^{[2]}(\text{col}) \rangle / F(F+1), \\ A_{1+}^{\text{col}} &= 2(3)^{-1/2} \langle T_{1+}^{[2]}(\text{col}) \rangle / F(F+1), \\ A_{2+}^{\text{col}} &= 2(3)^{-1/2} \langle T_{2+}^{[2]}(\text{col}) \rangle / F(F+1). \end{aligned} \tag{73}$$

The relation of the preceding state multipoles $\tau_q^{[k]}(L, \text{col})$ in the collision frame to the collision parameters of Hertel and Stoll (1974b) is given by

$$\tau_q^{[k]}(L, \text{col}) = \sum_M (-)^{k-L-M} (L - MLM + q \mid kq) \frac{q_{MM} + q}{u^q uu}. \tag{74}$$

For the case of an S → P transition the multipole components are summarized in Table IV.

TABLE IV

NONZERO MULTIPOLE MOMENT COMPONENTS $\langle \tau_0^{[0]} \rangle$, $\langle \tau_{1-}^{[1]} \rangle$, ..., IN THE COLLISION FRAME FOR THE CASE OF AN S → P TRANSITION (P BEING A LASER-EXCITED STATE) ACCORDING TO THE THEORY OF MACEK AND HERTEL (1974)

$$C = 1/\sum_\mu q_{\mu\mu} \qquad q_{M\bar{M}}(\theta) = \sum (2+1)/4 f_M f_M^*$$

$$\begin{aligned} \langle \tau_0^{[0]} \rangle &= \frac{1}{\sqrt{3}} C \sum_M q_{MM} = \frac{1}{\sqrt{3}} \\ \langle \tau_1^{[1]} \rangle &= 2 \cdot \sqrt{2} C(1011 \mid 11) \text{ Im } q_{01} = -2C \text{ Im } q_{01} = \sqrt{2} O_{1-}^{\text{col}} \\ \langle \tau_{0+}^{[2]} \rangle &= -C\{(1010 \mid 20) q_{00} - (111-1 \mid 20) q_{11}\} = -\sqrt{\tfrac{2}{3}} C(q_{00} - q_{11}) = \sqrt{\tfrac{2}{3}} A_0^{\text{col}} \\ \langle \tau_1^{[2]} \rangle &= -2\sqrt{2} C(1011 \mid 21) \text{ Re } q_{01} = -2C \text{ Re } q_{01} = \sqrt{2} A_{1+}^{\text{col}} \\ \langle \tau_2^{[2]} \rangle &= -\sqrt{2} C(1111 \mid 22) q_{11} = -\sqrt{2} C q_{11} = \sqrt{2} A_{2+}^{\text{col}} \end{aligned}$$

B. Experimental Scheme and Results

Hertel and Stoll (1974b) applied the standard crossed-beam techniques whereby the laser and the initial electron beam crossed an alkali atomic beam (see Fig. 38). They used 180° electrostatic analyzers with the primary electron beam and in front of the electron multiplier. The fluorescent light emitted from the alkali beam is continuously monitored as reference signal for the laser stabilization. The scattered electrons can be detected separately according to their energy loss or gain after the collision. A typical energy loss/gain spectra from sodium is shown in Fig. 39. The most pronounced loss peak results from the excitation of the 3^2P fine structure state. After the laser beam tuned to the 3S–3P transition is switched on, a number of further loss/gain peaks occur; those from the 3^2D and 4^2S excitation are the most prominent ones. The $3^2P \rightarrow 3^2S$ peak on the scale with excess energy ("gain") is clearly the superelastic peak resulting from the deexcitation process $3^2P \rightarrow 3^2S$. The negative peak at the position of the $3^2S \rightarrow 3^2P$ inelastic process in the lower right part of Fig. 39 corresponds to a depopulation of the ground state (difference of scattering intensities between laser beam on and off). Even $3^2P \rightarrow 4^2P$ and $\rightarrow 5^2S$ processes are clearly resolvable.

Although theoretical analysis for the electron scattering from laser-excited atoms provides many new aspects for the collision dynamics and the properties of target atom, the present experimental results were mainly concerned with test measurements on different experimental parameters. Particularly measurements of the scattering angle dependence of the excited state cross sections lead to difficulties concerning the scattering volume. These difficulties appear to be enhanced in the laser-excited scattering experiment because three beams, namely the atomic, the initial electron, and the photon beam, have to cross in a most ideal way in a common collision region. Nevertheless, the initial experimental data already provided promising results. What, however, was easily measurable was the dependence of the superelastic deexcitation cross section of the $3^2P_{3/2}$ state ($F = 3$) of sodium as a function of the polarization angle δ with respect to the scattering plane. Figure 40 shows results that clearly demonstrate that the differential scattering cross sections no longer only depend on the scattering angle θ but also on the orientation of the atom with respect to incoming beam of electrons.

The observables in Fig. 40 obviously follow a linear law such as $c_1 + c_2 \cos 2\delta$ which is consistent, according to Macek and Hertel (1974), with the well-established assumption that during the collisional excitation the total spin and total orbital angular momenta are separately conserved (e.g., see Percival and Seaton, 1958; and Baranger and Gerjuoy, 1958).

Another easy variable parameter in the experiment of Hertel and Stoll (1974b) is the angle θ_n between the incident light and outgoing electrons. Figure 41 shows the ratio

$$r = I(\theta_n, \delta = 0°)/I(\delta = 90°)$$

for the forward scattering intensities with the polarization vector parallel ($\delta = 0°$) or perpendicular ($\delta = 90°$); note that the scattering intensity, in this case, is independent of θ_n to the scattering plane.

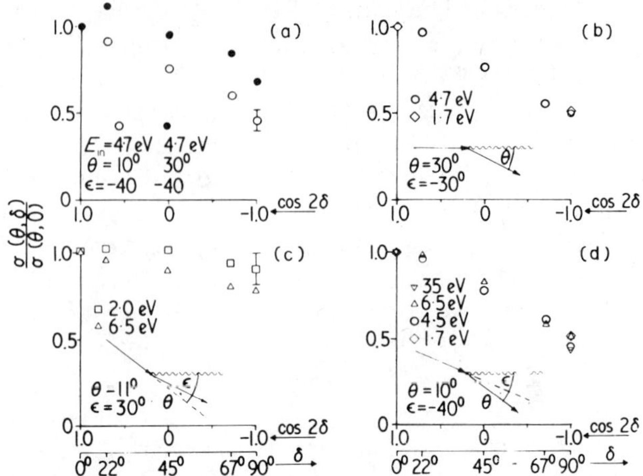

Fig. 40. Differential deexcitation cross section of the $3^2P_{3/2}$, $F = 3$ state of sodium as a function of the polarization angle δ of the exciting laser light for different electron energies and scattering parameters (see Fig. 37).

The experimental points match the theoretical ones for incomplete pumping reasonably well. Population densities of the magnetic substates derived from this curve can be used to apply the theoretical calculations of Moores and Norcross (1972) in order to calculate the preceding ratio for nonzero scattering angle. Figure 42 compares these theoretical data with experimental results.

The experimental data fit the theoretical curve very satisfactorily although the experimental output is not yet sufficient to separate scattering amplitudes and phase differences from each other.

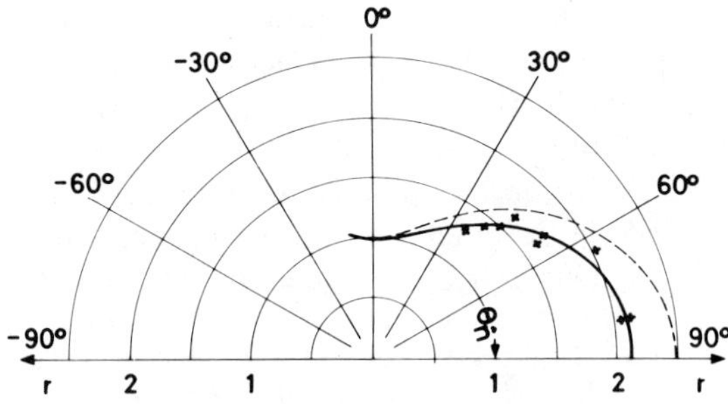

Fig. 41. Ratio $r = [I(\theta_n, \delta = 0°)]/[I(\delta = 90°)]$ of forward scattering for the Na, $3^2P \rightarrow 3^2S$ transition for initial electron energy $E = 3$ eV as a function of angle (θ_n) between the incident light and the outgoing electron. The dashed line is theoretical curve for stationary optical pumping; the full line, for incomplete optical pumping; experimental points of Hertel (1975).

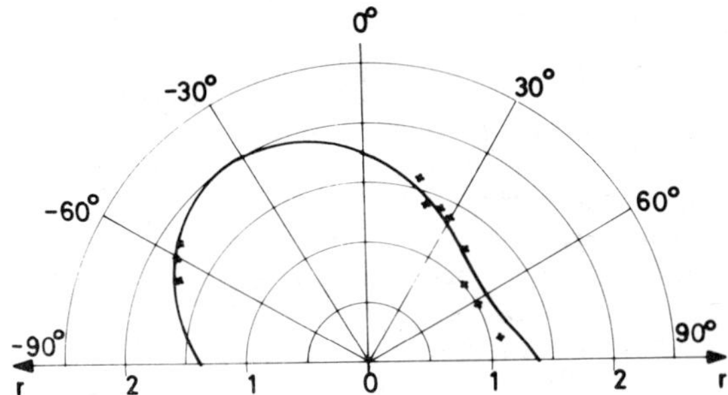

Fig. 42. Same ratio and quantities as in Fig. 41, except that the electron scattering angle is 20°. (×) Experimental data of Hertel (1975) with a theoretical prediction (full line) according to Moores and Norcross (1972).

Of course, as indicated in the theoretical interpretation of the electron scattering by laser-excited sodium atoms, a full analysis of the various types of measurements should result in many more details on the scattering and target parameters.

REFERENCES

Baranger, E., and Gerjuoy, E. (1958). *Proc. Phys. Soc., London* **73**, 326.

Barschall, H. H., and Haeberli, W., eds. (1971). *Polariz. Phenom. Nucl. React., Proc. Int. Symp., 3rd, 1970.*

Bates, D. R. (1968). *Adv. At. Mol. Phys.* **4**,

Bederson, B. (1969a). *Comments At. Mol. Phys.* **1**, 41.

Bederson, B. (1969b). *Comments At. Mol. Phys.* **2**, 65.

Berderson, B. (1973). *At. Phys., Proc. Int. Conf., 3rd, 1972* p. 401.

Bell, R. E. (1968). *In* "Alpha, Beta and Gamma Ray Spectroscopy," (K. Siegbahn, ed.), Vol. 2, p. 905. North-Holland Publ., Amsterdam.

Blum, K., and Kleinpoppen, H. (1974). *Phys. Rev. A* **9**, 1902.

Blum, K., and Kleinpoppen, H. (1975a). *J. Phys. B* **8**, 922.

Blum, K., and Kleinpoppen, H. (1975b). *Int. J. Quantum Chem. Symp.* **10**, 415.

Burke, P. G. (1969). *Lect. Theor. Phys.* **11c**, 1.

Burke, P. G., and Mitchell, J. F. B. (1974a). *J. Phys. B* **7**, 214.

Burke, P. G., and Mitchell, J. F. B. (1974b). *J. Phys. B* **7**, 229.

Burke, P. G., and Schey, H. M. (1962). *Phys. Rev.* **126**, 163.

Burke, P. G., and Taylor, A. J. (1969). *J. Phys. B* **2**, 869.

Byrne, J., and Farago, P. S. (1971). *J. Phys. B* **4**, 957.

Campbell, D. M., Brash, H. M., and Farago, P. S. (1971). *Phys. Rev. A* **36**, 449.

Collins, R. E., Bederson, B., and Goldstein, M. (1971). *Phys. Rev. A* **3**, 1976.

Crandall, D. H., Taylor, P. O., and Dunn, G. H. (1974). *Phys. Rev. A* **10**, 141.

Csanak, G., Taylor, H. S., and Tripethy, D. N. (1973). *J. Phys. B* **6**, 2040.

Drukarev, G. F., and Obedov, V. D. (1972). *Sov. Phys.—JETP, (Engl. Transl.)* **34**, 284.

Düweke, M., Kirchner, N., Reichert, E., and Staudt, E. (1973). *J. Phys. B* **6**, L208.

Eck, T. G. (1973). *Phys. Rev. Lett.* **31**, 270.
Ehlers, V. J., and Gallagher, A. C. (1973). *Phys. Rev. A* **7**, 1573.
Eminyan, M., MacAdam, K. B., Slevin, J., and Kleinpoppen, H. (1973). *Phys. Rev. Lett.* **31**, 575.
Eminyan, M., MacAdam, K. B., Slevin, J., and Kleinpoppen, H. (1974). *J. Phys. B* **7**, 1519.
Eminyan, M., MacAdam, K. B., Slevin, J., Standage, M. C., and Kleinpoppen, H. (1975). *J. Phys. B* (submitted for publication).
Enemark, E. A., and Gallagher, A. (1972). *Phys. Rev. A* **6**, 192.
Fano, U., and Copper, J. W. (1965). *Phys. Rev. A* **138**, 400.
Fano, U., and Macek, J. H. (1973). *Rev. Mod. Phys.* **45**, 553.
Farago, P. S. (1971). *Rep. Prog. Phys.* **34**, 1055.
Farago, P. S. (1974). *J. Phys. B* **7**, 128.
Feautrier, N. (1970). *J. Phys. B* **3**, L152.
Flannery, M. R., and McCann, K. J. (1975). *J. Phys. B* (in press).
Flower, D. R., and Seaton, M. J. (1967). *Proc. Phys. Soc., London* **91**, 59.
Geltman, S. (1969). "Topics in Atomic Collision Theory." Academic Press, New York.
Gould, G. (1970). Ph.D. Thesis, University of South Wales.
Hafner, H., and Kleinpoppen, H. (1967). *Z. Phys.* **198**, 315.
Hafner, H., Kleinpoppen, H., and Krüger, H. (1965). *Phys. Lett.* **18**, 270.
Hanne, G. F., and Kessler, J. (1974). *Phys. Rev. Lett.* **33**, 341.
Hanne, G. F., and Kessler, J. (1976a). *In* "Electron and Photon Interactions with Atoms" (H. Kleinpoppen and M. R. C. McDowell, eds.), p. 445. Plenum, New York.
Hanne, G. F., and Kessler, J. (1976b). *J. Phys. B* **9**, 791.
Heddle, D. W. O. (1975). *J. Phys. B* **8**, L33.
Heddle, D. W. O., Keesing, R. G. W., and Watkins, R. D. (1974). *Proc. R. Soc. London, Ser A* **337**, 443.
Heideman, H. G. M., Smit, C., and Smit, J. A. (1969). *Physica (Utrecht)* **45**, 305.
Hertel, I. V., (1976). *In* "Interactions of Electrons and Photons with Atoms" (H. Kleinpoppen and M. R. C. McDowell, eds.), p. 375. Plenum, New York.
Hertel, I. V., and Stoll, W. (1974a). *J. Phys. B* **7**, 570.
Hertel, I. V., and Stoll, W. (1974b). *J. Phys. B* **7**, 583.
Hils, D., McCusker, V., Kleinpoppen, H., and Smith, S. J. (1972). *Phys. Rev. Lett.* **29**, 398.
Imhof, R. E., and Read, F. H. (1971a). *J. Phys. B* **4**, 450.
Imhof, R. E., and Read, F. H. (1971b). *Chem. Phys. Lett.* **11**, 326.
Joachain, D. J., and Vanderpoorten, R. (1974). *J. Phys. B* **7**, L528.
Jost, K., and Kessler, J. (1966). *Z. Phys.* **195**, 1.
Karule, E. M. (1970). *Latv. PSR Zinat. Akad. Vestis* No. 3, p. 9.
Karule, E. M., and Peterkup, R. K. (1965). *In* "Atomic Collisions" (Y. Ia Veldre, ed.), Vol. 3. Larvian Acad. Sci., Riga, USSR.
Kessler, J. (1969). *Rev. Mod. Phys.* **41**, 3.
Kleinpoppen, H. (1971). *Phys. Rev. A* **3**, 2015.
Kleinpoppen, H. (1973). *In* "Fundamental Interactions in Physics" (B. Kursunoglu and A. Perlmutter., eds.), p. 229. Plenum, New York.
Lurio, A. (1965). *Phys. Rev. A* **140**, 1505.
McConnell, J. C., and Moiseiwitsch, B. L. (1968). *J. Phys. B* **1**, 406.
Macek, J. H., and Hertel, I. V. (1974). *J. Phys. B* **7**, 2173.
Macek, J. H., and Jaecks, D. H. (1971). *Phys. Rev. A* **4**, 2288.
Madison, D. H., and Shelton, W. N. (1973). *Phys. Rev. A* **7**, 449.
Mahan, A. H. (1974). Ph.D. Thesis, University of Colorado, Boulder.
Mahan, A. H., Krotkov, R. V., Gallagher, A. C., and Smith, S. J. (1973). *Bull. Am. Phys. Soc.* [2] **18**, 1506.

Massey, H. S. W. (1969). *In* "Physics of the One- and Two-Electron Atoms" (F. Bopp and H. Kleinpoppen, eds.), p. 511. North-Holland Publ., Amsterdam.
Massey, H. S. W., and Mohr, C. B. O. (1941). *Proc. Phys. Soc., London, Ser. A* **177**, 341.
Moiseiwitsch, B. L., and Smith, S. J. (1968). *Rev. Mod. Phys.* **40**, 238.
Moores, D. L., and Norcross, D. W. (1972). *J. Phys. B* **5**, 1482.
Morgan, L. A., and McDowell, M. R. C. (1975). *J. Phys. B* **8**, 1073.
Morgan, L. A., and Stauffer, A. D. (1975). In *J. Phys. B* **8**, 2342.
Mott, N. F. (1929). *Proc. R. Soc. London, Ser. A* **124**, 425.
Mott, N. F. (1932). *Proc. R. Soc. London, Ser. A* **135**, 429.
Mott, N. F., and Massey, H. S. W. (1965). "Theory of Atomic Collisions," 3rd ed. Oxford Univ. Press, London and New York.
Mumma, M. J., Misakian, M., Jackson, W. M., and Faris, J. L. (1974). *Phys. Rev. A* **9**, 203.
Oppenheimer, J. R. (1927a). *Z. Phys.* **43**, 27.
Oppenheimer, J. R. (1927b). *Proc. Natl. Acad. Sci. U.S.A.* **13**, 800.
Oppenheimer, J. R. (1928). *Phys. Rev.* **32**, 361.
Ott, W. R., Kauppila, W. E., and Fite, W. L. (1963). *Phys. Rev. Lett.* **19**, 1361.
Ottley, T. W. (1974). Ph.D. Thesis, University of Stirling.
Ottley, T. W., and Kleinpoppen, H. (1975). *J. Phys. B* **8**, 621.
Ottley, T. W., Denne, D. R., and Kleinpoppen, H. (1974). *J. Phys. B* **7**, L179.
Penney, W. G. (1932). *Proc. Natl. Acad. Sci. U.S.A.* **18**, 231.
Percival, I. C., and Seaton, M. J. (1958). *Philos. Trans. R. Soc. London, Ser. A* **251**, 113.
Rubin, K., Bederson, B., Goldstein, M., and Collins, R. E. (1969). *Phys. Rev.* **182**, 201.
Skinner, H. W. B., and Appleyard, E. T. (1927). *Proc. R. Soc. London, Ser. A* **117**, 224.
Smith, S. J. (1976). *In* "Interactions of Electrons and Photons with Atoms" (H. Kleinpoppen and M. R. C. McDowell, eds.), p. 365. Plenum, New York.
Tripathi, A. N., Mathur, K. C., and Joshi, S. K. (1973). *J. Phys. B* **6**, 1431.
Vriens, L., and Carriere, J. D. (1970). *Physica (Utrecht)* **49**, 517.
Wykes, J. (1972). *J. Phys. B* **5**, 1126.

Theoretical Interpretation of Hund's Rule

JACOB KATRIEL and
RUBEN PAUNCZ
Department of Chemistry
Technion Israel Institute of Technology
Haifa, Israel

I. Introduction . . . 143
A. Experimental Background . . . 143
B. Validity within the Independent Particle Model . . . 144
C. Recent Developments . . . 146
II. Scaled Atomic Orbitals . . . 147
A. General Consequences of Scaling . . . 147
B. Illustrative Computations . . . 151
III. Optimized Atomic Orbitals . . . 153
A. Minimal Basis Set . . . 153
B. Analysis of SCF Computations . . . 154
IV. Inclusion of the Correlation Correction . . . 156
A. Carbon Isoelectronic Sequence . . . 156
B. Helium Isoelectronic Sequence . . . 157
V. The $1/Z$ Perturbation Theory . . . 157
VI. Role of Inner Shells . . . 160
VII. The Pair Distribution Function . . . 162
A. A Qualitative Analysis . . . 162
B. Illustrative Computations . . . 167
C. Expectation Values of Two-Electron Operators . . . 174
VIII. An Interpretation of Hund's Rule . . . 176
Appendix A: Screening in $(n, l)^k$-Type Configurations . . . 180
Appendix B: The Pair Distribution Function . . . 181
References . . . 184

I. Introduction

A. Experimental Background

The original discovery of Hund's rule (Hund, 1925) was one of the last triumphs of atomic spectroscopy prior to the advent of modern quantum mechanics. It was by the analysis of experimental spectra that the importance of the interelectronic interactions was recognized. The first step toward the elucidation of complex spectra was the realization that the nature of the ground term, as well as the large intermultiplet splittings, could not be accounted for in terms of one-electron states, which had been sufficient for the understanding of the simpler spectra of alkali metals.

The role of the coupling of the angular momenta of the different electrons in determining the nature of the ground term was first noted by Russell and Saunders (1925). This was achieved through the analysis of alkaline earth spectra and constituted the first clear-cut demonstration of an effect due to interactions between the electrons.

The Pauli principle, formulated in the same year, pointed out the existence of an entirely different type of phenomenon, the essential significance of which was that electrons somehow felt each other so as to form a well-correlated entity; they were not completely independent particles as had been assumed in earlier work.

Hund's original formulation was a statement of the experimental situation in the spectra of the elements scandium to nickel. He noted the validity of the rule according to which the term of highest multiplicity, among those belonging to the lowest configuration, is the ground term. The experimental results were also found to be consistent with the rule according to which the terms of larger resultant orbital angular momentum are lower in energy for a given configuration and multiplicity.

In a later part of the same article the rule is referred to in a somewhat looser way. The formulation there given may be interpreted as a proposal of a more general ordering principle of the terms corresponding to a given configuration. It is not completely clear, however, if this interpretation is implied, since in the same section the ground terms of various atoms and positive ions are identified.

A subtle point concerning the formulation of Hund's rule is related to the range of its applicability. The minimal formulation would restrict the rule to the determination of the lowest term of the ground configuration. In a more general sense, one could consider it as an ordering principle for the ground configuration, or, at best, as a criterion for the lowest term of any configuration. The maximal formulation would attempt to use Hund's rule as a general ordering principle for any configuration. A loose rendering of this universal formulation seems to be the way in which the rule is generally understood.

Experimentally, Hund's rule is highly reliable so far as the ground state is concerned. It is, however, not nearly as dependable for the ordering of higher terms in the configuration. Experimental deviations from Hund's rule are sufficiently common. Some observed deviations in the spectrum of neutral cerium are mentioned by Martin (1963).

B. Validity within the Independent Particle Model

1. *Vector Model Interpretation*

Since Hund's rule had been suggested experimentally before the advent of quantum mechanics, it was first treated theoretically within the framework of the old quantum mechanical vector model. A model for complex atoms was set up by Slater (1926) consisting of electrons, the intrinsic and orbital angular momentum vectors of which act on each other with torques the energies of which are proportional to the scalar product of these vectors. Very satisfactory agreement with the available experimental results was obtained.

The vector model has thus provided the unifying framework for the understanding of atomic structure and spectra. Its fundamental drawback is that the types of interactions assumed between the electrons are of an *ad hoc* nature, having nothing to do with the realistic fundamental interactions between the constituent particles. The fundamental origin of the interactions introduced in the vector model has been touched upon by Slater in that early study. In particular, he realized that the torques assumed between the spin vectors are of such a magnitude as to exclude a magnetic origin.

2. *Formulation, Foundation, and Validity of Hund's Rule*

In attempting to assess the validity of Hund's rule, it is useful to distinguish between its limitations within the independent particle approximation, on the one hand, and violations of it associated with the breakdown of this approximation, on the other. Consequently, the electrostatic effects within the independent particle model can be separated in a well-defined way from other complications, due in part to configuration interaction, i.e., correlation, and, to a lesser extent, to the inadequacy of the electrostatic Hamiltonian upon which the independent particle model in the L–S coupling scheme is based.

In this restricted sense the energetic ordering of terms corresponding to a given configuration is first studied within the independent particle scheme; the results are then compared with the Hund's rule prediction, both from the point of view of the ordering and from the point of view of its energetic origin.

The most restrictive formulation of Hund's rule would state that in a configuration of equivalent electrons the deepest lying term corresponds to the highest possible value of the total spin. Generalizations of this rule to the ground state corresponding to configurations of nonequivalent electrons as well using it to predict the ordering of all the terms corresponding to a given configuration have been discussed.

The rule concerning the orbital angular momentum can similarly be stated in a restrictive way, according to which the lowest term of a given configuration of equivalent electrons is chosen from among the states with highest multiplicity by the criterion of highest resultant orbital angular momentum. Again, generalization to configurations of nonequivalent electrons as well as to a complete ordering of the terms has been discussed.

Slater's work on the theory of complex spectra is the first and main theoretical treatment of the ordering of atomic terms within a quantum mechanical context (Slater, 1929). Slater's work provides a foundation to Hund's rule in the restricted sense as well as plenty of counterexamples insofar as the general version is considered. Slater's treatment is a simplified independent particle model, in which the many electron wavefunctions in the various terms are constructed out of the same set of one-electron wavefunctions. Because the one-electron part of the Hamiltonian has the same expectation value for such many-electron functions, it follows that the ordering of levels is determined by the two-electron part of the Hamiltonian, the interelectronic repulsion. Slater's work provided the formal demonstration that the origin of term splitting is electrostatic and has nothing to do with the much smaller magnetic interactions. This fact had already been recognized by Slater in his earlier vector model work, on the basis of the magnitude of the energies involved, but there was nothing in that model to distinguish between the two types of interactions.

Hund's rule has been applied to molecules, the paramagnetism predicted in ground-state O_2 being the most popular example. The "breakdown" of Hund's rule in the case of a strong ligand field is completely understood and does not affect its validity in any fundamental sense.

From a theoretical point of view, within Slater's treatment, one could distinguish two types of situation. In certain sufficiently simple configurations, the expressions obtained for the energies of the various terms are simple enough to enable a unique ordering of levels without a numerical evaluation of the integrals involved. In more complicated configurations, the ordering depends on the two-electron integrals in such a way that their numerical evaluation for a properly chosen set of orbitals is necessary.

C. Recent Developments

Following Slater's successful interpretation of complex spectra, it was generally assumed that the first rule is intimately connected with the electronic correlation. Since the probability of finding two electrons with parallel spins at the same place is zero, it was concluded that in the triplet state the electrons are farther apart on the average than in the corresponding singlet state and the average electronic repulsion is smaller in the triplet state than in the singlet one.

Davidson (1965) was the first to observe that for excited states of He the electron repulsion in the triplet term is larger than in the corresponding singlet one. Similar observations have been made by Messmer and Birss (1969) again for excited states of He. Calculations using simple variational wavefunctions by Kohl (1972) and Killingbeck (1973) confirmed these findings. Another counterexample to the traditional explanation was found by Lemberger and Pauncz (1969) for the ^3P, ^1D, and ^1S states of the carbon atom. Katriel (1972a) extended these investigations to the isoelectronic series of carbon and analyzed the contributions of nuclear attraction and electron repulsion. He showed (1972b) that the usual interpretation of the electron correlation in the triplet state is incorrect and arrived (1972c) at a general interpretation of Hund's rule.

The Z expansion perturbation method was used for the analysis of energy differences by Colpa and Islip (1973). Colpa and Brown (1973) used these considerations and applied them to several isoelectronic sequences. Then Colpa (1974) reformulated the theory of Hund's rule in terms of inequalities (Colpa *et al.*, 1975). By using the local energy method, Harcourt and co-workers (Harcourt and Harcourt, 1973; Harcourt *et al.*, 1974) arrived at the same conclusions.

As a consequence of these investigations the role of Hund's rule in elementary quantum chemistry is now revised, as exemplified by Snow and Bills (1974) in the *Journal of Chemical Education.*

In the present review article, we intend to present a comprehensive treatment of some recent developments in the interpretation of Hund's rule.

II. Scaled Atomic Orbitals

A. General Consequences of Scaling

The use of the same atomic orbitals for the multiplets belonging to a certain configuration constitutes a basic assumption of conventional multiplet theory (Condon and Shortley, 1935). The one-electron energy components so obtained are identical for the different multiplets. The energetic differences between multiplets are due to differences in the two-electron terms, i.e., the interelectronic repulsion. In terms of the conventional approach, the lowest multiplet is the one in which the interelectronic repulsion is the smallest. An obvious flaw in this approach is that it does not satisfy the virial theorem, since the kinetic energy (a one-electron term) is the same in the different multiplets, but the total energy is not.

By independently scaling the wavefunction for each one of the multiplets, it is possible not only to satisfy the virial theorem but, at the same time, to improve the wavefunctions, in a variational sense (Löwdin, 1959).

The total energy before scaling can be written for each multiplet in the form

$$E = T + L + C, \tag{1}$$

where

$$T = \langle \psi | \sum_i -\tfrac{1}{2}\nabla_i^2 | \psi \rangle,$$

$$L = Z \cdot \langle \psi | \sum_i -\frac{1}{r_i} | \psi \rangle,$$

$$C = \langle \psi | \sum_{i<j} \frac{1}{r_{ij}} | \psi \rangle; \qquad \langle \psi | \psi \rangle = 1. \tag{2}$$

Scaling can formally be introduced by the operator

$$U = \prod_j \exp\left[i \frac{\ln \eta}{2\hbar} (\mathbf{r}_j \cdot \mathbf{p}_j + \mathbf{p}_j \cdot \mathbf{r}_j)\right] \tag{3}$$

so that $\tilde{\psi}(r) = U\psi(r) = \sqrt{\eta} \cdot \psi(\eta r)$. It is easily seen that because U is unitary,

$$\langle \tilde{\psi}(r) | \tilde{\psi}(r) \rangle = 1.$$

It follows that

$$\tilde{E}(\eta) = \langle \tilde{\psi} | H | \tilde{\psi} \rangle = \langle U\psi | H | U\psi \rangle = \langle \psi | U^{\dagger} H U | \psi \rangle \equiv \langle \psi | \tilde{H} | \psi \rangle. \tag{4}$$

From the definition of U, we obtain $U^{\dagger}(\eta) = U(1/\eta)$ and, therefore,

$$\tilde{H} = \eta^2 \cdot \sum_i -\frac{1}{2}\nabla_i^2 + \eta \cdot \left(\sum_i -\frac{Z}{r_i} + \sum_{i<j} \frac{1}{r_{ij}}\right). \tag{5}$$

Hence,

$$\tilde{E}(\eta) = \eta^2 \cdot T + \eta \cdot (L + C). \tag{6}$$

From the requirement,

$$\partial \tilde{E}/\partial \eta = 2\eta \cdot T + (L + C) = 0, \tag{7}$$

it follows that

$$\eta = -\frac{L + C}{2T} = -\frac{E - T}{2T} = -\frac{E}{2T} + \frac{1}{2}. \tag{8}$$

As $E < 0$ (the state is a bound one) and $T > 0$, it follows that $\eta > \frac{1}{2}$, i.e., certainly positive. Finally,

$$\begin{aligned} \tilde{E} &= -(L + C)^2/4T = -T \cdot \eta^2, \\ \tilde{T} &= T \cdot \eta^2 = -\tilde{E}, \\ \tilde{L} &= L \cdot \eta; \qquad \tilde{C} = C \cdot \eta. \end{aligned} \tag{9}$$

It should be emphasized that, from the condition $\partial \tilde{E}/\partial \eta = 0$, it follows that the energy correction will be of second order in $\delta\eta$, so that it can be expected to be small. However, $\partial \tilde{T}/\partial \eta = -(L + C)$; $\partial \tilde{L}/\partial \eta = L$ and $\partial \tilde{C}/\partial \eta = C$ and, therefore, for each one of the energy components, one may expect a meaningful change as a result of the scaling. It is important to note that this behavior, resulting from the variational principle, is typical of the total energy: The use of an inadequate wavefunction sharply affects the expectation value of different operators, but the total energy is only mildly dependent on it.

For the two multiplets belonging to the same configuration, we construct wavefunctions ψ_A and ψ_B, with the same atomic orbitals. We get

$$E_A = T + L + C_A; \qquad E_B = T + L + C_B. \tag{10}$$

Let us assume that $E_A < E_B$, i.e., $C_A < C_B$. The scaling parameters will be

$$\eta_A = -(L + C_A)/2T; \qquad \eta_B = -(L + C_B)/2T. \tag{11}$$

The C_A and C_B are expectation values of a positive definite operator and are, therefore, both positive. As T is also positive, $\eta_A > \eta_B$. For the energies, we get

$$\tilde{E}_A = -T \cdot \eta_A^2; \qquad \tilde{E}_B = -T \cdot \eta_B^2 \tag{12}$$

and therefore $\tilde{E}_A < \tilde{E}_B$, i.e., scaling does not affect the energetic ordering. The importance of this result derives from the fact that, by using the same orbitals for the different multiplets, one gets an energetic ordering that agrees with Hund's rule. Such an improvement of the wavefunction cannot change the level ordering. From the point of view of the theoretical confirmation of Hund's rule, this result is already an improvement on the result obtained when we assume that the orbitals are the same. It should be stressed that this argument is completely general and was not derived from a computation, which is necessarily associated with some specific system.

The result $\eta_A > \eta_B$ reflects a contraction of the wavefunction toward the nucleus in the lower energy multiplet. This contraction agrees with the results of SCF and configuration interaction (CI) computations for the multiplets of carbon and sulfur (Lemberger and Pauncz, 1969) and with further results to be discussed in the following sections. As a result of the contraction one should expect not only an increase in the nuclear attraction but also an increase in the kinetic energy and in the interelectronic repulsion. Formally the change in the nuclear attraction results from the fact that $L < 0$ and, therefore,

$$\tilde{L}_A = L \cdot \eta_A < L \cdot \eta_B = \tilde{L}_B. \tag{13}$$

From the viral theorem and the fact that $\tilde{E}_A < \tilde{E}_B$, it follows that $\tilde{T}_A > \tilde{T}_B$. In order to evaluate the relative contributions of the interelectronic repulsion and the nuclear attraction to the energy separation between the terms, we shall investigate the expression

$$\frac{\tilde{C}_A - \tilde{C}_B}{\tilde{L}_A - \tilde{L}_B} = \frac{C_A \cdot \eta_A - C_B \cdot \eta_B}{L \cdot (\eta_A - \eta_B)} = \frac{C_A \cdot (L + C_A) - C_B \cdot (L + C_B)}{L(C_A - C_B)}$$

$$= 1 - \frac{C_A + C_B}{-L}. \tag{14}$$

As E_A and E_B are negative, whereas T is positive, $L + C_A = E_A - T$ and $L + C_B = E_B - T$ are negative. Hence $C_A < -L$, $-L > 0$ and, therefore, $0 < C_A/-L < 1$, a condition that also holds for $C_B/-L$. Therefore

$$0 < (C_A + C_B)/-L < 2 \tag{15}$$

or

$$-1 < (\tilde{C}_A - \tilde{C}_B)/(\tilde{L}_A - \tilde{L}_B) < 1, \tag{16}$$

that is,

$$|\tilde{C}_A - \tilde{C}_B| < |\tilde{L}_A - \tilde{L}_B|. \tag{17}$$

From these results it follows that the main contribution to the energy differences between the different multiplets is associated with the difference in the nuclear attractions and not with the difference in the interelectronic repulsions. This fact is in accordance with the results of the SCF and CI computations mentioned earlier and, of course, contradicts the interpretation of classical multiplet theory that assigns the energetic differences to the interelectronic repulsion differences.

For an isoelectronic sequence the relative importance of the nuclear attraction increases with the increase in nuclear charge and, therefore, $(C_A + C_B)/-L$ decreases so that

$$\lim_{Z\to\infty} \frac{C_A + C_B}{-L} = 0. \tag{18}$$

We then find that

$$\frac{\tilde{C}_A - \tilde{C}_B}{\tilde{L}_A - \tilde{L}_B} \tag{19}$$

increases with the increase in the nuclear charge, so that

$$\lim_{Z\to\infty} \frac{\tilde{C}_A - \tilde{C}_B}{\tilde{L}_A - \tilde{L}_B} = 1. \tag{20}$$

The relative importance of the interelectronic repulsion differences in determining the energetic differences of the multiplets thus increases with the increase in the nuclear charge and approaches that of the nuclear attraction for a sufficiently large nuclear charge. This behavior is certainly surprising. We will discuss its origin in the forthcoming sections.

Since $\eta_A > \eta_B$ whereas $C_A < C_B$, the relative value of $\tilde{C}_A$ and $\tilde{C}_B$ must be carefully examined. From the relations

$$\tilde{C}_A - \tilde{C}_B = C_A \cdot \eta_A - C_B \cdot \eta_B = -\frac{1}{2T}[C_A \cdot (L + C_A) - C_B \cdot (L + C_B)]$$

$$= \frac{C_A - C_B}{-2T} \cdot (L + C_A + C_B) \tag{21}$$

and

$$(C_A - C_B)/-2T > 0,$$

it follows that $\tilde{C}_A - \tilde{C}_B < 0$ provided that $L + C_A + C_B < 0$. A sufficient condition is $C_B < -L/2$. This condition is highly conservative because if we express L in terms of an average interaction of a single electron with a proton, l, and C in terms of an average interaction of two electrons, c, in their respective orbitals, we obtain $L = Z \cdot N \cdot l$ and $C = c[N(N-1)]/2$. From electrostatic considerations it is clear that $|l| > c$ and, therefore, at least for $Z > N - 1$, there exists $C < -L/2$.

We see that throughout the isoelectronic series, starting from the negative ion, scaling does not change the ordering of the interelectronic repulsion values; i.e., if $C_A < C_B$ then $\tilde{C}_A < \tilde{C}_B$. This does not agree with the results of the SCF and CI computations for neutral atoms. It will be demonstrated in Section III how the correspondence with the results of the numerical computations can be improved. The significance of the gradual improvement of the description obtained will be discussed in detail in a later stage.

B. Illustrative Computations

In order to illustrate quantitatively the results of scaling with reference to the interpretation of Hund's rule, we will discuss the ground configuration of the carbon atom using Slater-type orbitals.

By carrying out an optimization of the parameters for the ^{3}P state and by using the orbitals thus obtained for the other states as well, we obtain the values for the energy and its components shown in Table I. By scaling each one of these states, we get the results appearing in Table II.

TABLE I

UNSCALED ENERGY COMPONENTS FOR CARBON USING A SLATER BASIS OPTIMIZED FOR ^{3}P [a]

State	T	L	C	E
^{3}P	37.622	−88.106	12.861	−37.622
^{1}D			12.927	−37.556
^{1}S			13.026	−37.457

[a] Atomic units are used throughout.

Since the deviation from the virial theorem is sufficiently small, the scaling parameter is close to the value $\eta = 1$, and therefore scaling does not affect the total energy at all, within the accuracy limits presented in Table II. However, the changes in the energy components are significant as concerns the interpretation of Hund's rule. It is particularly important that, as a result of scaling, the nuclear attraction is greater (in its absolute value) in the lower energy states than in the higher ones. The differences in the nuclear attraction values are considerably greater than the differences in the interelectronic repulsion values. As expected, scaling decreased the differences in the interelectronic repulsion values, although it did not change their ordering.

TABLE II

SCALED ENERGY COMPONENTS FOR CARBON IN ^{3}P OPTIMIZED SLATER BASIS

State	η	$T' = -E'$	L'	C'
^{3}P	1.00000	37.622	−88.106	12.861
^{1}D	0.99913	37.556	−88.028	12.916
^{1}S	0.99781	37.457	−87.912	12.998

Curve a in Fig. 1 represents the values of

$$R = \frac{C(^3\mathrm{P}) - C(^1\mathrm{D})}{L(^3\mathrm{P}) - L(^1\mathrm{D})}$$

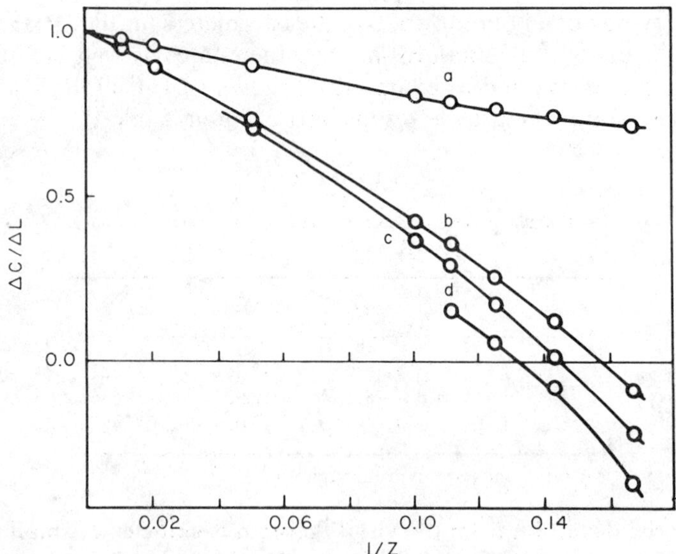

Fig. 1. Relative contribution of the interelectronic repulsion and nuclear attraction to the ^{3}P–^{1}D separation for the carbon isoelectronic series: (a) scaled orbitals; (b) optimized orbitals; (c) SCF; (d) exact.

obtained with the scaled functions for the isoelectronic sequence of carbon. The ratio R is smaller than unity throughout the isoelectronic sequence, i.e., the main energetic contribution to the multiplet splitting is connected with the nuclear attraction rather than with the interelectronic repulsion. This is in agreement with the general discussion in the previous section.

As is clear from the extrapolation in Fig. 1, the computational results satisfy $\lim_{Z \to \infty} (\Delta C/\Delta L) = 1$ in agreement with the general discussion.

III. Optimized Atomic Orbitals

A. Minimal Basis Set

When the atomic orbitals are optimized independently for every state the results obtained are, of course, better than those obtained by scaling alone. The results obtained for the carbon atom are presented in Table III. The optimal parameter values are given in Table IV.

TABLE III

ENERGY COMPONENTS FOR AN OPTIMIZED SLATER BASIS APPROXIMATION OF CARBON

State	L	C	$E = -T$
^{3}P	−88.106	12.861	−37.622
^{1}D	−87.963	12.849	−37.557
^{1}S	−87.746	12.824	−37.461

According to Hund's rule $E(^3\text{P}) < E(^1\text{D}) < E(^1\text{S})$, but, in contrast with the interpretation of classical multiplet theory, $C(^3\text{P}) > C(^1\text{D}) > C(^1\text{S})$. This result agrees with the results of the SCF and CI computations of Lemberger and Pauncz (1969).

Thus, the term responsible for the level ordering is the nuclear attraction. *Scaling is, therefore, the minimal improvement of the classical multiplet theory required to point out the central role of the nuclear attraction in determining the energetic differences between the multiplets. Independent optimization is the minimal improvement required in order to obtain a correct qualitative picture concerning the relative values of the interelectronic repulsion as well.*

The optimal values of the parameters point to a shrinking of the outer orbital, 2p, in low energy states but to an expansion of orbitals 1s and 2s. This behavior is necessarily different from the one obtained through simple scaling which makes all the orbitals shrink homogeneously in the lower energy state in relation to the higher one. The shrinking of the outer orbital as opposed to the expansion of the inner orbitals is precisely like

TABLE IV

SLATER FUNCTION OPTIMIZED EXPONENTIAL PARAMETERS FOR CARBON

State	1s	2s	2p
^{3}P	5.6726	1.608	1.568
^{1}D	5.6730	1.614	1.536
^{1}S	5.6735	1.623	1.487

the behavior of SCF functions for the same states. The differences in the 1s orbital between the various states are, as expected, very small, but the differences in orbitals 2s and 2p are considerable.

Analogous computations have been carried out for the isoelectronic ions, and the ratio $[C(^3\mathrm{P}) - C(^1\mathrm{D})]/[L(^3\mathrm{P}) - L(^1\mathrm{D})]$ for states $^3\mathrm{P}$–$^1\mathrm{D}$ is presented as curve b in Fig. 1.

It is clear from the drawing that the condition $\lim_{Z\to\infty} (\Delta C/\Delta L) = 1$ holds also for this approximation.

B. Analysis of SCF Computations

Self-consistent field computations that correspond to the basic configurations of light and medium atoms were carried out in various ways, beginning with the first works of Hartree in the early years of quantum mechanics. The most extensive calculations are due to Clementi (1965) whose functions were used by Lemberger and Pauncz (1969) to compute the nuclear attraction and the interelectronic repulsion values for the low multiplets of carbon, nitrogen, oxygen, silicon, phosphorus, and sulfur. Similar computations were performed by Davidson (1965) and by Messmer and Birss (1969) for a number of excited-state configurations of helium. Hund's rule was confirmed in all cases, but it turned out that the lower energy is associated with greater nuclear attraction rather than with lower interelectronic repulsion. On the contrary, it was found that the interelectronic repulsion is in all cases greater in the lower multiplet, which is in conflict with the conventional interpretation of Hund's rule.

We extend here the analysis by examining the behavior of isoelectronic positive ions of the atoms mentioned above.

The results of the previous paragraphs point out clearly the interest that lies in this extension, both from the point of view of the relative value of the interelectronic repulsions and from that of the various contributions to the total energy. The nuclear attraction and the interelectronic repulsion can be obtained from the results of Clementi in three different ways: (*a*) a direct computation of the expectation values of the corresponding operators with the given wavefunctions; (*b*) by using the orbital energies ε_i and the total energy in order to obtain the interelectronic repulsion from the expression,

$$\sum_i n_i \varepsilon_i = E + C, \tag{22}$$

in which n_i is the population of the orbital i, and the nuclear attraction is obtained from the virial theorem

$$L + C = 2E; \tag{23}$$

(*c*) by using the Hellmann–Feynman theorem to evaluate L,

$$L = Z \cdot \partial E/\partial Z, \tag{24}$$

where $\partial E/\partial Z$ is obtained numerically from the values of E against Z (for integral values of Z) and again using the virial theorem to determine C.

It can be shown that the three approaches yield identical results; the first one is undoubtedly the most difficult to carry out as it requires the greatest amount of work, whereas the third is the least trustworthy, because of the need for numerical differentiation. The results we will present from now on have been obtained by the second approach and in some cases have been checked and found identical with those obtained by the first.

The results of the computations for the isoelectronic sequence of carbon are given in Table V.

The corresponding $[C(^3\mathrm{P}) - C(^1\mathrm{D})]/[L(^3\mathrm{P}) - L(^1\mathrm{D})]$ values are given in curve c of Fig. 1. It is readily observed that the nuclear attraction is dominant along the isoelectronic sequence, but the ordering of the interelectronic repulsions is reversed. Although the interelectronic repulsion for the neutral atom is greater in the lower multiplet than in the higher one, it becomes lower in the lower multiplet for positive ions. This behavior is similar to that obtained with a minimal basis set and, roughly, to the one obtained by scaling.

TABLE V

SELF-CONSISTENT FIELD ENERGY COMPONENTS FOR THE CARBON ISOELECTRONIC SEQUENCE

	$E = -T$			L			C		
Z	^{3}P	^{1}D	^{1}S	^{3}P	^{1}D	^{1}S	^{3}P	^{1}D	^{1}S
6	−37.689	−37.631	−37.549	−88.138	−87.990	−87.765	12.760	12.728	12.667
7	−53.888	−53.807	−53.690	−123.951	−123.792	−123.550	16.175	16.178	16.170
8	−73.100	−72.997	−72.846	−165.731	−165.556	−165.286	19.531	19.562	19.594
9	−95.320	−95.195	−95.011	−213.498	−213.304	−213.007	22.858	22.914	22.985
10	−120.544	−120.397	−120.180	−267.257	−267.042	−266.715	26.169	26.248	26.355
20	−537.868	−537.508	−536.972	−1134.709	−1134.285	−1133.650	58.973	59.269	59.706

The most important fact, as all these findings would seem to indicate, is that nuclear attraction differences are dominant in determining the energetic ordering of the multiplets whereas the relative value of the interelectronic repulsions varies and is only of secondary significance.

IV. Inclusion of the Correlation Correction

A. Carbon Isoelectronic Sequence

In order to obtain the exact values of the energy and its components, correlation should be taken into consideration. Correlation can be introduced by *ab initio* computation as a superposition of configurations. In effect, this correlation is necessarily partial, very expensive, and requires a great deal of computational work. It was done for the carbon atom by Lemberger and Pauncz (1969) whose relevant results are in qualitative agreement with those of the corresponding Hartree–Fock computations.

One can obtain reasonably precise estimates of the correlation correction for the various energetic components by using the Hellmann–Feynman theorem and the semiempirical values of the correlation energies for the corresponding isoelectronic sequences (Katriel, 1972a). The correlation energy components thus computed, and the SCF energy components obtained in the previous paragraph, yield the exact components presented in Table VI. These values were used in computing curve d in Fig. 1, from which it is clear that the behavior of $\Delta C/\Delta L$ versus the nuclear charge is very similar to that obtained from the sequence of approximations presented in the foregoing. The exact and SCF results should become even closer for higher nuclear charges.

TABLE VI

Exact Energy Components for the Carbon Isoelectronic Series

Z	$C(^3P)$	$C(^1D)$	$C(^1S)$	$L(^3P)$	$L(^1D)$	$L(^1S)$
6	12.50	12.44	12.41	−88.20	−88.04	−87.91
7	15.91	15.90	15.85	−124.02	−123.88	−123.67
8	19.24	19.25	19.26	−165.79	−165.62	−165.42
9	22.55	22.58	22.63	−213.55	−213.36	−213.16

Correlation, therefore, introduces quantitative corrections with respect to the results of the Hartree–Fock computation but does not affect the qualitative picture associated with the interpretation of Hund's rule.

B. Helium Isoelectronic Sequence

Similar computations have been carried out for a few other systems, and the results have led to exactly the same conclusion as that drawn from the isoelectronic sequence of carbon. The (1s, 2p), ^{1,3}P states of the helium isoelectronic sequence are particularly significant because, for this sequence, there exist exact results due to Pekeris. This sequence will be used later as a model for further analysis. The exact results were used to compute $[C(^3\mathrm{P}) - C(^1\mathrm{P})]/[L(^3\mathrm{P}) - L(^1\mathrm{P})]$ values presented in curve E of Fig. 2. Also, the values for this ratio computed on the basis of a Gaussian approximation for the 1s and 2p orbitals with optimization of the two exponential parameters are presented in curve G of Fig. 2. In both cases, $\Delta C/\Delta L < 0$ for a sufficiently small nuclear charge and $\lim_{Z\to\infty} (\Delta C/\Delta L) = 1$.

The results are in perfect agreement with the conclusions of the previous discussion. The Gaussian approximation describes the qualitative behavior quite satisfactorily, thus reinforcing the conclusion that an optimized minimal basis set is sufficient to reproduce the crucial features of interest.

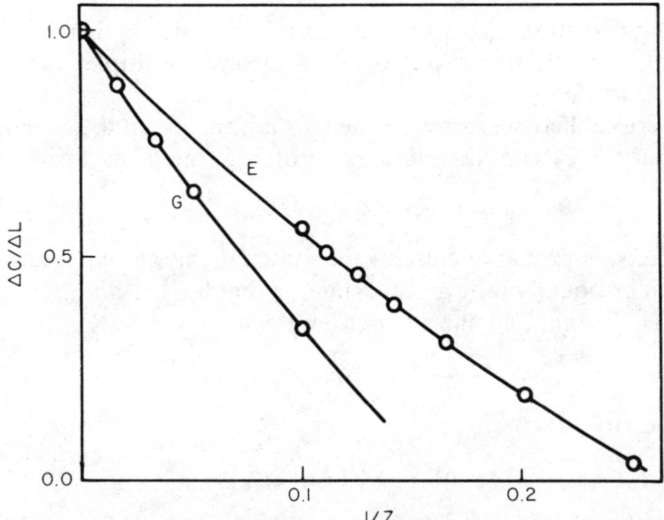

Fig. 2. Behavior of $[C(^3\mathrm{P}) - C(^1\mathrm{P})]/[L(^3\mathrm{P}) - L(^1\mathrm{P})]$ for the He isoelectronic series: (G) Gaussian approximation; (E) exact.

V. The $1/Z$ Perturbation Theory

Following the procedure of $1/Z$ perturbation theory (Hylleraas, 1930), we write the Hamiltonian in the form

$$H = H_0 + \frac{1}{Z} \cdot V, \tag{25}$$

where

$$H_0 = \sum_{i=1}^{N} \left(-\frac{1}{2}\nabla_i^2 - \frac{1}{r_i} \right) \quad \text{and} \quad V = \sum_{i<j} \frac{1}{r_{ij}}. \tag{26}$$

The corresponding eigenvalue ε is related to the energy in atomic units by $E = Z^2 \cdot \varepsilon$.

The zero-order energy is the sum of the hydrogenic energies for the occupied orbitals. This energy is equal for all the multiplets that belong to the same configuration. The first-order correction removes the degeneracy between the different multiplets. As we know, the energy up to first order satisfies Hund's rule so that for two multiplets A and B that belong to the same configuration, the first-order corrections are

$$\varepsilon_{\mathrm{A}}^{(1)} = \langle \psi_{\mathrm{A}} | V | \psi_{\mathrm{A}} \rangle \tag{27}$$

and

$$\varepsilon_{\mathrm{B}}^{(1)} = \langle \psi_{\mathrm{B}} | V | \psi_{\mathrm{B}} \rangle. \tag{28}$$

According to this treatment, the energy difference is therefore only the result of the interelectronic repulsion. The lower multiplet is the one for which $\varepsilon^{(1)}$ is smaller.

The energy difference between the two multiplets up to second order in perturbation theory can, therefore, be written in the form

$$\Delta E = \alpha \cdot (Z - Z_0) \tag{29}$$

with $\alpha < 0$. Z_0 is probably close to the minimal charge required for all the electrons to be bound in the ground state, so that it is certain that $Z_0 > 0$. By using the Hellmann–Feyman theorem, we get

$$\Delta L = Z \cdot \partial\, \Delta E / \partial Z = \alpha \cdot Z, \tag{30}$$

whereas the virial theorem leads to

$$\Delta C = \alpha \cdot (Z - 2Z_0). \tag{31}$$

Therefore $\Delta L < 0$ for any value of the nuclear charge, whereas $\Delta C < 0$ for $Z > 2Z_0$, but positive for $Z < 2Z_0$. Hence

$$\Delta C / \Delta L = 1 - (2Z_0 / Z), \tag{32}$$

and it follows that $\Delta C/\Delta L < 1$ for any Z, and $\lim_{Z \to \infty} (\Delta C/\Delta L) = 1$.

Thus, Slater's classical multiplet theory adequately describes the first order in the perturbation theory, which is the order that yields multiplet splittings. The picture obtained in second order is, of course, much more complete and closer to the full description. From the qualitative energetic point of view, the first-order picture is adequate even for very low nuclear charges ($Z \simeq Z_0$), but regarding the various energy components, the difference between the first and second orders is quite fundamental, as we have seen above.

It has not been sufficiently appreciated that perturbation theory does allow a complete discussion of intermultiplet spacing and their origins in terms of the energy components involved. All that is needed is a consistent evaluation of all the energy components to the same order in perturbation theory, bearing in mind the fact that both the Hellmann–Feynman theorem and the virial theorem, which have been shown to play such a central role in the discussion based on the variation principle, are satisfied to any order in perturbation theory. This is a straightforward consequence of the uniqueness of the perturbation theory expansion.

In $1/Z$ perturbation theory,

$$E = \sum_{p=0}^{\infty} \varepsilon_p \cdot Z^{2-p} \tag{33}$$

$$L = \sum_{p=0}^{\infty} L_p \cdot Z^{2-p} \tag{34}$$

$$C = \sum_{p=0}^{\infty} C_p Z^{1-p}. \tag{35}$$

By using the Hellmann–Feynman theorem, we get

$$L = Z \cdot \frac{\partial E}{\partial Z} = \sum_{p=0}^{\infty} (2-p)\varepsilon_p \cdot Z^{2-p}, \tag{36}$$

and, by using the virial theorem,

$$C = 2E - L = \sum_{p=1}^{\infty} p \cdot \varepsilon_p \cdot Z^{2-p}. \tag{37}$$

Therefore

$$L_p = (2-p)\varepsilon_p, \tag{38}$$

$$C_p = p \cdot \varepsilon_p. \tag{39}$$

These relations between the pth-order corrections to the nuclear attraction and the interelectronic repulsion and the energy were derived by Scherr and Knight (1963). Explicitly, the energy, correct to first order, is

$$E^{(0)} + \frac{1}{Z} \cdot E^{(1)} = \langle \psi^{(0)} | H | \psi^{(0)} \rangle = \langle \psi^{(0)} | H^0 | \psi^{(0)} \rangle + \frac{1}{Z} \langle \psi^{(0)} | V | \psi^{(0)} \rangle. \tag{40}$$

The kinetic energy, correct to first order is not the expectation value with respect to $\psi^{(0)}$ but rather

$$T^{(0)} + \frac{1}{Z} \cdot T^{(1)} = \langle \psi^{(0)} | \sum_i -\frac{1}{2} \nabla_i^2 | \psi^{(0)} \rangle + \frac{1}{Z} \left\{ \langle \psi^{(0)} | \sum_i -\frac{1}{2} \nabla_i^2 | \psi^{(1)} \rangle + \text{h.c.} \right\}. \tag{41}$$

The nuclear attraction, correct to first order, is

$$L_{(0)} + L_{(1)} = \langle \psi^{(0)} | L | \psi^{(0)} \rangle + \frac{1}{Z} \{ \langle \psi^{(0)} | L | \psi^{(1)} \rangle + \text{h.c.} \}$$

and the interelectronic repulsion

$$C_{(0)} + C_{(1)} = \frac{1}{Z} \langle \psi^{(0)} | \sum_{i<j} \frac{1}{r_{ij}} | \psi^{(0)} \rangle.$$

Because $C_{(0)} = 0$, then, from the virial theorem,

$$L_{(1)} + C_{(1)} = E_{(1)}$$

and, therefore,

$$L_{(1)} = 2E_{(1)} - C_{(1)}.$$

But $E_{(1)} = C_{(1)}$, so that $L_{(1)} = E_{(1)}$.

A somewhat different treatment using the $1/Z$ expansion was given by Colpa and Islip (1973).

VI. Role of Inner Shells

In all the systems discussed thus far, there are at least two occupied major shells. Things are different in the configurations (2s, 2p) and $2p^2$ for the isoelectronic sequence of helium. There are some fundamental difficulties in the application of the variation principle to these configurations because they are embedded in the continuum of (1s, free), but a few remarks can be made about them based on qualitative considerations and approximate results.

We have seen that independent scaling of the wavefunction for the different multiplets is sufficient to assign the greater part of the energetic difference between them to the nuclear attraction and not to the interelectronic repulsion. We have also seen that scaling alone is not sufficient to reverse the order of the interelectronic repulsions. Independent optimization of the parameters, even in a minimal basis of the Slater or Gauss type, is sufficient to bring about a reversal of the interelectronic repulsions in those cases for which such reversal is observed in more precise computations. It is clear from the discussion of the previous paragraph that the reversal of the interelectronic repulsions originates from the first-order corrections to the wavefunctions obtained with a perturbation theory approach. This correction is asymptotically similar to the variational correction, obtained by optimization of a wavefunction that is made up of hydrogenic functions, for, at the limit $Z \to \infty$, it coincides with the exact solution, which is the zero-order function in the perturbation theory. The claim that the results of minimal basis set computations, with optimization of the parameters, point to the correct qualitative picture concerning the relative values of the interelectronic repulsions in the different multiplets is, therefore, substantive enough.

For configurations with only one occupied shell, there is a single variational parameter in the minimal basis, and therefore complete optimization is equivalent to simple scaling. For such configurations there will be no reversal of the interelectronic repulsion (with reference to classical multiplet theory) even by optimization of the parameters for each multiplet independently, in the minimal basis scheme. Optimized minimal basis set computations have been shown to describe the exact behavior, at least qualitatively, in the cases previously considered. Hence, the fact that reversal of the ordering of interelectronic repulsions has not been observed in this scheme for the one-shell configurations seems sufficient to exclude the likelihood of such a reversal for precise computations as well. The scaling for these configurations is discussed in detail in Appendix A.

Eliezer and Moualem (1971) worked out perturbation theory approximations of the Hartree–Fock solutions for the multiplets derived from the $2p^2$ configuration of the isoelectronic sequence of helium. By applying the Hellmann–Feynman theorem, we obtain the contribution of the interelectronic repulsion. It is clear from the results given in Table VII that for $Z > 1$ there is no reversal of the interelectronic repulsion.

The reversal of the interelectronic repulsions is, therefore, closely connected with the interaction between the inner and outer shell. This conclusion, the significance of which will be discussed later, agrees with the results of the analysis of the interelectronic repulsion in the lower multiplets of carbon as carried out by Lemberger and Pauncz (1969). The interesting fact pointed out by this analysis is that the increase in the interelectronic repulsion in the higher multiplets over that in the lower ones is due to the increase in the repulsion between the closed shell $1s^2\ 2s^2$ and the open shell $2p^2$. The interelectronic repulsion within both the open and the closed shells is almost constant, and the small changes in it are in the conventional direction.

TABLE VII

ENERGY AND INTERELECTRONIC REPULSION FOR THE CONFIGURATION $2p^2$: HARTREE–FOCK PERTURBATION THEORY

	E			C		
Z	$(2p^2)\,^3P$	$(2p^2)\,^1D$	$(2p^2)\,^1S$	$(2p^2)\,^3P$	$(2p^2)\,^1D$	$(2p^2)\,^1S$
2	−0.70135	−0.66869	−0.62257	0.26903	0.29203	0.32075
3	−1.78724	−1.73344	−1.65560	0.43323	0.47746	0.53807
4	−3.37316	−3.29824	−3.18871	0.59736	0.66276	0.75513
5	−5.45908	−5.36306	−5.22186	0.76147	0.84800	0.97208
6	−8.04501	−7.92788	−7.75503	0.92556	1.03321	1.18898
7	−11.1309	−10.9927	−10.7882	1.08964	1.21841	1.40586
8	−14.7169	−14.5575	−14.3214	1.25372	1.40359	1.62271

VII. The Pair Distribution Function

A. A Qualitative Analysis

The results obtained for the relative value of the interelectronic repulsion in the various multiplets point to a behavior that is fundamentally different from what it is believed to be within the framework of classical multiplet theory. The claim that the average distance between electrons of parallel spins is greater than the average distance between electrons with reversed spins is not borne out by the computational results.

In order to understand better the origin of this contradiction between the qualitative claim, based on Pauli's principle, and the computational results, let us consider the pair-distribution function.

For a two-electron atom this function is

$$\rho(r'_{12}) = \langle \psi | \delta(r_{12} - r'_{12}) | \psi \rangle. \tag{42}$$

$\rho(r_{12}) \cdot dr_{12}$ is the probability that the distance between the two electrons will be between r_{12} and $r_{12} + dr_{12}$, whereas all the other coordinates receive any value whatsoever. For every operator that is a function of the distance between the electrons only, it is, of course, true that

$$\langle \psi | \mathbf{O}(r_{12}) | \psi \rangle = \int_0^\infty \mathbf{O}(r_{12}) \rho(r_{12})\, dr_{12}. \tag{43}$$

Let us consider the configuration (1s, 2p) for the isoelectronic sequence of helium. Within the independent particle approximation, the wavefunctions for the states ^{3}P and ^{1}P are

$$\psi(^3P) = \frac{1}{\sqrt{2}}[\phi_{1s}(1) \cdot \phi_{2p}(2) - \phi_{1s}(2) \cdot \phi_{2p}(1)] \cdot T(1, 2), \tag{44}$$

$$\psi(^3P) = \frac{1}{\sqrt{2}}[\phi'_{1s}(1) \cdot \phi'_{2p}(2) + \phi'_{1s}(2) \cdot \phi'_{2p}(1)] \cdot S(1, 2), \tag{45}$$

where T and S are spin functions of the triplet and singlet type, respectively. The orbitals in the singlet function are primed so as to emphasize the fact that, in contrast to the convention of classical multiplet theory, we do not assume here that the orbitals are identical in both multiplets.

The pair distribution function can be written for the triplet and singlet, respectively, as follows:

$$^3\rho(r_{12}) = D(\phi_{1s}, \phi_{2p}) - E(\phi_{1s}, \phi_{2p}) \tag{46}$$

$$^1\rho(r_{12}) = D(\phi'_{1s}, \phi'_{2p}) + E(\phi_{1s}, \phi'_{2p}) \tag{47}$$

where

$$D_{s,p} \equiv D(\phi_{1s}, \phi_{2p}) = \iint \phi^2_{1s}(1) \cdot \phi^2_{2p}(2) \cdot \delta(r_{12} - r'_{12})\, dV_1\, dV_2 \tag{48}$$

and

$$E_{s,p} \equiv E(\phi_{1s}, \phi_{2p}) = \iint \phi_{1s}(1) \cdot \phi_{2p}(1) \cdot \phi_{1s}(2) \cdot \phi_{2p}(2) \cdot \delta(r_{12} - r'_{12})\, dV_1\, dV_2\,. \tag{49}$$

For a minimal basis set of Gaussians

$$\phi_{1s} = \left(\frac{2\alpha}{\pi}\right)^{3/4} \cdot e^{-\alpha \cdot r^2}$$

$$\phi_{2p} = \sqrt{4\beta} \cdot \left(\frac{2\beta}{\pi}\right)^{3/4} \cdot x \cdot e^{-\beta \cdot r^2}$$

the integrals $D_{s,p}$ and $E_{s,p}$ are evaluated in Appendix B. The results are

$$D_{s,p}(r) = C \cdot r^2 \cdot \exp\left[-\frac{2\alpha\beta}{\alpha+\beta} r^2\right] \cdot \left[1 + \frac{4}{3} \cdot \frac{\alpha^2}{\alpha+\beta} r^2\right], \tag{50}$$

$$E_{s,p}(r) = C \cdot r^2 \cdot \exp\left[\frac{-(\alpha+\beta)}{2} r^2\right] \cdot \left[1 - \frac{\alpha+\beta}{3} r^2\right], \tag{51}$$

where

$$C = \frac{64}{\sqrt{\pi}} \cdot \frac{\alpha^{3/2} \cdot \beta^{5/2}}{[2(\alpha + \beta)]^{5/2}} \tag{52}$$

If r_{12} is very small, these relations reduce to

$$\begin{aligned} D(r) &\simeq C \cdot r^2 \cdot \left(1 - \frac{2\alpha\beta}{\alpha + \beta} r^2\right) \cdot \left[1 + \frac{4\alpha^2}{3(\alpha + \beta)} r^2\right] \\ &\simeq C \cdot r^2 \left[1 + \frac{2\alpha(2\alpha - 3\beta)}{3(\alpha + \beta)} r^2\right] \end{aligned} \tag{53}$$

and

$$\begin{aligned} E(r) &\simeq C \cdot r^2 \cdot \left(1 - \frac{\alpha + \beta}{2} r^2\right) \cdot \left(1 - \frac{\alpha + \beta}{3} r^2\right) \\ &\simeq C \cdot r^2 \cdot \left[1 - \frac{5(\alpha + \beta)}{6} r^2\right]. \end{aligned} \tag{54}$$

Therefore

$$^1\rho(r) = (D + E) \propto r^2$$

whereas

$$^3\rho(r) = (D - E) \propto r^4.$$

It is important to stress that this behavior is not dependent on the assumption of the orbitals being equal for the different multiplets and is, therefore, equally true for classical multiplet theory and for an approach that allows for independent optimization of the orbitals of each multiplet. From the asymptotic behavior obtained, it therefore follows that, for small distances between the two electrons,

$$^3\rho(r_{12}) < {}^1\rho(r_{12})$$

i.e., it is more probable that the electrons will be found to be closer to each other for the triplet than for the singlet. This result is a manifestation of Pauli's principle.

For very large r_{12},

$$D(r) \simeq C \cdot \frac{4}{3} \cdot \frac{\alpha^2}{\alpha + \beta} \cdot r^4 \cdot \exp\left[-\frac{2\alpha\beta}{\alpha + \beta} r^2\right] \tag{55}$$

and

$$E(r) \simeq -C \cdot \frac{\alpha + \beta}{3} \cdot r^4 \cdot \exp\left[\frac{-(\alpha + \beta)}{2} r^2\right], \tag{56}$$

hence

$$\frac{E(r)}{D(r)} \simeq -\left(\frac{\alpha+\beta}{2\alpha}\right)^2 \cdot \exp\left[-\left(\frac{\alpha+\beta}{2} - \frac{2\alpha\beta}{\alpha+\beta}\right)r^2\right]. \tag{57}$$

If the exponential parameter satisfies the condition

$$\frac{\alpha+\beta}{2} - \frac{2\alpha\beta}{\alpha+\beta} > 0,$$

then for a large enough value of r, $E(r)$ will be very small as compared to $D(r)$. The above condition is equivalent to the condition $\alpha \neq \beta$ which is bound to be true since α and β are connected with effective charges for an inner and outer electron, respectively.

In contrast to the behavior for very small r_{12}, the relative value of $^3\rho$ and $^1\rho$ for large interelectronic distances obtained from the approximation of classical multiplet theory is fundamentally different from the result obtained when independent optimization of the orbitals is allowed.

The interesting value is the difference in the pair distribution function for the two multiplets:

$$\Delta(r_{12}) = {}^3\rho(r_{12}) - {}^1\rho(r_{12}). \tag{58}$$

Meanwhile, we know that, for very small values of the interelectronic distance,

$$\Delta(r_{12}) < 0.$$

From the normalization of the wavefunctions, it follows that

$$\int_0^\infty {}^1\rho(r)\, dr = \int_0^\infty {}^3\rho(r)\, dr = 1 \tag{59}$$

and, therefore,

$$\int_0^\infty \Delta(r_{12}) \cdot dr_{12} = 0. \tag{60}$$

For the approximation of identical orbitals in the two multiplets, there exists

$$\Delta(r) = (D - E) - (D + E) = -2E(r). \tag{61}$$

From the expression for $E(r)$ for a minimal basis of Gaussian functions, it is clear that $E(r) > 0$ for $r < \sqrt{2/(\alpha+\beta)}$ and $E(r) < 0$ for larger values of the interelectronic distance. The probability of finding the two electrons at small distances is lower in the triplet, whereas the probability of finding them at great distances is lower in the singlet. The conclusion concerning the relative value of the interelectronic repulsion within the framework of the classical multiplet theory is a direct consequence from this.

When independent optimization of the parameters is carried out, then

$$\Delta(r) = ({}^3D - {}^3E) - ({}^1D + {}^1E) = ({}^3D - {}^1D) - ({}^3E + {}^1E). \tag{62}$$

Since at large interelectronic distances, $E \ll D$ for both multiplets, the value of $\Delta(r)$ will be mainly dependent on ${}^3D - {}^1D$.

We must therefore compare the values of ${}^3D$ and ${}^1D$:

$$\frac{{}^3D(r)}{{}^1D(r)} \simeq \exp\left\{-2 \cdot \left[\frac{{}^3\alpha \cdot {}^3\beta}{{}^3\alpha + {}^3\beta} - \frac{{}^1\alpha \cdot {}^1\beta}{{}^1\alpha + {}^1\beta}\right] \cdot r^2\right\}. \tag{63}$$

Since ${}^{1,3}\alpha$ are parameters of the inner orbital, which is only slightly affected by the presence of the outer orbital, it may be expected that ${}^3\alpha \simeq {}^1\alpha$. On the other hand, it is expected that ${}^3\beta > {}^1\beta$ so that the nuclear attraction is greater in the triplet.

Consequently,

$$\frac{{}^3\alpha \cdot {}^3\beta}{{}^3\alpha + {}^3\beta} > \frac{{}^1\alpha \cdot {}^1\beta}{{}^1\alpha + {}^1\beta} \tag{64}$$

and, therefore, when the interelectronic distances are sufficiently large,

$${}^3D < {}^1D.$$

For large interelectronic distances, we expect that

$$\Delta(r_{12}) = {}^3\rho(r_{12}) - {}^1\rho(r_{12}) < 0 \tag{65}$$

follows as a result of the difference in the $D(r)$ functions between the triplet and the singlet, whereas the exchange terms are negligible.

In order to maintain the three requirements,

$$\Delta(r_{12}) < 0 \qquad r_{12} \to 0,$$

$$\Delta(r_{12}) < 0 \qquad r_{12} \to \infty,$$

$$\int_0^\infty \Delta(r_{12}) \cdot dr_{12} = 0,$$

there must exist a range of medium interelectronic distances, in which $\Delta(r_{12}) > 0$. Contrary to the simplified picture of classical multiplet theory according to which electrons with parallel spins are likely to be farther apart than electrons with reversed spins, we have here a more complex situation.

It is found that both for very small and for very large interelectronic distances, electrons with reversed spins are more likely to be found, whereas in the intermediate range we are likely to find electrons with parallel spins. The qualitative use of Pauli's principle, therefore, yields an adequate qualitative picture in the range $r_{12} \simeq 0$, but the extrapolation from this local behavior to the whole space is both unjustified and misleading.

B. Illustrative Computations

By performing independent optimization of the parameters for the two multiplets, we obtain the results presented in Fig. 3.

In accordance with the results of the previous paragraphs, we obtain here too, a shrinking of the outer orbital, 2p, in the triplet, as well as an expansion of the inner orbital, 1s.

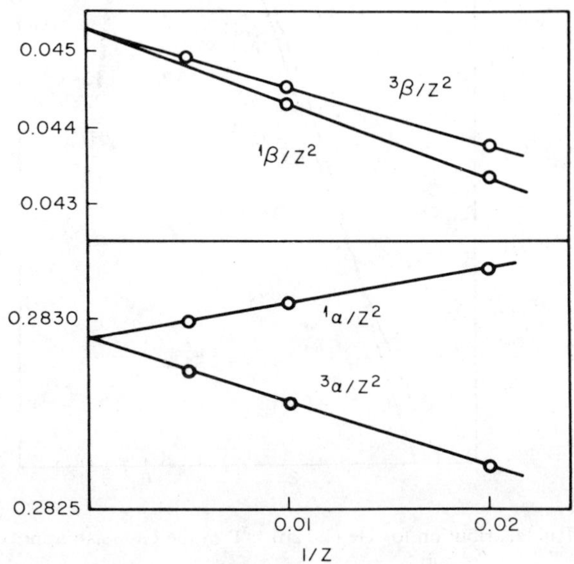

Fig. 3. Asymptotic behavior of the exponential parameters for the ^{1,3}P states of the He isoelectronic sequence in the Gaussian approximation.

It is clear from Fig. 3 that the asymptotic behavior in the limit $Z \to \infty$ is, for each parameter,

$$\eta/Z^2 = \eta_0 + \eta_1 \cdot (1/Z) \tag{66}$$

so that

$$^3\alpha_0 = {^1\alpha_0} = 0.28295,$$

$$^3\beta_0 = {^1\beta_0} = 0.0453.$$

These values correspond to the limit at which the interelectronic repulsion can be neglected and, therefore, they must be equal to the parameters for the 1s and 2p Gaussian orbitals for a hydrogen atom.

The values given by Reeves (1963) for the hydrogenic parameters are

$$\alpha_{1s} = 0.28294,$$

$$\alpha_{2p} = 0.04527,$$

and the correspondence is, therefore, excellent.

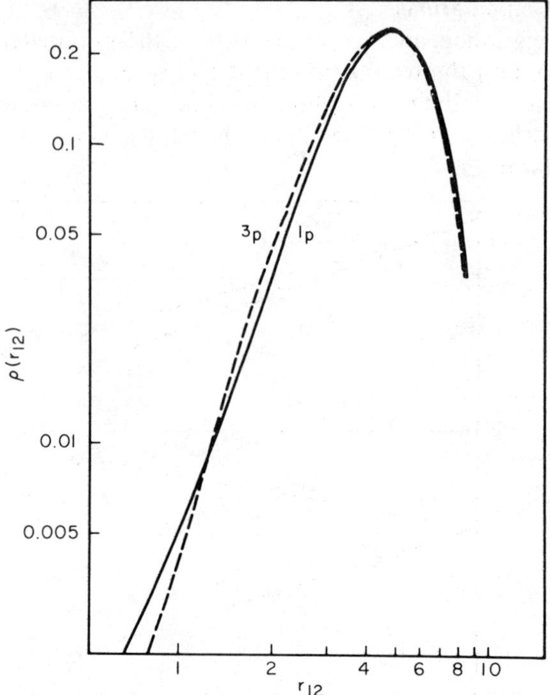

Fig. 4. Pair distribution for He (1s, 2p) ^{1,3}P in the Gaussian approximation.

The pair distribution functions for the states $^{1,\,3}$P of helium, in the Gaussian approximation are presented in Fig. 4. The results presented in Fig. 5 were obtained by computing the difference between the pair distribution functions. It is clear from the figure that, although for $Z = 200$, the behavior is quite close to the one expected according to classical multiplet theory, the behavior for $Z = 2$ is quite different and corresponds very well to the qualitative results obtained from the asymptotic behavior of the pair distribution function.

In order to obtain a more complete and more trustworthy picture, the behavior for orbitals made up as linear combinations of N even-tempered, Gaussian functions,

$$\phi_{1s} = \sum_{i=1}^{N} C_i \cdot e^{-\alpha_i r^2} \qquad \alpha_i = \alpha \cdot \varepsilon^{i-1}$$

$$\phi_{2p} = x \cdot \sum_{i=1}^{N} D_i \cdot e^{-\beta_i r^2} \qquad \beta_i = \beta \cdot \delta^{i-1}.$$

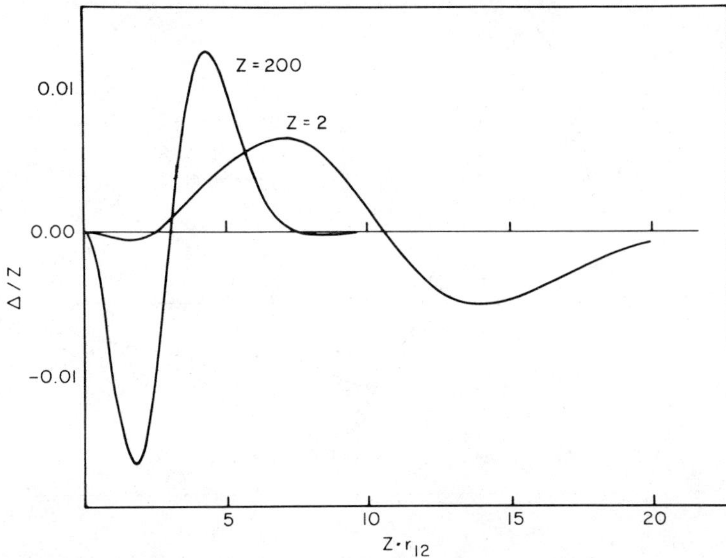

Fig. 5. The pair distribution difference for the $^3P - {}^1P$ states of the He sequence in the Gaussian approximation.

was investigated (Katriel, 1972a). For $N \to \infty$, it converges to the Hartree–Fock solution. The four nonlinear parameters were determined my minimization of the energy. The $2N$ linear parameters were determined by an iterative solution of SCF-type equations. The results for helium are presented in Table VIII.

The pair distribution functions for the two multiplets of helium were computed by applying the expressions developed in Appendix B; the difference between them is presented in Fig. 6. The qualitative behavior is similar

TABLE VIII

INDEPENDENT PARTICLE MODEL RESULTS FOR $^{1,3}P$ He

	Energy		Nuclear attraction		Electronic repulsion	
N	Triplet	Singlet	Triplet	Singlet	Triplet	Singlet
2	−2.06991	−2.06491	−4.39798	−4.37138	0.25817	0.24156
3	−2.11317	−2.10577	−4.49061	−4.45446	0.26428	0.24293
4	−2.12591	−2.11750	−4.51737	−4.47843	0.26554	0.24343
5	−2.12947	−2.12074	−4.52492	−4.48502	0.26598	0.24355
∞[a]	−2.13143	−2.12246	—	—	0.2661	0.2436

[a] Davidson (1965).

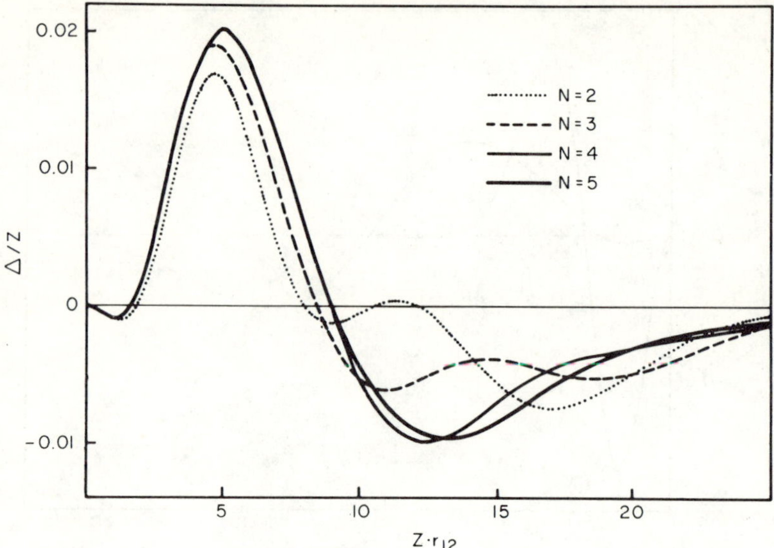

Fig. 6. The pair distribution difference for the $^3P - {}^1P$ states of helium.

to that obtained for a Gaussian approximation ($N = 1$) as presented in Fig. 5. Pauli's principle is manifested in the behavior at very small interelectronic distances, but the energetic effect of this region is, of course, secondary. It would be closer to reality to claim that electrons of parallel spins are more likely to be found at a small distance from each other, whereas electrons with reversed spins are more likely to be found at a considerable distance from each other, rather than the other way around.

The behavior of the pair distribution difference is in perfect agreement with the results of the computations for the interelectronic repulsion values in the singlet and triplet states, respectively. These findings do not argue against the Pauli principle but merely against an oversimplified interpretation of its consequences.

From the asymptotic behavior of the variational parameters as well as from the perturbation theory development of the wavefunction, it follows that the asymptotic behavior of the difference between the pair distribution functions, for $Z \to \infty$, is

$$\Delta(r_{12})/Z = \Delta_0(\tilde{r}_{12}) + (1/Z) \cdot \Delta_1(\tilde{r}_{12}), \tag{67}$$

where

$$\tilde{r}_{12} = Z \cdot r_{12}.$$

$\Delta_0(r_{12})$ is the difference between the pair distribution functions obtained from identical and unscreened orbitals, for the two multiplets, whereas $\Delta_1(r_{12})$ is the first-order correction.

By comparing the expression,

$$\Delta C = \int_0^\infty \Delta(r_{12}) \cdot \frac{dr_{12}}{r_{12}}$$

$$= Z \cdot \int_0^\infty \Delta_0(r_{12}) \frac{dr_{12}}{r_{12}} + \int_0^\infty \Delta_1(r_{12}) \frac{dr_{12}}{r_{12}}, \tag{68}$$

with the expression obtained from perturbation theory,

$$\Delta C = \alpha(Z - 2Z_0), \tag{69}$$

we obtain

$$\int_0^\infty \Delta_0(r_{12}) \frac{dr_{12}}{r_{12}} = \alpha < 0 \tag{70}$$

whereas

$$\int_0^\infty \Delta_1(r_{12}) \frac{dr_{12}}{r_{12}} = -2\alpha Z_0 > 0. \tag{71}$$

From the normalization of the pair distribution functions, it follows that

$$\int_0^\infty \Delta(r_{12})\, dr_{12} = 0 \tag{72}$$

and, therefore,

$$\int_0^\infty \Delta_0(r_{12})\, dr_{12} = 0 \tag{73}$$

and also

$$\int_0^\infty \Delta_1(r_{12})\, dr_{12} = 0. \tag{74}$$

$\Delta_0(r_{12})$ is, therefore, necessarily negative for small interelectronic distances and positive for large ones, in agreement with Pauli's principle. This is exactly the behavior of the difference between the pair distribution functions computed with the same orbitals for the two multiplets.

The behavior of $\Delta_1(r_{12})$ is exactly reversed; it is positive for small interelectronic distances and negative for large ones. For high nuclear charges the first-order correction is negligible and, therefore, we obtain the representation of the classical multiplet theory, which corresponds to zero order. For low nuclear charges the first-order correction affects the picture more significantly.

The functions $\Delta_0(r_{12})$ and $\Delta_1(r_{12})$ were obtained for the Gaussian approximation by a numerical analysis of $\Delta(r_{12})$ for high nuclear charges.

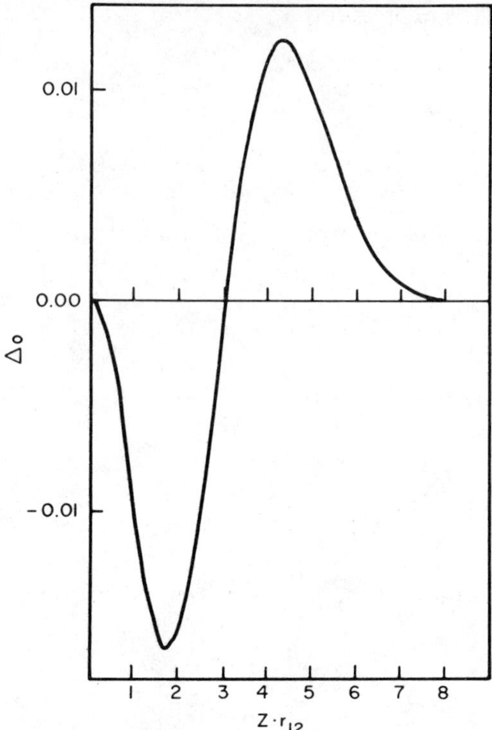

Fig. 7. Zero-order component of the pair distribution difference.

The results are presented in Figs. 7 and 8 and correspond well to the qualitative results of the above discussion.

The functions $D(r)$ and $E(r)$ are connected with the Coulomb integral and with the exchange integral,

$$J = \int_0^\infty D(r_{12}) \cdot \frac{dr_{12}}{r_{12}}, \tag{75}$$

$$K = \int_0^\infty E(r_{12}) \cdot \frac{dr_{12}}{r_{12}}, \tag{76}$$

so that in the approximation of the classical multiplet theory, there exists

$$^3C - {}^1C = (J - K) - (J + K) = -2K,$$

where ^3C and ^1C are the interelectronic repulsion terms in the two multiplets.

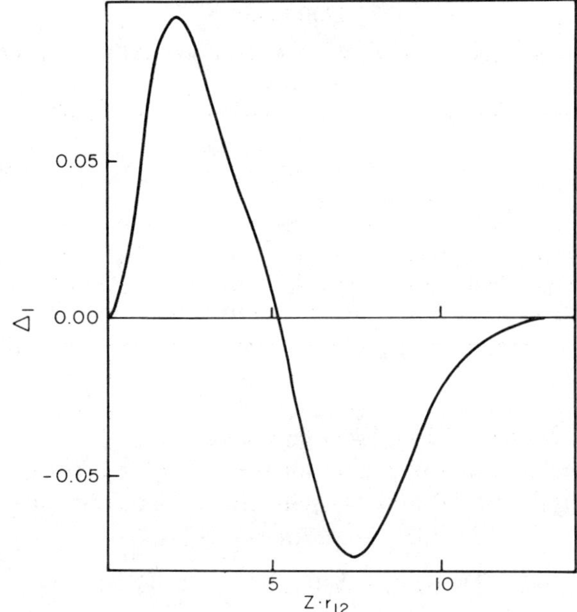

Fig. 8. First-order component of the pair distribution difference.

According to the classical multiplet theory, the splitting between the multiplets, therefore, follows from the fact that the exchange term appears in the energy expressions with reversed signs. By independent optimization for each term we get

$$^3C - {}^1C = ({}^3J - {}^3K) - ({}^1J + {}^1K) = ({}^3J - {}^1J) - ({}^3K + {}^1K), \quad (77)$$

and therefore the difference in the Coulomb integrals will also contribute to the difference in the interelectronic repulsions. Since in the triplet the wavefunction has shrunk more toward the nucleus than in the singlet, one may expect that $^3J > {}^1J$ and therefore in cases where the difference in the Coulomb integrals is big enough there will be $^3C - {}^1C > 0$, contradicting the expectations of the classical multiplet theory (Colpa, 1974).

The values obtained for the interelectronic repulsion components in the Gaussian approximation are given in Table IX.

It is clear from the values in Table IX that for a small nuclear charge the difference in the Coulomb terms is the dominant value and, therefore, $^3C - {}^1C > 0$, whereas for a sufficiently large nuclear charge the difference in the interelectronic repulsions is determined by the exchange term. This behavior corresponds to the qualitative discussion presented above.

TABLE IX

COMPONENTS OF INTERELECTRONIC REPULSION FOR He SEQUENCE IN THE GAUSSIAN APPROXIMATION

	J		K		C	
Z	^3P	^1P	^3P	^1P	^3P	^1P
2	0.23016	0.22404	0.000687	0.000602	0.22947	0.22464
4	0.69318	0.66886	0.00911	0.00765	0.68407	0.67652
10	2.05343	2.01278	0.04759	0.04320	2.00584	2.05598

C. Expectation Values of Two-Electron Operators

From the asymptotic expression for the difference between the pair distribution functions for ^1P and ^3P states in the isoelectronic sequence of helium,

$$\Delta(r_{12})/Z = \Delta_0(\tilde{r}_{12}) + \frac{1}{Z} \cdot \Delta_1(\tilde{r}_{12}), \tag{78}$$

where $\tilde{r}_{12} = Z \cdot r_{12}$, we get

$$\Delta(r_{12})\, dr_{12} = \left[\Delta_0(\tilde{r}_{12}) + \frac{1}{Z} \cdot \Delta_1(\tilde{r}_{12})\right] d\tilde{r}_{12}, \tag{79}$$

Hence,

$$\int_0^\infty \frac{1}{r_{12}} \cdot \Delta(r_{12})\, dr_{12} = Z \cdot \int_0^\infty \frac{1}{r_{12}} \cdot \Delta_0(r_{12})\, dr_{12} + \int_0^\infty \frac{1}{r_{12}} \cdot \Delta_1(r_{12})\, dr_{12}, \tag{80}$$

$$Z^2 \cdot \int_0^\infty r_{12} \cdot \Delta(r_{12})\, dr_{12} = Z \cdot \int_0^\infty r_{12} \cdot \Delta_0(r_{12})\, dr_{12} + \int_0^\infty r_{12} \cdot \Delta_1(r_{12})\, dr_{12}, \tag{81}$$

$$Z^3 \cdot \int_0^\infty r_{12}^2 \cdot \Delta(r_{12})\, dr_{12} = Z \cdot \int_0^\infty r_{12}^2 \cdot \Delta_0(r_{12})\, dr_{12} + \int_0^\infty r_{12}^2 \cdot \Delta_1(r_{12})\, dr_{12}. \tag{82}$$

TABLE X

DIFFERENCE BETWEEN EXPECTATION VALUES OF TWO-ELECTRON OPERATORS FOR THE 2 ^3P–^1P STATES IN THE He SEQUENCE[a]

Z	$\int \Delta \cdot \frac{dr_{12}}{r_{12}}$	$\int \Delta \cdot r_{12} \cdot dr_{12}$	$\int \Delta \cdot r_{12}^2 \cdot dr_{12}$
2	0.021617	−0.438373	−4.95572
3	0.016108	−0.164216	−1.02758
4	−0.004462	−0.071344	−0.3343
5	−0.030808	−0.032776	−0.135499
6	−0.05991	−0.014177	−0.061672
7	−0.09053	−0.004292	−0.029588
8	−0.12208	0.0013061	−0.014136
9	−0.15424	0.0046070	−0.0061645
10	−0.18680	0.0065964	−0.0018634

[a] Computed from values given by Accad *et al.* (1971).

By a numerical or graphical analysis of the exact expectation values, which were computed for the isoelectronic sequence of helium by Accad *et al.* (1971), one can obtain approximate values of the zero- and first-order components. Qualitative conclusions concerning the behavior of Δ_0 and Δ_1, which correspond the exact computation, can then be drawn from these.

This analysis was carried out for the values given in Table X. The results obtained are

$$\int \frac{1}{r_{12}} \cdot \Delta_0 \cdot dr_{12} = 0.031 \qquad \int \frac{1}{r_{12}} \cdot \Delta_1 \cdot dr_{12} = 0.127,$$

$$\int \Delta_0 \cdot dr_{12} = 0 \qquad \int \Delta_1 \cdot dr_{12} = 0,$$

$$\int r_{12} \cdot \Delta_0 \cdot dr_{12} = 0.292 \qquad \int r_{12} \cdot \Delta_1 \cdot dr_{12} = -2.26,$$

$$\int r_{12}^2 \cdot \Delta_0 \cdot dr_{12} = 2.74 \qquad \int r_{12}^2 \cdot \Delta_1 \cdot dr_{12} = -29.3.$$

Δ_0 is, therefore, negative for small r_{12} and positive for larger r_{12} and, conversely, for Δ_1. This qualitative confirmation of the preceding conclusions is, of course, important since it is based on exact values of the corresponding expectation values. Qualitatively, at least correlation does not affect the differences between the pair distribution functions to any significant extent.

VIII. An Interpretation of Hund's Rule

It is clear from the results presented earlier that the main energetic contribution to the energy difference between different multiplets derived from the same configuration is that of the nuclear attraction. This contribution, unlike that of the interelectronic repulsion, is always greater in the lower energy state. It is therefore clear that the classical interpretation of Hund's rule, which assigns the energy difference to the differences in the interelectronic repulsion, does not present a correct description.

The picture we get from these results is more complex than that of classical multiplet theory. However, it has a number of both coherent and consistent qualities that point to a possibility of achieving a general qualitative interpretation.

A theoretical interpretation of Hund's rule must account for the energetic ordering of the different levels belonging to a given configuration. It should also account for the roles of the various energy components and their relative values and for the qualitative features of the wavefunctions that correspond to the different levels.

The energetic aspect of Hund's rule is explained within the framework of the independent particle model as follows.

Let us consider two multiplets belonging to a given configuration. For the multiplet with the lower multiplicity (multiplet A), we will find the best wavefunction within the framework of the independent particle approximation, i.e., Hartree–Fock function. The atomic orbitals obtained will be used to construct the corresponding wavefunction for the state with the high multiplicity (multiplet B). The energy corresponding to this wavefunction differs from that of the multiplet A only in the interelectronic repulsion. As a result of the Pauli principle, the interelectronic repulsion will be lower for multiplet B when the same atomic orbitals are used for both states. These orbitals are already optimal (Hartree–Fock) for multiplet A but can be improved for B.

As a result of the variation principle this improvement of the orbitals can only decrease the total energy and, therefore, increase the multiplet splitting. Thus, we have shown that within the framework of the independent particle model (Hartree–Fock approximation), the high multiplicity state is energetically lower, which is in accordance with Hund's rule.

The energetic ordering of the levels is, therefore, a direct consequence of the difference in the interelectronic repulsion predicted by the classical interpretation of Hund's rule. The conventional approach constitutes an indispensable component of the suggested rationale. It is incorporated in the argument in such a way as to yield, together with the variation principle, a general prediction of the level ordering within the framework of the

Hartree–Fock approximation. This step would be an improvement in comparison with classical multiplet theory even without the fundamental difficulties encountered in the latter, since a Hartree–Fock approximation is undoubtedly much better than the approximation for which the same atomic orbitals are used for the different states. It is clear that the relaxation of the orbital forms in the high multiplicity state toward the Hartree–Fock orbitals may lead to extreme changes in various energy component values, but so long as we are interested only in the total energy, this is of no significance. It is enough to know that it may only decrease as a result of relaxation. The philosophy of using the variation principle is similar to that of Feinberg and Ruedenberg (1971) in a completely different context. A graphical description of the argument is presented in Fig. 9.

The only difficulty that remains with reference to the energetic ordering of the multiplets is connected to the effect of the correlation energy. We have

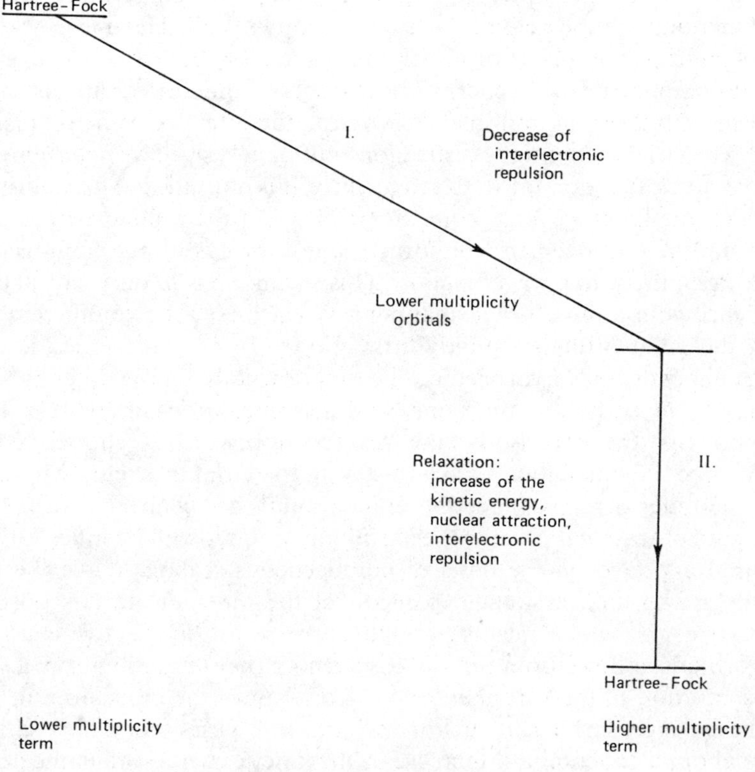

Fig. 9. The interpretation of Hund's rule.

already seen that the correlation energy is greater in the lower multiplicity state and, therefore, tends to diminish the energy difference between the different states.

We do not have a satisfactory general argument that would assure that the correlation can never reverse the level ordering which is obtained from a Hartree–Fock approximation and which corresponds to Hund's rule. We can, at any rate, point out that from the point of view of $1/Z$ perturbation theory the multiplet splitting is a first-order effect. Correlation, on the other hand, is a second-order effect since both in the Hartree–Fock scheme and for the exact solution the zero-order wavefunction is the same antisymmetrical product of hydrogenic functions. Thus, it is highly probable—although we do not have an exact proof—that the correlation effect will be always of secondary importance with regards to the qualitative features of the multiplet ordering.

In order to understand the contributions of the different energy components, let us consider more closely the two steps entailed in constructing the wavefunction for multiplet B. At a first step, we constructed for it a wavefunction with the help of the atomic orbitals of the Hartree–Fock function of multiplet A. As a result of this choice of the orbitals, the single-particle components of the energy (kinetic energy and nuclear attraction) are identical with those of multiplet A, whereas the interelectronic repulsion is lower. The virial theorem is satisfied for multiplet A, since we determined the Hartree–Fock function for it. Consequently, it is not satisfied for multiplet B for which the kinetic energy equals that of A but the total energy is lower. The simplest improvement, uniform scaling, brings about homogeneous shrinking of the whole wavefunction. This shrinking is in only partial correspondence with a more exact treatment. When there are two different main shells, the inner orbitals are only little affected by the outer ones, i.e., they remain almost simple hydrogenic. The outer orbitals, on the other hand, are strongly screened by the inner ones and deformed accordingly. It is, therefore, clear that the relaxation of the wavefunction, which is required to make it a Hartree–Fock solution (while satisfying the virial theorem), will mainly affect the outer orbitals. Since the outer orbitals are connected with only a small part of the total energy, the shrinking required will be much stronger than that expected on the basis of homogeneous scaling. While the outer orbitals are shrunk, a slight expansion of the inner orbitals is observed. This expansion, which means a small decrease in the effective charge for these orbitals, follows from the slight screening that originates from a stronger penetration of the shrunken outer orbitals into the inner orbital.

The balance, formed in such a way that the virial theorem is satisfied, brings about a concomitant increase in the kinetic energy and in the nuclear attraction, as a result of the shrinking of the outer orbitals and their approach to the nucleus. The kinetic energy and the nuclear attraction (in its absolute value) are higher in B than in multiplet A. The approach between the outer and inner orbitals, resulting from the shrinking of the former and the expansion of the latter causes an increase in the interelectronic repulsion which is much greater than when uniform scaling is assumed. Thus, the interelectronic repulsion decreases in the first stage of wavefunction construction, but increases in its relaxation stage.

It is quite possible that the increase in the second stage will be greater than the decrease in the first, so that, on the whole, we will get stronger repulsion just in multiplet B. The total energy increases more or less quadratically with nuclear charge, whereas the interelectronic repulsion increases only linearly. It may, therefore, be expected that with the increase of the nuclear charge the extent of the shrinking of the outer orbitals required by the virial theorem would be smaller, so that the increase in the interelectronic repulsion would be progressively smaller. On the other hand, it is clear that the increase predicted by the uniform scaling model is a lower bound to the increase obtained in a Hartree–Fock approximation, in which the considerable shrinking in the outer orbitals and the expansion of the inner ones are much more effective in increasing the interelectronic repulsion.

Since we have shown for the uniform scaling model that the differences in nuclear attraction are the dominant differences between the multiplets, it follows from the lower bound claim that the same state of affairs is true for the Hartree–Fock approximation as well.

This conclusion is in close agreement with the computational results presented above, but it is important here to emphasize the general nature of the argument and its independence of any specific computations.

We must keep in mind, of course, the fact that the picture is incomplete without a discussion of the contribution of correlation. We have already seen that this contribution is not expected to affect the qualitative description of the energetic ordering of the multiplets, although the energetic distance between them is expected to become smaller as a result of its introduction. The major effect of correlation is the decrease in the interelectronic repulsion, accompanied by a secondary increase in the nuclear attraction.

Correlation is expected to have a greater effect on the interelectronic repulsion in multiplet A and, therefore, the interelectronic repulsion in multiplet B will be still higher in relation to A than it is within the framework of the Hartree–Fock approximation.

The fact that there is no reversal of the interelectronic repulsions for configurations with only one main shell agrees well with the argument presented above. The only origin for the increase in the interelectronic repulsion in multiplet B is the shrinking of the single shell, which is needed to increase the kinetic energy so that the virial theorem is satisfied.

This shrinking, of course, also increases the nuclear attraction (in its absolute value) and the interelectronic repulsion. Clearly, the increase in the two one-electron elements will be steeper than the increase in the interelectronic repulsion.

The change needed in the kinetic energy so that the virial theorem is satisfied is of the order of magnitude of the difference in the interelectronic repulsions before the shrinking, which then equals the total energy difference. This change in the kinetic energy is achieved along with a considerably smaller change in the interelectronic repulsion. A measure of the relative sensitivity of the interelectronic repulsion and the nuclear attraction to the orbital shrinking is the ratio between the screening that originates from the other electrons and the charge of the nucleus as a whole. It is clear that the ratio is very small when the screening originates from electrons of the same shell, and, in any case, it is smaller than unity. It is impossible, therefore, to expect a reversal of the interelectronic repulsion without expansion of the inner shell toward the shrinking outer shell.

Finally, we wish to bring the characterization of the difficulties encountered in the interpretation of Hund's rule into sharper focus. It is quite clear that the term in the Hamiltonian that is responsible for the energetic differences between the different multiplets of a given configuration is the interelectronic repulsion. It is found that the introduction of interelectronic repulsion into the Hamiltonian brings about a change in the wavefunction in such a way that the energetic effect of the multiplet splitting appears in the single-electron part. This is, of course, a result of the noncommutativity between the one-electron Hamiltonian and the interelectronic repulsion operator. The quantum mechanical origin of Hund's rule is expressed in this behavior much more conspicuously than in the electrostatic interpretation of classical multiplet theory, an interpretation that takes into account the quantum mechanical nature of the problem by making use of the Pauli principle but disregards the surprising results of noncommutativity just mentioned.

Appendix A: Screening in $(n, l)^k$-Type Configurations

For an atom with charge Z and k electrons in a configuration $(n, l)^k$, let

$$T = k \cdot \langle \phi | -\frac{1}{2} \nabla^2 | \phi \rangle, \qquad L = k \cdot \langle \phi | -\frac{1}{r} | \phi \rangle,$$

$$C = \langle \psi | \sum_{i<j} \frac{1}{r_{ij}} | \psi \rangle,$$

where $\phi = \phi_{n,l}(r, \theta, \varphi)$. By introducing scaling we get

$$E = \eta^2 \cdot T + \eta(Z \cdot L + C)$$

and from $\theta E/\partial\eta = 0$, it follows as usual that

$$\eta = -(1/2T)(Z \cdot L + C).$$

If hydrogenic orbitals are used, then $T = k \cdot (1/2n^2)$, $L = -k \cdot (1/n^2)$ so that $\eta = Z - (n^2 \cdot C/k)$. The screening is given by $\sigma = n^2 \cdot C/k$ and is in this approximation constant along the isoelectronic sequence and proportional to the interelectronic repulsion. The simplest special case is the ground-state configuration of the helium sequence, for which $n = 1$, $k = 2$, and $C = 5/8$ so that $\sigma = 5/16$.

Appendix B: The Pair Distribution Function

If the pair distribution function is defined as

$$\rho(r_0) = \langle\psi| \sum_{i<j} \delta(r_{ij} - r_0) |\psi\rangle,$$

then one immediately notices that for a wavefunction constructed out of Slater determinants the typical matrix elements are

$$\langle\phi_i(1)\phi_j(2)|\delta(r_{12} - r_0)|\phi_{i'}(1)\phi_{j'}(2)\rangle.$$

For Gaussian basis functions, we use the Talmi (1952) transformation to center-of-mass and relative coordinates, express the relative coordinates in a spherical polar framework and integrate over all coordinates but r_{12}.

We write $\mathbf{R} = \delta \cdot \mathbf{r}_1 + \varepsilon \cdot \mathbf{r}_2$ and $\mathbf{r} = -\mathbf{r}_1 + \mathbf{r}_2$ and require $\delta + \varepsilon = 1$ to obtain a Jacobian which is equal to unity.

The typical exponential expression is

$$\alpha r_1^2 + \beta r_2^2 = (\alpha + \beta)R^2 + [\delta(\alpha + \beta) - \alpha]2\mathbf{R} \cdot \mathbf{r} + [\alpha(1 - \delta^2) + \beta\delta^2]r^2.$$

We now choose δ so as to cancel the mixed term, i.e., $\delta = \alpha/(\alpha + \beta)$, getting

$$\mathbf{r}_1 = \mathbf{R} - \frac{\beta}{\alpha + \beta}\mathbf{r}, \qquad \mathbf{r}_2 = \mathbf{R} + \frac{\alpha}{\alpha + \beta}\mathbf{r}$$

and

$$\alpha r_1^2 + \beta r_2^2 = (\alpha + \beta)R^2 + \frac{\alpha\beta}{\alpha + \beta}r^2.$$

For simple Gaussians,

$$\rho(r_0) = \langle \exp(-\alpha r_1^2) \exp(-\beta r_2^2) | \delta(r_{12} - r_0) | \exp(-\alpha' r_1^2) \exp(-\beta' r_2^2) \rangle$$

$$= \int d^3R \cdot \exp[-(\alpha + \alpha' + \beta + \beta')R^2] \cdot \int d^3r \cdot \delta(r - r_0)$$

$$\cdot \exp\left[-\frac{(\alpha + \alpha')(\beta + \beta')}{\alpha + \alpha' + \beta + \beta'} r^2\right]$$

$$= \left(\frac{\pi}{\alpha + \alpha' + \beta + \beta'}\right)^{3/2} \cdot \exp\left[-\frac{(\alpha + \alpha')(\beta + \beta')}{\alpha + \alpha' + \beta + \beta'} r_0^2\right] \cdot 4\pi r_0^2$$

which is identical with the expression given by Lester and Krauss (1964).

We will now evaluate the pair distribution functions,

$$D_{s,p}(r) = \langle \phi_s(\alpha; 1) \cdot \phi_p(\beta; 2) | \delta(r_{12} - r) | \phi_s(\alpha'; 1) \cdot \phi_p(\beta'; 2) \rangle \qquad (83)$$

and

$$E_{s,p}(r) = \langle \phi_s(\alpha; 1) \cdot \phi_p(\beta; 2) | \delta(r_{12} - r) | \phi_p(\beta'; 1) \cdot \phi_s(\alpha'; 2) \rangle \qquad (84)$$

in which

$$\phi_s(\alpha; i) = \left(\frac{2\alpha}{\pi}\right)^{3/4} \cdot \exp(-\alpha r_i^2), \qquad (85)$$

$$\phi_p(\beta; 1) = \sqrt{4\beta} \cdot \left(\frac{2\beta}{\pi}\right)^{3/4} \cdot x_i \cdot \exp(-\beta r_i^2). \qquad (86)$$

We get

$$D_{s,p}(r') = 4\sqrt{\beta \cdot \beta'} \cdot \left(\frac{2}{\pi}\right)^3 (\alpha\alpha'\beta\beta')^{3/4}$$

$$\cdot \int \exp[-(\alpha + \alpha')r_1^2] \cdot x_2^2 \cdot \exp[-(\beta + \beta')r_2^2]$$

$$\cdot \delta(r_{12} - r') \, d^3r_1 \, d^3r_2$$

$$= \frac{32}{\pi^3} (\alpha\alpha')^{3/4} \cdot (\beta\beta')^{5/4}$$

$$\cdot \int \exp[-(\alpha + \alpha' + \beta + \beta')R^2]$$

$$\cdot \exp\left[-\frac{(\alpha + \alpha')(\beta + \beta')}{\alpha + \alpha' + \beta + \beta'} r^2\right]$$

$$\cdot \left(X + \frac{\alpha + \alpha'}{\alpha + \alpha' + \beta + \beta'} \cdot x\right)^2$$

$$\cdot \delta(r_{12} - r') \cdot d^3R \cdot d^3r$$

$$= \frac{32}{\pi^3} (\alpha \cdot \alpha')^{3/4} \cdot (\beta\beta')^{5/4}$$

$$\cdot \left\{ \int d^3R \cdot X^2 \cdot \exp[-(\alpha + \alpha' + \beta + \beta')R^2] \cdot \int d^3r \right.$$

$$\cdot \delta(r_{12} - r') \cdot \exp\left[-\frac{(\alpha + \alpha')(\beta + \beta')}{\alpha + \alpha' + \beta + \beta'} r^2 \right]$$

$$+ \left(\frac{\alpha + \alpha'}{\alpha + \alpha' + \beta + \beta'} \right)^2 \cdot \int d^3R$$

$$\cdot \exp[-(\alpha + \alpha' + \beta + \beta')R^2] \cdot \int d^3r \cdot \delta(r_{12} - r')$$

$$\left. \cdot x^2 \cdot \exp\left[-\frac{(\alpha + \alpha')(\beta + \beta')}{\alpha + \alpha' + \beta + \beta'} r^2 \right] \right\}$$

$$= \frac{32}{\pi^3} (\alpha\alpha')^{3/4} \cdot (\beta\beta')^{5/4}$$

$$\cdot \left\{ \frac{\pi^{3/2}}{2(\alpha + \alpha' + \beta + \beta')^{5/2}} \right.$$

$$\cdot \exp\left[-\frac{(\alpha + \alpha')(\beta + \beta')}{\alpha + \alpha' + \beta + \beta'} r'^2 \right] \cdot 4\pi r'^2$$

$$+ \left(\frac{\alpha + \alpha'}{\alpha + \alpha' + \beta + \beta'} \right)^2 \cdot \left(\frac{\pi}{\alpha + \alpha' + \beta + \beta'} \right)^{3/2}$$

$$\left. \cdot \exp\left[-\frac{(\alpha + \alpha')(\beta + \beta')}{\alpha + \alpha' + \beta + \beta'} r'^2 \right] \cdot r'^4 \cdot \frac{4\pi}{3} \right\}.$$

Finally,

$$D_{s,p}(r) = \frac{64}{\sqrt{\pi}} \cdot \frac{(\alpha \cdot \alpha')^{3/4} (\beta \cdot \beta')^{5/4}}{(\alpha + \alpha' + \beta + \beta')^{5/2}} \cdot r^2$$

$$\cdot \exp\left[-\frac{(\alpha + \alpha')(\beta + \beta')}{\alpha + \alpha' + \beta + \beta'} \cdot r^2 \right]$$

$$\cdot \left[1 + \frac{2}{3} \frac{(\alpha + \alpha')^2}{\alpha + \alpha' + \beta + \beta'} \cdot r^2 \right].$$

In a similar manner,

$$E_{s,p}(r) = \frac{64}{\sqrt{\pi}} \frac{(\alpha \cdot \alpha')^{3/4} \cdot (\beta \cdot \beta')^{5/4}}{(\alpha + \alpha' + \beta + \beta')^{5/2}} \cdot r^2 \cdot \exp\left[-\frac{(\alpha + \beta')(\alpha' + \beta)}{\alpha + \alpha' + \beta + \beta'} \cdot r^2\right] \cdot \left[1 - \frac{2}{3} \cdot \frac{(\alpha + \beta')(\alpha' + \beta)}{\alpha + \alpha' + \beta + \beta'} \cdot r^2\right].$$

For $\alpha = \alpha'$ and $\beta = \beta'$,

$$D_{s,p}(r) = \frac{64}{\sqrt{\pi}} \cdot \frac{\alpha^{3/2} \cdot \beta^{5/2}}{[2(\alpha + \beta)]^{5/2}} \cdot r^2 \cdot \exp\left(-\frac{2\alpha\beta}{\alpha + \beta} r^2\right) \cdot \left(1 + \frac{4}{3} \cdot \frac{\alpha^2}{\alpha + \beta} r^2\right),$$

$$E_{s,p}(r) = \frac{64}{\sqrt{\pi}} \cdot \frac{\alpha^{3/2} \cdot \beta^{5/2}}{[2(\alpha + \beta)]^{5/2}} \cdot r^2 \cdot \exp\left(-\frac{\alpha + \beta}{2} r^2\right)\left(1 - \frac{\alpha + \beta}{3} r^2\right).$$

Acknowledgment

The work reported here is part of a research project supported by the U.S.-Israel Binational Science Foundation.

References

Accad, Y., Pekeris, C. L., and Schiff, B. (1971). *Phys. Rev. A* **4**, 516.
Clementi, E. (1965). *IBM J. Res. Dev.* **9**, 2.
Colpa, J. P. (1974). *Mol. Phys.* **28**, 581.
Colpa, J. P., and Brown, R. E. (1973). *Mol. Phys.* **26**, 1453.
Colpa, J. P., and Islip, M. F. J. (1973). *Mol. Phys.* **25**, 701.
Colpa, J. P., Thakkar, A. J., Smith, V. H. Jr., and Randle, P. (1975). *Mol. Phys.* **29**, 1861.
Condon, E. U., and Shortley, G. H., (1935). "The Theory of Atomic Spectra." Cambridge Univ. Press, London and New York.
Davidson, E. R. (1965). *J. Chem. Phys.* **42**, 4199.
Eliezer, L., and Moualem, A. (1971). *Theor. Chim. Acta* **23**, 39.
Feinberg, M. J., and Ruedenberg, K. (1971). *J. Chem. Phys.* **54**, 1495.
Harcourt, R. D., and Harcourt, A. (1973). *Chem. Phys.* **1**, 238.

Harcourt, R. D., Beckworth, J., and Feigen, R. (1974). *Chem. Phys. Lett.* **26**, 126.
Hund, F. (1925). *Z. Phys.* **33**, 345.
Hylleraas, E. A. (1930). *Z. Phys.* **65**, 209.
Katriel, J. (1972a). *Theor. Chim. Acta* **23**, 309.
Katriel, J. (1972b). *Phys. Rev. A* **5**, 1990.
Katriel, J. (1972c). *Theor. Chim. Acta* **26**, 163.
Killingbeck, J. (1973). *J. Mol. Phys.* **25**, 455.
Kohl, J. (1972). *J. Chem. Phys.* **56**, 4236.
Lemberger, A., and Pauncz, R. (1969). *Acta Phys. Acad. Sci. Hung.* **27**, 169.
Lester, W. A., and Krauss, M. (1964). *J. Chem. Phys.* **41**, 1407.
Löwdin, P.-O. (1959). *J. Mol. Spectrosc.* **3**, 46.
Martin, W. C. (1963). *J. Opt. Soc. Am.* **53**, 1047.
Messmer, R. P., and Birss, F. W. (1969). *J. Phys. Chem.* **73**, 2085.
Reeves, C. M. (1963). *J. Chem. Phys.* **39**, 1.
Russel, H. N., and Saunders, F. A. (1925). *Astrophys. J.* **61**, 38.
Scherr, C. W., and Knight, R. E. (1963). *Rev. Mod. Phys.* **35**, 436.
Slater, J. C. (1926). *Phys. Rev.* **28**, 291.
Slater, J. C. (1929). *Phys. Rev.* **34**, 1293.
Snow, R. L., and Bills, J. L. (1974). *J. Chem. Educ.* **51**, 585.
Talmi, I. (1952). *Helv. Phys. Acta* **25**, 185.

Linked-Cluster Perturbation Theory for Closed- and Open-Shell Systems*

B. H. BRANDOW
Theoretical Division
Los Alamos Scientific Laboratory
University of California
Los Alamos, New Mexico

I. Introduction . . . 188
II. Degenerate Brillouin–Wigner Perturbation Theory . . . 190
III. The Linked-Cluster Expansion for Closed-Shell Systems . . . 192
A. Diagrammatic Notation . . . 192
B. The Unlinked-Cluster Problem . . . 195
C. Cancellation of Unlinked Diagrams . . . 196
D. The Factorization Theorem . . . 197
E. Exclusion-Violating Diagrams . . . 199
IV. Physical Interpretation and General Philosophy of Applications . . . 201
V. Further Closed-Shell Results . . . 204
A. Expectation Values . . . 204
B. The Ground-State Wavefunction . . . 205
C. Norm of the Wavefunction . . . 207
D. The Coester–Kümmel–Čižek Formalism . . . 209
E. Single-Particle Potentials . . . 213
F. Outline of the Adiabatic Derivation of Goldstone . . . 215
VI. Open-Shell Systems—A General Formalism for Model Hamiltonians and Model Operators . . . 220
A. Diagrammatic Conventions . . . 222
B. Separation of Core and Valence Contributions . . . 223
C. Need for a Fully Linked Valence Expansion . . . 225
D. Degenerate Rayleigh–Schrödinger Perturbation Theory . . . 225
E. The Folded-Diagram Expansion . . . 226
F. Reduction of the Bloch–Horowitz Diagrams . . . 229
G. Hermiticity and Effective Operators . . . 231
H. Relation to Other Many-Body Theories and Results . . . 233
I. Potential Applications . . . 236
VII. Formal Derivation of Effective Spin Hamiltonians . . . 237
A. Two Nonorthogonality Catastrophes . . . 237
B. Choice of H_0 . . . 239
C. Results and Discussion . . . 242
D. Effective π-Electron Hamiltonians . . . 244
References . . . 246

* Work performed under the auspices of the U.S. Energy Research and Development Administration.

I. Introduction

For *ab initio* energy calculations, the standard tools of the quantum chemist are the Hartree–Fock or SCF approximation, the variational method, configuration mixing, and perturbation theory. The Hartree–Fock approximation usually describes gross features quite well, but the resulting total energy is frequently unsatisfactory. This error, known as the correlation energy, can be calculated to very high accuracy for two- or three-electron systems by the variational method and to somewhat less accuracy by configuration mixing. Beyond three or four electrons, however, the straightforward application of these methods becomes prohibitively difficult, and one therefore needs more powerful many-body techniques. My personal conviction, which is shared by a number of others, is that the diagrammatic perturbation theory à la Brueckner (1955, 1959) and Goldstone (1957) constitutes the most flexible and general approach now available. The pioneering application in quantum chemistry is, of course, the work of Kelly (1963, 1964b, 1969). The reward for learning this formalism is quite significant. Not only does this provide a very useful computational tool, it also offers much general insight into the nature of many-body systems.

Unfortunately, papers on this subject have tended to be forbiddingly formal and complicated. Many competent researchers have been repelled by the esoteric opacity of much of the literature. When I realized that this was a common reaction among nuclear physicists, I felt challenged to try to remedy this situation. The first result was to develop what I consider to be the simplest and most elementary derivation of the basic linked-cluster energy expansion. In this approach it is not necessary to be familiar with Wick's theorem, the adiabatic theorem, or even the language of second quantization. This derivation is presented and discussed in Sections III and IV, starting with the Brillouin–Wigner form of perturbation theory, which is reviewed in Section II. Besides conveying the flavor of the subject, I hope that this presentation may serve to convince the reader that this formalism can be mastered, to a useful degree of proficiency, without an extraordinary effort.

In Section V we present the corresponding linked-cluster results for the expectation value of a general operator, for the total wavefunction of a closed-shell system, and for the norm of this wavefunction. In contrast to the energy and expectation value expansions, the wavefunction result requires some understanding of second quantization. The connection between this wavefunction result and the nonperturbative formalism of Coester and Kümmel (1960) and Čižek (1966) is also discussed in this section. After some comments about the choice of single-particle potentials, we conclude the discussion of the closed-shell case by outlining the manner in which these linked-cluster results emerge from the time-dependent adiabatic approach of Goldstone.

While working on the pedagogy of Goldstone's famous result, I happened upon the important and beautiful paper of Bloch and Horowitz (1958), concerning the extension of Goldstone's result to open-shell systems. This opened up the vista of a general formalism for effective Hamiltonians and effective operators, for which the nuclear shell model is merely one field of application. The open-shell diagrams of Bloch and Horowitz were not fully linked, however, which meant that in practice their formalism is limited to systems with only a few valence particles, say two or three. A hint in the direction of a corresponding fully linked result was provided in a paper by Morita (1963), which we subsequently developed into a formally satisfactory general scheme (Brandow, 1966a, 1967). The resulting open-shell formalism is described in Section VI. (Quite fortunately, the elementary time-independent approach we had developed for the closed-shell case turned out to have an objective advantage for the open-shell case, as compared to the adiabatic technique used by Goldstone and Morita. This remark is explained in Section VI,F.) This open-shell formalism is much more general and flexible than that for closed-shell systems, in the sense that it can deal with a far greater variety of physical problems. It is, however, considerably more complicated, thus the reader will appreciate that more homework is required to understand this fully. Nevertheless, the basic features should be clear enough from reading Section VI.

The most obvious advantage of a fully linked version of the Bloch–Horowitz result is that the number of valence particles need not be small. This number can, in fact, be macroscopically large. A specific example is presented in Section VII, where we use this formalism to provide a general derivation for effective spin Hamiltonians (à la Heisenberg) for macroscopic samples of magnetic insulator materials. A less obvious but equally important point is that, even for a system with only two valence particles, the fully linked feature is necessary to provide a clean distinction between the one-body aspect (the single-particle potential) and the two-body aspect, i.e., the "true" effective interaction.

Most of the material in this article is discussed more fully in our cited publications. Those with a scholarly interest in the basic linked-cluster formalism may wish to consult the introductory sections of Brandow (1967) and Paldus and Čižek (1975) for references to alternative derivations and some of the many previous reviews. A general review of degenerate perturbation theory and the existing alternatives to the present open-shell perturbative formulation is also available (Brandow, 1975a). An excellent source of information on quantum-chemical applications of a variety of linked-cluster techniques is the proceedings of the Advanced Summer Institute on Correlation Effects in Atoms and Molecules (Lefebvre and Moser, 1969), where most of the active investigators reported on their work as of 1967. A more recent descriptive survey by Freed (1971) is also recommended.

II. Degenerate Brillouin–Wigner Perturbation Theory

There are basically just two types of perturbation theory, namely the Brillouin–Wigner (BW) and Rayleigh–Schrödinger (RS) forms. We shall begin with the BW form, in order to keep the later developments within a simple overall perspective.

Consider a very general quantum system, not necessarily a "many-body" system. Starting with the usual formulas

$$H = H_0 + V, \tag{1}$$

$$\Psi = \sum_i a_i \Phi_i, \tag{2}$$

$$H\Psi = E\Psi, \tag{3}$$

$$H_0 \Phi_i = E_i \Phi_i, \tag{4}$$

we immediately obtain

$$(E - E_i)a_i = \langle \Phi_i | V | \Psi \rangle. \tag{5}$$

We then select a certain number d of the Φ_i's to span a degenerate subspace or "model subspace" D. The corresponding E_i's may be only quasi-degenerate; exact degeneracy is not required. With this choice of D we associate a degenerate projection operator P, a degenerate or "model" wavefunction

$$\Psi_D = P\Psi = \sum_{i \in D} a_i \Phi_i, \tag{6}$$

and a resolvent or Green's function

$$\mathcal{G} = \sum_{i \notin D} \frac{|\Phi_i\rangle\langle\Phi_i|}{E - E_i} = \frac{Q}{E - H_0}, \tag{7}$$

where $P + Q = 1$. Then Eqs. (2) and (5) can be combined in the form

$$\begin{aligned} \Psi &= \Psi_D + \mathcal{G}V\Psi \\ &= \sum_{n=0}^{\infty} (\mathcal{G}V)^n \Psi_D. \end{aligned} \tag{8}$$

It is convenient to define a wave operator or "model operator" Ω such that

$$\Psi = \Omega \Psi_D, \tag{9}$$

and also a reaction matrix, effective interaction, or "model interaction"

$$\mathscr{V} = V\Omega. \tag{10}$$

Substitution in Eq. (8) yields

$$\begin{aligned}\Omega &= P + \mathscr{G}V\Omega \\ &= \sum_{n=0}^{\infty} (\mathscr{G}V)^n P,\end{aligned} \tag{11}$$

$$\begin{aligned}\mathscr{V} &= VP + V\mathscr{G}\mathscr{V} \\ &= V\sum_{n=0}^{\infty} (\mathscr{G}V)^n P.\end{aligned} \tag{12}$$

Now Eq. (5) can be rewritten as

$$\begin{aligned}(E - E_i)a_i &= \langle \Phi_i | V\Omega | \Psi_D \rangle \\ &= \sum_{j \in D} \langle \Phi_i | \mathscr{V} | \Phi_j \rangle a_j .\end{aligned} \tag{13}$$

This is valid for *all* amplitudes a_i, whether in D or not. In particular it holds for the a_i's in D, and these d equations can be expressed in the form of a d-dimensional secular equation

$$[H_0 + P\mathscr{V}(E) - EI]A = 0, \tag{14}$$

where A is a column vector composed of the amplitudes a_i in Eq. (6).

We have now obtained a formally exact version of degenerate perturbation theory (Bloch and Horowitz, 1958; Löwdin, 1962). This is expressed in the convenient form of a "model Hamiltonian," about which we shall say much more in Sections VI and VII. At this stage it is sufficient to recognize that all of the "nondegenerate" basis states Φ_i are concealed within $\mathscr{V}$.

Iteration of the first lines of Eqs. (8), (11), and (12) leads to perturbation expansions of the Brillouin–Wigner form. The two identifying characteristics of the BW form of perturbation theory should be noted at this point: (1) it has the very simple formal structure of a geometric series in the operator $\mathscr{G}V$; and (2) the desired (and presumably unknown) energy E appears in all of the energy denominators.

III. The Linked-Cluster Expansion for Closed-Shell Systems

By specializing to the nondegenerate case where $d = 1$, $P = |\Phi_0\rangle\langle\Phi_0|$, we immediately obtain the usual BW expressions

$$\Psi = \sum_{n=0}^{\infty} (\mathscr{G}V)^n \Phi_0 , \tag{15}$$

$$\begin{aligned} \Delta E \equiv E - E_0 &= \langle \Phi_0 | V | \Psi \rangle = \langle \Phi_0 | \mathscr{V} | \Phi_0 \rangle \\ &= \langle \Phi_0 | V + V\mathscr{G}V + V\mathscr{G}V\mathscr{G}V + \cdots | \Phi_0 \rangle. \end{aligned} \tag{16}$$

It should be noted that the "intermediate normalization" convention

$$\langle \Phi_0 | \Psi \rangle = \langle \Phi_0 | \Omega | \Phi_0 \rangle = \langle \Phi_0 | \Phi_0 \rangle = 1 \tag{17}$$

is implicit in these formulas. We now want to apply Eq. (16) to a many-body system where Φ_0 represents a closed-shell Slater determinant, for example the Hartree–Fock ground-state wavefunction.

A. Diagrammatic Notation

The first problem one encounters is that, beyond the first few orders in V, so many different types of intermediate-configuration sequences (sequences of Φ_i's) are possible that one needs an efficient shorthand notation. This problem is neatly solved by the diagrams introduced by Goldstone (1957), some low-order examples of which are shown in Fig. 1. Here Figs. 1a and c

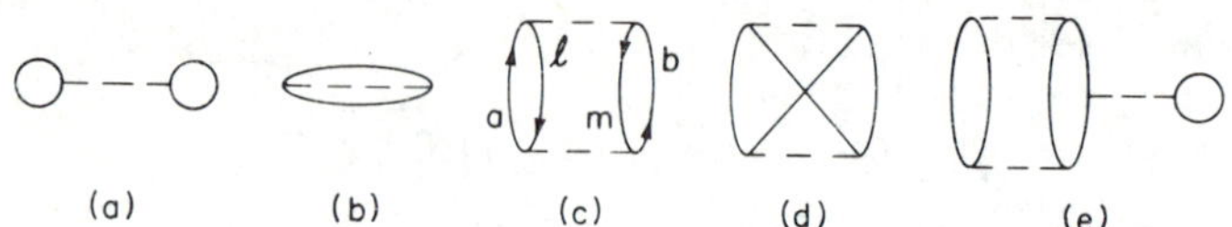

Fig. 1. Some low-order Goldstone diagrams.

represent "direct" first- and second-order terms from Eq. (16), whereas Figs. 1b and d show the corresponding exchange terms. The combination of diagrams a + b represents the potential energy of the Hartree–Fock approximation. The connection with the various terms $\langle V \rangle$, $\langle V\mathscr{G}V \rangle$, etc., in Eq. (16) follows from the rule that the successive horizontal levels, proceeding from bottom to top, correspond to the successive V's and $\mathscr{G}$'s in the various terms of Eq. (16), proceeding from right to left. Thus in diagrams c and d, the bottom (top) dotted lines represent matrix elements of the two-body parts of the right-hand (left-hand) interactions V in $\langle \Phi_0 | V\mathscr{G}V | \Phi_0 \rangle$.

Although our present procedures are obviously time-independent, this correspondence is often described by saying that "time runs upward" in the Goldstone diagrams.

The two downgoing lines in diagram c represent "hole" orbitals, namely orbitals that were occupied in Φ_0 but are vacant in a particular intermediate-state determinant Φ_i, the latter coming from the sum in Eq. (7). The two upgoing lines represent "particle" orbitals, namely, orbitals occupied in Φ_i but not in Φ_0. Line segments that begin and end at the same horizontal level, as seen in diagrams a, b, and e, represent orbitals in Φ_0 ("passive hole orbitals") that remain occupied before and after the associated interaction V. Although not always shown explicitly in Fig. 1, each line segment should be thought of as carrying the label of an orbital from the basis generated by the one-body operator H_0. These labels are eventually summed over, the hole (particle) labels running over all orbitals contained within (or outside of) Φ_0. These orbital summations are equivalent to the Φ_i summation in Eq. (7). Momentum or angular momentum conservation enters, as usual, via the vanishing of all the V-matrix elements that violate the appropriate conservation condition. The solid lines always form closed loops (with the arrow directions proceeding in a continuous manner), which simply means that the total number of particles is conserved by each and every matrix element of V.

A very convenient feature of these diagrams is that the appropriate energy denominators [from Eq. (7)] can be written down immediately. The general rule is

$$(E_0 + \Delta E - H_0) = \Delta E + \Sigma \text{ (all downgoing line energies)} - \Sigma \text{ (all upgoing line energies)}. \tag{18}$$

Thus for diagram c, for example,

$$e(\text{diagram c}) = \Delta E + (\varepsilon_l + \varepsilon_m) - (\varepsilon_a + \varepsilon_b), \tag{19}$$

where the ε_i's are the orbital eigenvalues generated by H_0.

Once these conventions are understood, the algebraic expression corresponding to each diagram may be written down immediately. Thus

$$\Delta E(\text{diagram c}) = \frac{1}{2} \sum_{ab} \sum_{lm} \frac{\langle lm | V | ab \rangle \langle ab | V | lm \rangle}{\Delta E + (\varepsilon_l + \varepsilon_m) - (\varepsilon_a + \varepsilon_b)}, \tag{20}$$

and similarly for all of the other diagrams. Another valuable feature is that there are simple and completely mechanical rules, involving the topological features of the diagram, which make it easy to keep track of the various factors of $\frac{1}{2}$, and similarly for the minus signs associated with the exchange terms. These rules become indispensable for complicated diagrams, where elementary ideas become too cumbersome to work out explicitly. For a particularly convenient form of these rules, see Brandow (1967) Appendix B. This version effectively cuts through the annoying business of "topological weighting factors," which have burdened a number of linked-cluster derivations to be found in the literature. The reader should be reassured, at this point, that when he begins to feel comfortable with these diagrammatic conventions and the associated rules, the struggle to master this form of perturbation theory is already half-way over.

There is an alternative diagrammatic convention which comes readily to mind. One could simply start off with N vertical lines arranged side by side, each of which is to prepresent the "history" of one of the N particles in the system, and then add dashed horizontal lines to represent the various V interactions. This scheme is as elementary as possible, and it has been found useful for some specialized purposes. It does have some obvious drawbacks, however. The recipes for the overall sign, the number of factors of $\frac{1}{2}$, and the energy denominators are a bit more cumbersome. However, the chief disadvantage becomes apparent when N is large and one considers perturbation terms of order V^r, where $r \ll N$. Then *most* of the N lines remain completely passive, and thus represent "excess baggage." Given the initial state Φ_0, these passive lines are redundant since they convey no further useful information. There is no such redundancy in the Goldstone convention; all of the information presented is "new" and therefore useful. The Goldstone diagrams can be thought of as being the result of subtracting the "zeroth-order" N-line diagram (which simply describes Φ_0) from each of the present N-line diagrams. This correspondence is illustrated in Fig. 2. By analogy to the diagrammatic scheme for quantum electrodynamics (Feynman, 1949), the reference state Φ_0 is often referred to as the "vacuum state."

There is another diagrammatic convention, due to Hugenholtz (1957), which is quite similar to that of Goldstone. In this convention, direct and

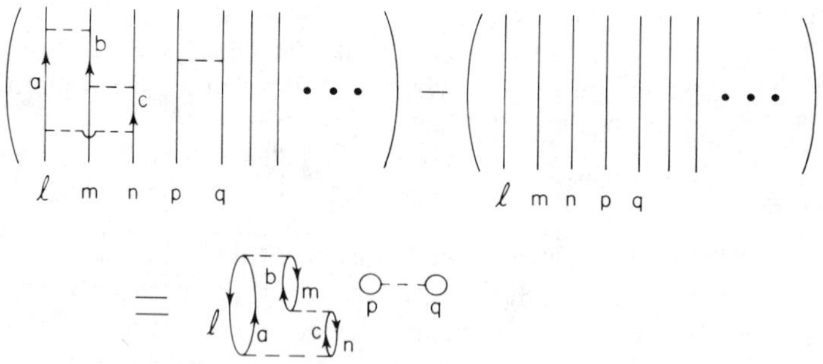

Fig. 2. Use of Φ_0 as a reference state—the "vacuum convention."

exchange matrix elements of V are always combined into "black dot" vertices, thus preserving antisymmetry at every stage. We have argued that it is most convenient, in practice, to work with a combination of both of these conventions (Brandow, 1967, Appendix B).

B. The Unlinked-Cluster Problem

We now consider a macroscopic sample of some quantum fluid (liquid or gas) such as the electron gas, liquid ^3He, or nuclear matter, where the basis orbitals ϕ_l, ϕ_a, etc., are just plane waves. The "direct" diagrams from the first three orders in Eq. (16) are shown in Fig. 3, together with a few of the

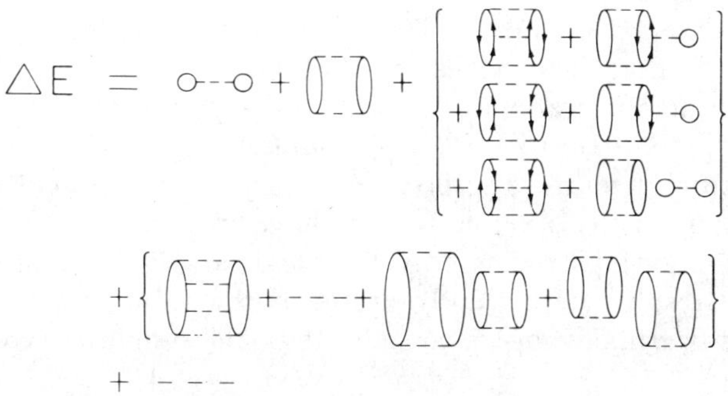

Fig. 3. The first few orders of "direct" diagrams from the BW energy expansion [Eq. (16)].

many fourth-order diagrams. For a fixed bulk density $\rho = N/\text{volume}$, it is physically obvious that ΔE and E_0 must be proportional to the total number of particles N. Unfortunately, however, this property is not at all obvious from the Brillouin–Wigner algebraic expressions corresponding to the various diagrams shown here. From elementary counting arguments (Brandow, 1967, Appendix C), one readily finds that each linked (topologically connected) piece of a diagram contributes an overall factor of N, arising from the various summations and orbital normalization factors. On the other hand, the ΔE term in each of the BW energy denominators contributes a further factor of N^{-1}. Thus diagrams a and b of Fig. 1 are $\mathcal{O}(N)$, diagrams c and d are $\mathcal{O}(1)$, diagram e is $\mathcal{O}(N^{-1})$, whereas the unlinked third-order diagram in Fig. 3 is $\mathcal{O}(1)$. Similarly, the unlinked diagram in Fig. 2 is $\mathcal{O}(N^{-1})$. It turns out that the first-order diagrams a and b, corresponding to $\langle\Phi_0|V|\Phi_0\rangle$, are the *only* diagrams of the BW expansion whose contribution is $\mathcal{O}(N)$; all of the other BW contributions are $\mathcal{O}(1)$ or smaller [$\mathcal{O}(N^{-1})$, $\mathcal{O}(N^{-2})$, etc.]. [This smallness of the individual diagrams is compensated, of course, by the huge number of unlinked diagrams that are generated by the high-order terms in Eq. (16).] The BW expansion is therefore quite impractical for macroscopic systems. On the other hand, if we were to go over to the Rayleigh–Schrödinger form of perturbation theory in a naive manner, say by simply ignoring the ΔE's in all of the BW energy denominators, we would find terms proportional to arbitrarily large positive powers of N(N^2, N^3, etc.), due to the unlinked diagrams. This is equally unsatisfactory.

It was Brueckner (1955, 1959) who first drew attention to this problem. He proceeded to show, however, that when the RS expansion is worked out with proper care, all of the unlinked-diagram terms mutually cancel and one is simply left with linked (fully connected) diagram terms, all of which scale properly with N. This is the famous linked-cluster result. (At one time it was thought that this cancellation might hold only to relative order N^{-1}, which could have serious consequences for small-N systems such as atoms, but it is now recognized that the cancellation is exact.)

Brueckner demonstrated this cancellation explicitly in low orders (eventually up to sixth order in V), but he did not have a general proof to cover all of the higher order terms. The first general proof was presented by Goldstone (1957), and many other derivations have appeared since (see Brandow, 1967, Introduction). [It seems only fair to remark that Hugenholtz (1957) obtained this result independently, at almost the same time as Goldstone.] We shall now outline what we consider to be the simplest derivation of all.

C. Cancellation of Unlinked Diagrams

The BW expansion [Eq. (16)] can be converted into the RS expansion by the following procedure. The first step is to formally expand ΔE out of all the energy denominators of Eq. (16), each BW denominator being replaced by a geometric series involving RS-type denominators:

$$\frac{1}{E_0 + \Delta E - H_0} = \frac{1}{E_0 - H_0} + \frac{1}{E_0 - H_0}(-\Delta E)\frac{1}{E_0 - H_0} + \frac{1}{E_0 - H_0}(-\Delta E)\frac{1}{E_0 - H_0}(-\Delta E)\frac{1}{E_0 - H_0} + \cdots. \tag{21}$$

It follows from Eq. (18) that these RS denominators are determined *entirely* by the upgoing and downgoing orbital eigenvalues at the various intermediate-state levels within the diagrams. Expansion (21) can be represented diagrammatically as in Fig. 4, where each solid horizontal bar represents an “insertion” (a fictitious diagonal interaction) of numerical value $(-\Delta E)$. We now have an expansion of ΔE in terms of diagrams, some of which themselves contain ΔE. This expansion may thus be substituted back

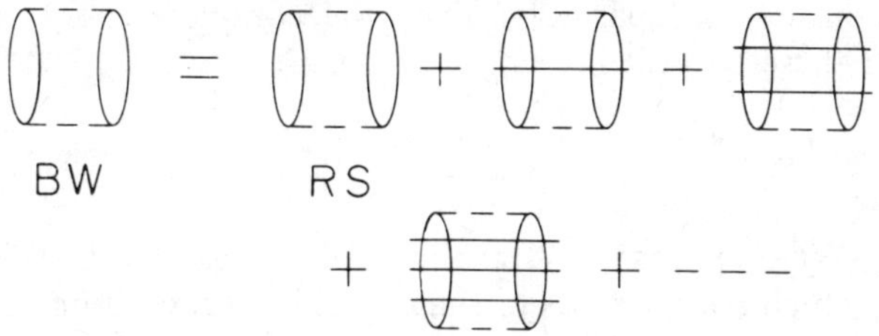

BW RS

Fig. 4. Diagrammatic application of the denominator expansion [Eq. (21)].

into itself to eliminate the ΔE's in the right-hand side of Eq. (21) and Fig. 4, as illustrated in Fig. 5. At this point we observe that the first term on the right-hand side of Fig. 5 is just equal and opposite to the leading (third-order) unlinked term seen in Fig. 3, so these terms cancel identically. It can easily be shown (Brandow, 1966a, 1967) that this type of cancellation is completely general—all of the unlinked terms from the original BW series are canceled by terms arising from the $(-\Delta E)$'s generated in Eq. (21), and vice versa. The only surviving terms are the fully linked diagrams with no $(-\Delta E)$ insertions, such as the first diagram to the right in Fig. 4. This is just Goldstone's result. We shall not present the full proof here, although this is not at all difficult. There are, however, two further ingredients in this proof that deserve some discussion.

D. The Factorization Theorem

To go beyond the simple example of cancellation just described, we must introduce the *factorization theorem.* Consider the RS-type denominators [from the first term of the series Eq. (21)] for the two different unlinked fourth-order diagrams shown in Fig. 3. Let us assume the same set of orbital labels (unsummed) for both diagrams, so that these diagrams differ only in the relative "time" orders of their lower interactions. Let e_L and e_R be the RS denominators appropriate for the left-hand and right-hand linked parts,

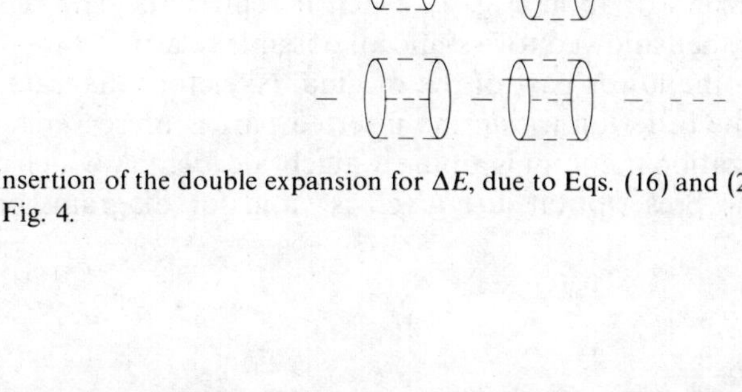

$(-\Delta E)$

Fig. 5. Insertion of the double expansion for ΔE, due to Eqs. (16) and (21), into one of the diagrams of Fig. 4.

each part considered separately. Then the complete energy denominator products for these two diagrams are

$$\frac{1}{e_L}\left(\frac{1}{e_L+e_R}\right)\frac{1}{e_L}, \qquad \frac{1}{e_L}\left(\frac{1}{e_L+e_R}\right)\frac{1}{e_R},$$

respectively. We now observe that the *sum* of these two products is simply $(e_L e_R e_L)^{-1}$, which corresponds to the negative of the second diagram on the right-hand side of Fig. 5. This is summarized in Fig. 6.

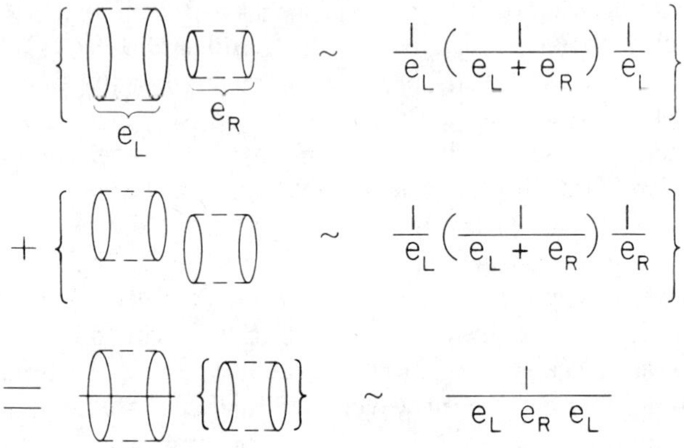

Fig. 6. Illustration of the factorization theorem. The first two diagrams contain "off shell" insertions, whereas in the last diagram the insertion is "on shell."

Let us now rephrase what we have just done. The denominators of the last-mentioned diagram are all of the most elementary RS form, since the second-order "insertion" is evaluated "on the energy shell." In other words, the denominator (e_R) of the insertion part is not influenced by the excitation energy (e_L) of the rest of the diagram (the "skeleton") to which it belongs. We have just seen that this RS diagram is equivalent to a sum of diagrams where the insertion part is "off the energy shell," meaning that at least one of its denominators now contains the excitation energy (e_L) of the rest of the diagram, in addition to its own excitation energy e_R. In the general case, the corresponding sum of diagrams is characterized as follows. The *top* of the inserted part is placed at the same intermediate-state level as the $(-\Delta E)$ insertion (horizontal bar in Fig. 4) which it represents. The remainder of the insertion is then allowed to assume all possible relative "time" orderings with respect to the lower part of the original (skeleton) diagram, such that the original time ordering *within* the inserted part is preserved.

The factorization theorem is a purely algebraic identity which shows that the form of the prescription just given is valid for diagrams of arbitrary complexity. That is, each of the summations just described, of diagrams with off-energy-shell denominators, leads to the corresponding on-shell diagram that is needed to complete the cancellation. This theorem can be proved by induction (Hugenholtz, 1957; Frantz and Mills, 1960), building upon the example just given. Brueckner's demonstration of cancellation was limited to low orders because he did not recognize the general form of this identity. Once this is recognized, however, his method of demonstration (which begins with the RS form of perturbation theory) can easily be made general, as noted by Goldstone (1957) and shown in detail by Baker (1971). In Goldstone's derivation, the corresponding identities are generated automatically by relaxing the relative time-order restrictions for interactions occurring within different linked parts, when the time integrations of his adiabatic formalism are carried out. Given this factorization theorem, however, it seems simpler and more straightforward to avoid the use of a time-dependent formalism to derive the final time-independent result.

E. Exclusion-Violating Diagrams

The other important ingredient of the general proof is the matter of "ignoring the exclusion principle in intermediate states," an idea dating back to the first paper on the use of diagrams in quantum electrodynamics (Feynman, 1949). The orbital summations for the various "insertions" in Fig. 5 must obviously be independent of the orbitals in the "skeleton" diagram on the left side of this figure, since these insertions arise from the original expansion [Eq. (16)]. This contrasts with the corresponding diagrams in Fig. 3, where, according to the exclusion principle for Slater determinants, the orbital labels at each horizontal level must all be distinct. It thus appears that the cancellation of unlinked diagrams cannot be complete. Goldstone has shown how to resolve this problem by a judicious addition and subtraction of exclusion-violating diagrams. An example is shown in Fig. 7, where the normally occupied orbital m has been emptied twice. Diagram 7a is a physically unallowed contribution to one of the fourth-order diagrams of Fig. 3, whereas diagram 7b has, according to the standard

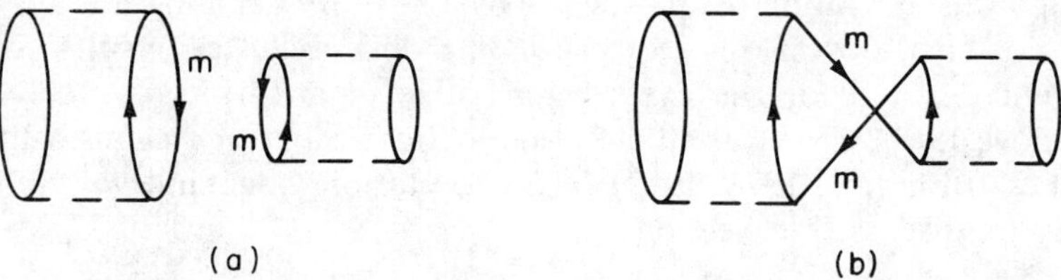

Fig. 7. Effect of ignoring the exclusion principle in intermediate states: (a) a nonphysical unlinked diagram; (b) an "exclusion-violating" linked diagram that exactly compensates for (a).

diagrammatic rules (Goldstone, 1957; Brandow, 1967), a precisely equal and opposite numerical value. Thus no harm is done by adding *both* of these diagrams to the terms originally in Fig. 3. Thanks to diagram 7a and its generalizations, we now obtain a *precise* cancellation of all the unlinked diagrams, but only at the cost of including the extra diagrams 7b, etc., among the set of surviving fully linked diagrams. These "extra" linked diagrams are all $\mathcal{O}(N)$ for macroscopic systems, as desired.

One of the marvels of the Goldstone formalism is that the required compensation diagrams, 7b, etc., can all be generated quite automatically, by the simple recipe that (1) linked diagrams of all possible topologies must be included, and (2) all of the orbital-label summations must be carried out independently. This is what is meant by "ignoring the exclusion principle in intermediate states." The formal meaning of this addition and subtraction is that one has abandoned the original Slater-determinant description of the intermediate configurations Φ_i, replacing this by a "second-quantized" description. The consistent use of Fermion creation and annihilation operators automatically generates the diagrams of Fig. 7. This also accounts for the remarkable simplicity of the diagrammatic rules.

If the reader has not learned the basic language and formalism of second-quantization (the occupation-number representation), we would encourage him to do so. The most simple and straightforward exposition, in our opinion, is an unpublished treatment by G. C. Wick; fortunately, two abbreviated accounts of this approach are now available (Ambegaokar, 1969; Brown, 1972).[1] (We reiterate, though, that this formalism is *not* essential for understanding the basic linked-cluster result.) Conversely, it is also very helpful in practice to understand the present connection with the more elementary Slater-determinant description.

To summarize this section, we have seen that the essential ingredients of the linked-cluster result are (1) the use of diagrams, (2) the RS form of perturbation theory, (3) the factorization theorem, and (4) ignoring the exclusion principle in intermediate states. We wish to emphasize that although the present discussion has been brief, this approach does lead to a complete and rigorous derivation (Brandow, 1966a, 1967). By pursuing this approach one can also obtain the corresponding linked-cluster results for the wavefunction and for the expectation value of an arbitrary operator $\mathcal{O}$ (Brandow, 1967), as well as all of the analogous results for systems with Bose statistics (Brandow, 1971, 1972). The expectation value and wavefunction results are discussed in Section V.

[1] Another good treatment has recently been presented by Manne (1977), who also provides simple derivations for the diagram rules.

IV. Physical Interpretation and General Philosophy of Applications

Since the original motivation for this formalism was to deal with systems having a macroscopic number of particles, a quantum chemist will naturally ask what advantages, if any, this offers for relatively few-electron systems such as atoms and small molecules. The answer involves the physical interpretation of this formalism (Brandow, 1967), which we now discuss.

For a macroscopic system one believes intuitively that particles should not interact to any significant extent when they are farther apart than some appropriate correlation length. This notion leads to the argument, familiar in statistical mechanics, that a macroscopic system can be divided into many comparatively small subsystems, each of which is still large enough to be considered approximately independent. That is, the "boundary" contributions are still small compared to the "bulk" contributions. Let us pursue this idea by considering a fictitious "cellular" system, where the various subsystems are rigorously isolated from each other by means of physical barriers. This guarantees complete independence for the various subsystems s, whereby the energy is now rigorously just the sum of the subsystem energies,

$$E_0 = \sum_s E_{0s} \sim N, \qquad \Delta E = \sum_s \Delta E_s \sim N. \tag{22}$$

The wavefunction has a product form

$$\Psi = \prod_s \Psi_s, \tag{23}$$

and thus the (normalized) overlap of Φ_0 and Ψ is exponentially small,

$$\langle \Phi_0 | \Psi \rangle = \prod_s \langle \Phi_{0s} | \Psi_s \rangle \sim e^{-\alpha N}. \tag{24}$$

(In terms of the intermediate normalization convention [Eq. (17)], however, one finds that $\langle \Psi | \Psi \rangle \sim e^{+2\alpha N}$.) But this near-vanishing overlap is nothing to be concerned about, since the individual overlaps $\langle \Phi_{0s} | \Psi_s \rangle$ need not be small, and all expectation values are simply additive:

$$\frac{\langle \Psi | \mathcal{O} | \Psi \rangle}{\langle \Psi | \Psi \rangle} = \sum_s \frac{\langle \Psi_s | \mathcal{O}_s | \Psi_s \rangle}{\langle \Psi_s | \Psi_s \rangle} \sim N. \tag{25}$$

When we consider the BW expansion for this cellular system, we see immediately that it is physically inappropriate to have all of the "other" $\Delta E_{s'}$'s ($s' \neq s$) occurring within the energy denominators for subsystem s. Going over to the RS form of perturbation theory is clearly a reasonable way to cure this problem. Furthermore, the factorization theorem also follows very simply from the notion of physically independent subsystems, as Hugenholtz (1957) has pointed out. It is therefore quite reasonable to expect the RS perturbation theory to give a simple additive result for the energy, with no cross-terms involving more than one subsystem s.

Our linked-cluster derivation is obviously very closely related to this elementary picture of independent subsystems. The comparison shows that each linked-cluster diagram is behaving formally as if it represents an independent subsystem. This holds true for each fixed set of orbital labels within a given diagram, as well as for the "complete" diagram resulting from the orbital summations.

At first sight this result seems somewhat too strong, since one's physical intuition says that the various clusters or correlations cannot be fully independent within any realistic system. The "hole" or normally occupied orbitals extend throughout the entire volume of the system (except perhaps for a "quantum solid"), thus the various cluster terms are spatially interpenetrating. The clusters must therefore interact, both dynamically (through V) and statistically (via the exclusion principle). This is indeed true. However, the Brueckner–Goldstone formalism has the very nice feature that the expected "cluster–cluster interactions" are always represented by *higher order* linked clusters, whereby the "original" clusters *must* be regarded as completely independent. An example of a statistical interaction between two simple clusters is shown in Fig. 7b.

The practical value of this formalism lies in the fact that one can usually find a judicious way of grouping the various cluster terms, which corresponds physically to decomposing the many-body system into an appropriate set of subsystems, such that the residual interactions between these subsystems turn out to be relatively small. Since the computational effort generally increases very rapidly with the size of the system (or subsystem), such a "cluster decomposition" can be of great value even for relatively small systems such as atoms.

In practice, the problem of finding an appropriate cluster decomposition scheme is as much art as science. One experiments with various possibilities, searching for a scheme in which the residual interactions (higher cluster corrections) are suitably small. Two well-known examples are the independent-pair approximation of Sinanoğlu (1962, 1964), Kelly (1963, 1964b), and Nesbet (1965, 1967), for atoms and small molecules (inspired by nuclear-matter theory), and the random-phase approximation for the high-density electron gas (see Pines, 1961). In the former case, each subsystem consists of all perturbative diagrams involving a given pair of hole orbitals, whereas for the latter one collects together all "ring" diagrams characterized by a given momentum transfer $\mathbf{q}$. Such groupings or "partial summations" of physically similar terms are of great value computationally, since they are chosen such that the sum can be evaluated in closed form. Kelly has carried out such diagrammatic summations explicitly, for ladders of hole–hole, particle–hole, and particle–particle interactions, since all of these have the character of simple geometric series, at least approximately. Although entirely different calculational procedures were used by Sinanoğlu and co-workers (restricted variations) and Nesbet (restricted configuration mixing), their calculations are *formally* equivalent to those of Kelly. (This formal equivalence is discussed in Brandow, 1972, Sections VI, VII, and pp. 169–174.) Note also that although the independent-pair approximation is equivalent to a very simple type of restricted configuration mixing, this is not true in general. The random-phase approximation represents a selection of certain matrix elements within a very complicated configuration mixing.

Let us return for a moment to the cellular model of Eqs. (22)–(25). For an ordinary (non-many-body) quantum system, one expects the BW form of perturbation theory to be more rapidly converging than the RS form, since, in view of Eq. (21), the BW form is "less expanded." Therefore, although the RS form was needed to "uncouple" the various subsystems, it would seem desirable to identify and sum out the higher order cluster terms that correspond to putting the interaction energy ΔE_s of subsystem s back into all of the energy denominators of this subsystem. One can readily see that all of these diagrams must be of the exclusion-violating type (although the converse is not true). Kelly's partial summation of ladders of hole–hole interactions is an example of such a ΔE_s-type summation. This refinement is generally of negligible importance for macroscopic fluid systems, but it can be significant for small systems and also for "quantum solids" where the basis orbitals are spatially localized (Brandow, 1972).

The various formulations of the independent-pair approximation illustrate the fact that the appropriate "leading cluster approximation" can usually be derived and expressed equally well from several different viewpoints. This is just what one should expect, of course, since the leading approximation should be based on the most prominent and therefore "intuitive" features of the particular system under consideration. By contrast, however, one usually faces a much more difficult problem when one seeks to know "what to do next." In our opinion, it is at *this* stage that the diagrammatic perturbation theory displays its greatest virtue as compared to rival formalisms. This formalism constitutes the most flexible and conceptually unbiased bookkeeping system known. No matter what one's "first approximation" may be, and regardless of how it was derived, it is usually a fairly straightforward matter to express this approximation in terms of a set of perturbation diagrams and then to identify the leading perturbative corrections to this approximation. These various corrections can then be calculated individually, if necessary, to obtain objective evidence upon which to base one's "second approximation." This philosophy has been responsible for most of the developments in the nuclear many-body problem during the past 15 years.

In specific cases, of course, other methods of studying the correction terms may also prove satisfactory, provided they are sufficiently systematic. The method developed by Coester and Kümmel (1960) and Čižek (1966) appears to be especially promising for electronic systems. This method is discussed in the following section.

V. Further Closed-Shell Results

In this section we shall derive the corresponding linked-cluster results for the expectation value $\langle \mathcal{O} \rangle$ of a general operator $\mathcal{O}$, for the correlated ground-state wavefunction Ψ, and for the norm of this wavefunction. The wavefunction result will then be used to provide a simple derivation for the nonperturbative coupled-cluster formalism of Coester and Kümmel (1960) and Čižek (1966). After some comments about the choice of single-particle potentials, we shall conclude the discussion of the closed-shell case by indicating the way these results emerge from the adiabatic approach of Goldstone.

A. Expectation Values

Given the linked-cluster expansion for the total energy, there is a very simple device that converts this into a recipe for general expectation values. The basic idea actually has nothing to do with the many-body nature of the system.

Consider a generalized Hamiltonian operator,

$$H_\lambda \equiv H + \lambda \mathcal{O} = H_0 + (V + \lambda \mathcal{O}), \tag{26}$$

where λ is assumed real so that H_λ is Hermitian. The $\mathcal{O}$ here is an arbitrary Hermitian operator. Let the corresponding ground-state wavefunction and energy be Ψ_λ and E_λ. Comparing the Schrödinger equation for finite λ with the Hermitian conjugate equation for $\lambda = 0$, we obtain

$$\begin{aligned} \lambda \langle \Psi_{\lambda=0} | \mathcal{O} | \Psi_\lambda \rangle &= \langle \Psi_{\lambda=0} | H_\lambda - H_{\lambda=0} | \Psi_\lambda \rangle \\ &= (E_\lambda - E_{\lambda=0}) \langle \Psi_{\lambda=0} | \Psi_\lambda \rangle, \end{aligned} \tag{27}$$

and thus

$$\frac{\langle \Psi_{\lambda=0} | \mathcal{O} | \Psi_\lambda \rangle}{\langle \Psi_{\lambda=0} | \Psi_\lambda \rangle} = \frac{E_\lambda - E_{\lambda=0}}{\lambda}. \tag{28}$$

The desired expectation value is therefore given by the $\lambda \to 0$ limit of the quantity $(E_\lambda - E_{\lambda=0})/\lambda$. Consequently, it is only necessary to evaluate

$(E_\lambda - E_{\lambda=0})$ to *first order* in $\lambda\mathcal{O}$, and then discard the λ factor. Comparison with the last expression in Eq. (26) shows that $\langle\mathcal{O}\rangle$ must therefore be given by the sum of all of the "energy" linked-cluster perturbation diagrams in which the special operator $\mathcal{O}$ appears just once. Of course this $\mathcal{O}$ can appear at any position within the diagram; all such possibilities must be included.

This result was first obtained by Thouless (1958, 1961), although the present elementary argument is due to Glassgold *et al.* (1959). Formally, this argument amounts to an application of the Hellmann–Feynman theorem to a suitably generalized "force." The latter arises from a fictitious external field, of strength λ, which couples to the system via the operator $\mathcal{O}$.

B. The Ground-State Wavefunction

For the total (correlated) ground-state wavefunction, it is again useful to employ diagrams to keep track of the multitude of terms generated by the perturbation expansion Eq. (8). These diagrams differ from those of Section III,A in one respect, namely, they always involve some linked pieces that are "open" at the top. A wavefunction analog of Fig. 2 is shown in Fig. 8, where

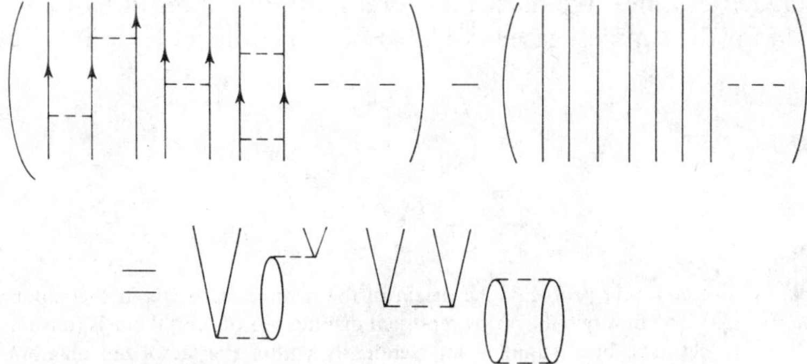

Fig. 8. The vacuum convention for wavefunction diagrams.

the upgoing arrows again indicate occupation of "particle" orbitals (orbitals not found in Φ_0). The most general wavefunction diagram will therefore consist of a number of open (one or more) and closed (none or more) linked pieces. This is the BW perturbation result. By means of the BW → RS expansion [Eq. (21)], the factorization theorem, and ignoring the exclusion principle in intermediate states, one can easily eliminate all of the diagrams containing closed ("vacuum fluctuation") linked parts. [Some low-order examples have been worked out in detail by Kelly (1964a), following exactly this approach.] The result is a form of the RS expansion for Ψ (see Goldstone, 1957), but this has not yet been cast into its most simple and elegant form.

The factorization theorem can be further exploited to simplify the RS diagrams containing more than one open linked part. By combining all possible relative time orders for interactions in *different* linked parts, subject to preserving the original order *within* each linked part, one converts the total energy-denominator product to an on-energy-shell form. This means that each linked piece now contributes a factor that is independent of the presence of the other linked pieces. We then sum over all of the orbital labels, for the external as well as the internal line segments, again ignoring the exclusion principle to the extent that these summations are to be carried out independently of each other. The compensating exclusion-violating diagrams must also be included, of course [see Kelly (1964a) and Brandow (1967) for examples].

We must now be careful not to overcount the various "original" RS contributions. Consider a diagram consisting of n topologically identical linked parts. Thanks to the various time orders included by the factorization, together with the independent orbital summations, it turns out that each of the original RS contributions from this type of diagram has now been included $n!$ times. We must therefore multiply the result by $(n!)^{-1}$, as illustrated for the simplest case in Fig. 9. For a diagram with n_{r1} linked parts

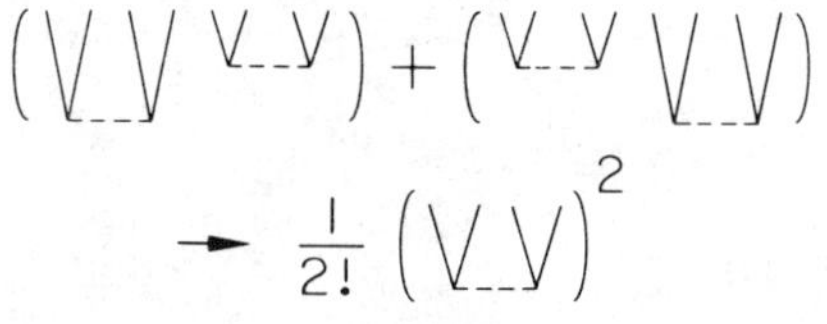

Fig. 9. An overcounting problem: the origin of the inverse factorials in the exponential expression Eq. (29). The first two diagrams represent distinct sets of orbital labels (unsummed), whereas these labels have been summed independently within the factorized diagram. The arrow implies orbital summation.

of topological form r_1, n_{r2} parts with topology r_2, etc., the required compensation factor is $[\Pi_r(n_r!)]^{-1}$. The net result of these factors is that the total wavefunction can be expressed in the form

$$\Psi = e^W \Phi_0 , \tag{29}$$

where W represents the sum of all open diagrams consisting of just a *single* linked piece. This elegant result is due to Hubbard (1957), Hugenholtz (1957), and Bloch (1958). (We have chosen the symbol W because of the form of the leading linked diagram, as seen in Fig. 9. Other authors use the symbol S or T instead.) Note that this expression obeys the intermediate normalization convention [Eq. (17)]. For this and other reasons, the operator e^W is not unitary. In any application, of course, one will naturally be led to decompose W in the form

$$W = \sum_{n=1}^{N} W_n , \tag{30}$$

where W_n is the fully linked amplitude for exciting n particles out of the "Fermi sea" Φ_0.

It is very important to recognize that this W is a *second-quantized* operator. Each upgoing (downgoing) external line carries a Fermion creation (annihilation) operator. This operator character of W plays a crucial role, in the following manner. Consider an "ordinary" RS diagram, such as shown at the top of Fig. 9. When all of the external orbital labels on the two linked pieces are distinct, one obtains the obvious contribution and the operator character plays no role. As soon as two of the external labels coincide, however, the corresponding Ψ amplitude vanishes, thanks to the Fermion operator identities $a_i^\dagger a_i^\dagger = 0$, $a_i a_i = 0$. We see, therefore, that this second-quantized aspect is essential when one uses the form Eq. (29), since this expression necessarily implies that the external-line orbital summations for each linked part are independent of those in any other linked part that may happen to be present simultaneously. This feature constitutes the analog, for the external lines, of the addition of exclusion-violating diagrams such as Fig. 7b. Consequently, although the formalism of second quantization was not essential for understanding the linked-cluster energy expansion, it *is* essential for the present wavefunction result [Eq. (29)].

C. Norm of the Wavefunction

Because the wavefunction expression (29) satisfies the intermediate normalization condition Eq. (17), it is of some interest to evaluate the normalization overlap $\langle\Psi|\Psi\rangle$. This is most easily done, within our present framework, by returning to the "primitive" RS expression for Ψ mentioned in the first paragraph of the previous subsection. This is the stage where the closed-linked parts have been eliminated, but the factorization shown in Fig. 9 has not yet been carried out.

Now imagine that one has drawn a solid horizontal line, which is to represent the boundary between the Ψ and $\Psi^\dagger$ amplitude contributions in $\langle\Psi|\Psi\rangle$. Below this line place any number of linked open W pieces (for various W_n's), and above this place any number of linked $W^\dagger$ pieces. (The latter have the same topology as the W's, but are drawn "upside down" in order to represent $\Psi^\dagger$.) These $W^\dagger$ pieces must be selected such that the total number of their "external" particle and hole lines (vertical lines at the level of the horizontal boundary line) is the same as that for the chosen set of W

pieces. Now join these external lines together (particle to particle and hole to hole), across the boundary line, to form a closed diagram consisting of one or more closed linked pieces. Suppose, furthermore, that these steps have been repeated in all possible ways. The resulting set of diagrams (plus unity for the $\langle\Phi_0|\Phi_0\rangle$ contribution) represents $\langle\Psi|\Psi\rangle$.

We now cast this set of closed (but generally unlinked) diagrams into a form analogous to Eq. (29), namely, an exponential of a sum of fully linked diagrams. This is done by applying the factorization theorem twice, once to the set of W's and once to the $W^\dagger$'s, in both cases regarding the boundary line as the fixed point of reference. In cases where two or more W's (or $W^\dagger$'s) belong to a common closed linked piece, these "common W's" ($W^\dagger$'s) are not to be factorized with respect to each other; the present goal is merely to disentangle the various closed linked pieces from each other. The result is that each unlinked diagram is reduced to the product of the contributions from its various linked pieces, with each piece evaluated separately from the others. In cases where two or more of the closed linked pieces have exactly the same topology (including vertical position with respect to the boundary), it is necessary to include inverse-factorial weighting factors to maintain correct counting. The argument is similar to that of Fig. 9, but applied now to the *entire* closed linked pieces instead of their individual W and $W^\dagger$ components. The result of these factorizations and inverse factorial weights is that $\langle\Psi|\Psi\rangle$ becomes the exponential of the sum of all distinct diagrams that are fully linked.

This result can be expressed in an analytic form, since the logarithm of $\langle\Psi|\Psi\rangle$ is closely related to the linked expansion for ΔE. Consider the function

$$F_L(x) = \langle\Phi_0|V\sum_{n=0}^{\infty}\left(\frac{Q}{E_0 + x - H_0}V\right)^n|\Phi_0\rangle_L, \tag{31}$$

where the subscript L means that only the perturbative contributions from fully linked diagrams are included here. Note that when evaluating these diagrams, quantity x plays exactly the same role as the ΔE appearing in Eq. (18). When x is set equal to zero, this function becomes identical to Goldstone's linked-cluster expression for ΔE. It is now easy to confirm that

$$\langle\Psi|\Psi\rangle = \exp[-F_L'(0)], \tag{32}$$

where $F_L' = dF_L/dx$. To see this, focus attention on the energy denominator of a particular intermediate state within a particular linked diagram for ΔE. Noting that

$$-\frac{d}{dx}\left(\frac{Q}{E_0 - x - H_0}\right)\bigg|_{x=0} = +\left(\frac{Q}{E_0 - H_0}\right)^2, \tag{33}$$

it is clear that this differentiation plays exactly the same role as the solid horizontal boundary line used above, when this line passes through the same intermediate state. [This line is also similar to the horizontal bars in Fig. 4, except that its assigned numerical value is unity instead of $(-\Delta E)$.] Since the differentiation acts successively on each intermediate state of every ΔE diagram, according to the product rule for derivatives, all diagrams in the logarithm of $\langle \Psi | \Psi \rangle$ are eventually reproduced. This elegant result is due to Hugenholtz (1957); see also Bloch (1958) and Brandow (1967).

In the case of a macroscopic system one finds that $F'_{\mathrm{L}} \sim N$, according to the same counting analysis used to demonstrate that $F_{\mathrm{L}}(0) = \Delta E \sim N$. This agrees with the comment following Eq. (24). In practical applications, however, one rarely if ever needs to know $\langle \Psi | \Psi \rangle$. Our main reason for presenting this result is a pedagogical one which will become apparent at the end of this section.

D. The Coester–Kümmel–Čižek Formalism

If one simply *assumes* that the wavefunction can be expressed in the form Eqs. (29) and (30), a proposition that can be verified independently of perturbation theory (da Providencia, 1963), then the Schrödinger equation takes the form

$$He^{W}\Phi_0 = Ee^{W}\Phi_0 \,, \tag{34}$$

which is equivalent to

$$e^{-W}He^{W}\Phi_0 = E\Phi_0 \,. \tag{35}$$

Left-multiplying this successively by each of the Φ_i's, for $i \neq 0$, provides an infinite set of constraints on W. These constraints must suffice to determine W completely, since the original Schrödinger equation suffices for Ψ. The result has the form of a set of coupled nonlinear equations linking the various n-body amplitudes W_n to each other. Moreover, the equations for W_1 and W_2 are inhomogeneous; this feature serves to fix the norms for *all* of the W_n's. By suitably truncating this set of equations, one can then obtain computationally feasible approximations for a subset of the W_n's. After W (and therefore Ψ) has been calculated by such an approximation, one can obtain the corresponding E from the left-multiplication of Eq. (35) by Φ_0. We have just summarized the essence of the nonperturbative formalism of Coester and Kümmel (1960; also Kümmel, 1961; Coester, 1958, 1969; Luhrmann, 1977; Kümmel *et al.*, 1977) and Čižek (1966; also Paldus *et al.*, 1972; Paldus and Čižek, 1975).

The purely algebraic derivation of these coupled equations is tedious and unenlightening. In particular, this does not provide a satisfying physical explanation for the fact that the coupled equations for the W_n's (and thus the

W_n's themselves) are independent of the desired eigenvalue E. The combinatorial aspects of this approach can also be rather forbidding, in spite of the fact that the resulting equations have a reasonably simple form.

The derivation just outlined can be greatly streamlined by means of diagrams, where each W_n is represented by n particle and n hole lines emerging from the top of a blob. We break up H into $H_0 + V$, and express each of these operators in second-quantized form. We also replace e^W by the formal Taylor series $1 + W + W^2/2! + \cdots$ (which is of course the original definition of e^W). Now note the following features:

1. Since the various W_n operators can only *create* particle–hole pairs, they must all commute with each other.

2. When H_0 is "contracted" with the Taylor series for e^W, it can link itself diagrammatically to no more than one of the W_n's within this series. This is because H_0 is a one-body operator, chosen to be diagonal in the given orbital basis. The resulting diagrammatic object (another second-quantized operator), formed by linking H_0 to a W_n, also shares the feature of only creating (never annihilating) particle–hole pairs, thus it also commutes with all of the remaining (unattached) W_n's.

3. The V operator can interact with existing particles and/or holes, but it can also create or annihilate one or two pairs, since this is a (nondiagonal) two-body operator. One therefore encounters a variety of possibilities when contracting (linking up) V with the series for e^W. When V attacks the "1" of this series (which actually represents Φ_0), it can produce one particle-hole pair, or two pairs, or no pair at all. It can also be contracted with one, two, three, or four of the W_n's. In any event, the resulting linked diagram also commutes with all of the remaining W_n's.

The result of these considerations is that

$$(H_0 + V)e^W|\Phi_0\rangle = e^W(\tilde{H}_0 + \tilde{V})|\Phi_0\rangle, \tag{36}$$

where $\tilde{H}_0$ and $\tilde{V}$ represent the various linked-diagram operators generated in steps 2 and 3 above. From Eq. (35) we then find

$$(\tilde{H}_0 + \tilde{V})|\Phi_0\rangle = E|\Phi_0\rangle, \tag{37}$$

which says that the linked-operator sum $(\tilde{H}_0 + \tilde{V})$ is unable to create any particle–hole pairs whatsoever. The one-pair and two-pair components of this result are symbolized diagrammatically in Fig. 10. These are the basic Coester–Kümmel–Čižek equations.

In each of these equations, the black dot in the leading diagram stands for the H_0 operator. Its successive action on each of the external lines leads to factors resembling Rayleigh–Schrödinger energy denominators,

$$(\varepsilon_p - \varepsilon_h)W_1, \qquad (\varepsilon_p + \varepsilon_{p'} - \varepsilon_h - \varepsilon_{h'})W_2\,, \tag{38}$$

Fig. 10. Diagrammatic expression of the Coester–Kümmel–Čižek equations: the one-pair and two-pair components of Eq. (37).

where p, p′(h, h′) refer to the specific particle (hole) labels on the external lines. All the remaining diagrams come from $\tilde{V}$. Except where shown explicitly, the orbital lines are to be assigned upgoing (particle) and downgoing (hole) arrows in all possible ways consistent with particle conservation; the corresponding exchange diagrams are also implied. The analogous relations for all $n > 2$ can be generated diagrammatically by building upon these examples, although the number of $\tilde{V}$ diagrams increases rapidly with n. Note that there are no inhomogeneous ($V\Phi_0$) terms for $n > 2$.

Given the W_n's satisfying this hierarchy of equations, the interaction energy is determined from Eq. (37) to be

$$\Delta E = \langle \Phi_0 | \tilde{V} | \Phi_0 \rangle = \langle \Phi_0 | [V(1 + W_1 + \tfrac{1}{2}W_1^2 + W_2)]_L | \Phi_0 \rangle, \tag{39}$$

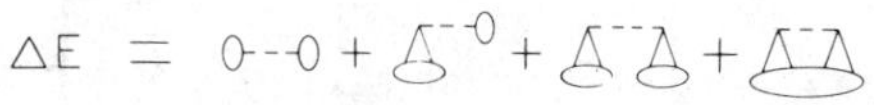

$\Delta E =$

Fig. 11. Illustration of the result [Eq. (39)] for the interaction energy.

where subscript L means that the "factors" V and W_1, W_2 are linked together. This result is shown diagrammatically in Fig. 11.

There is an alternative derivation which starts from the known RS diagrammatic perturbation expansions for the various W_n's. Since the basic form, Eq. (29), is intrinsically a Rayleigh–Schrödinger result, this approach makes it obvious why E (or ΔE) does not enter into the recipe for the W_n's. This derivation is extremely simple. Take any particular linked perturbation diagram, and erase or "snip off" the final (topmost) V interaction. The remaining diagram will be found to correspond to one of the general topologies symbolized by the various $\tilde{V}$ diagrams in Fig. 10 (or their analogs for $n > 2$), but with the explicit V removed. By considering specific examples, one soon recognizes that Fig. 10, etc., provide a concise summary of *all* of the possible topological structures for all of the W_n perturbation diagrams. In other words, the iterative solution of these Fig. 10 relations will generate *all* of the perturbation diagrams for the set of W_n's. This conclusion becomes transparent when, for each particular choice of external-line labels, the corresponding Fig. 10 relation is divided through by the energy-denominator factor seen in Eq. (38). This connection with the RS perturbation expansion was first pointed out by Coester (1958).

The most obvious virtue of this approach is that it provides a nonperturbative technique for calculating Ψ and E. A more important feature, however, is that this formalism appears to be very well suited for the physical characteristics of many electronic systems. It is now well established that the "independent pair" approximation provides an excellent starting point for atoms and small molecules, where, for converged basis calculations, it typically gives between 105 and 120% of the total correlation energy (Nesbet *et al.*, 1969; Barr and Davidson, 1970; Lee *et al.*, 1971; Miller and Kelly, 1971; Langhoff and Davidson, 1974), the results usually being somewhat worse for molecules than for atoms. However, the problem of how best to proceed beyond this approximation, to go after the remaining -10% or so of correlation energy, is not at all obvious. For example, it has been suggested that the next step should be to calculate all three-electron clusters (Nesbet, 1967), which is analogous to the approximation sequence followed in nuclear-matter theory (Bethe, 1965; Brandow, 1966b; Rajaraman and Bcthe, 1967; Day, 1969). On the other hand, electron-gas studies suggest that the ring diagrams of the random-phase approximation should play a prominent role. It is obvious from Fig. 10 that one can easily devise a truncated equation for W_2 which generates all of the ring diagrams, in addition to all of the diagrams that enter in the independent-pair approximation. This approach is far more convenient than a straightforward configuration-interaction scheme because, on the one hand, the rings introduce quite complicated (many pair) excited configurations, while on the other hand, one is able to treat selectively some physically dominant matrix elements while disregarding a multitude of less-important elements. The quantitative validity of this approach has been nicely demonstrated in a study by Paldus *et al.* (1972).

E. Single-Particle Potentials

Up to this point we have been consistently ignoring the possible presence of an auxiliary single-particle potential V_{sp}, such as the Hartree–Fock potential. One generally works with an orbital basis defined by

$$H_0 = T + V_{sp}, \tag{40}$$

where T is the basic one-body operator (kinetic energy plus any nuclear or ionic potential). The full perturbation operator is therefore

$$V = v - V_{sp}, \tag{41}$$

where v is the true two-body interaction. The extra diagrams resulting from the $(-V_{sp})$ term in Eq. (41) are quite straightforward and are discussed in most of the papers dealing with practical applications.

How should this V_{sp} be defined in practice? The most useful and general point of view is to say that V_{sp} is simply a "free parameter" in the formalism, to be specified in any manner that turns out to be most convenient for the problem at hand. In closed-shell electronic systems, the strong shell structure tends to produce fairly rapid convergence, thus computational simplicity may be the primary consideration. The Hartree–Fock potential is therefore an obvious and popular choice, at least for the normally occupied (hole) orbitals. Other choices may be preferred, however, for the "particle" orbitals (Kelly, 1964b; Silverstone and Yin, 1968; Huzinaga and Arnau, 1970; Miller and Kelly, 1971; Davidson, 1972). In strongly interacting homogeneous systems, such as nuclear matter or liquid helium (^{3}He or ^{4}He), the need for rapid convergence takes precedence. The choice of hole–orbital potentials is reasonably well settled (Brandow, 1966b; Davies *et al.*, 1974), but unfortunately there is still no consensus about the best choice for the particle orbitals,[2] especially for those orbitals just above the Fermi surface

[2] We now believe that the most satisfactory choice for ground-state calculations is a constant (orbital-independent) potential for the virtual orbitals, chosen to reproduce the effect of three-body clusters (Zabolitsky, 1976). However, this choice requires a separate treatment for the ring diagrams (zero-point fluctuations of the collective modes), a point we intend to discuss elsewhere.

(see, for example, the discussion of Brandow, 1967, pp. 803–6). (In homogeneous systems, of course, the orbitals are constrained to be plane waves, thus V_{sp} merely affects the single-particle eigenvalues ε_k.) When one adds to this problem the feature of inhomogeneity, as in the case of nuclei or solid ^{3}He, there are a number of additional subtleties to be dealt with.

One of these subtleties concerns the so-called generalized Brillouin condition. A familiar feature of the ordinary Hartree–Fock (HF) theory is that the first-order amplitude for creating a *single* particle–hole pair is identically zero when HF self-consistency is satisfied. In terms of the present diagrammatic formalism, this "Brillouin theorem" takes the form shown in

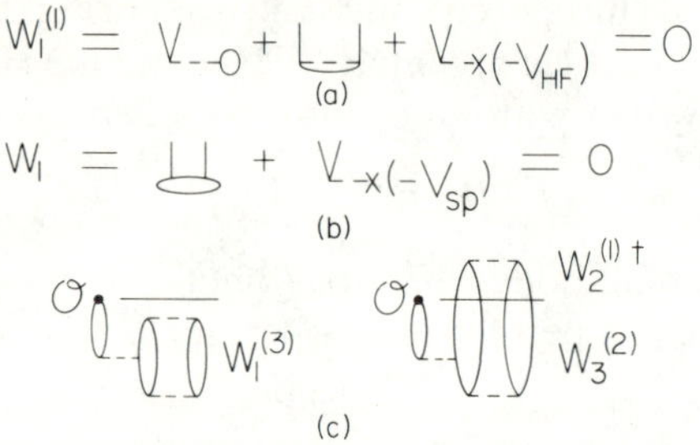

Fig. 12. (a) Diagrammatic form of the Brillouin theorem. (b) The "generalized Brillouin condition," Eq. (42). (c) Some diagrams included in the "extended Brillouin condition," Eq. (43).

Fig. 12a (where the exchange diagram is now shown explicitly). An obvious generalization of this relation is to insist that the *total* amplitude for exciting a particle b and hole m must vanish for all pairs bm,

$$\langle \Phi_0 | a_m^\dagger a_b | \Psi \rangle = W_1(b, m) = 0, \tag{42}$$

where Eq. (29) has been used. This is shown schematically in Fig. 12b, where the blob now represents all of W_1 except the first-order $(-V_{sp})$ contribution. This self-consistency requirement for V_{sp} is the original form of the "generalized Brillouin condition," as suggested by several investigators (Nesbet, 1958; Coester and Kümmel, 1960; Löwdin, 1962). This requirement is very appealing from the standpoint of the Coester–Kümmel–Čižek formalism, as it obviously eliminates a large number of the $\tilde{V}$ diagrams; more than half of the W_2 diagrams of Fig. 10, for example.

But this feature is not quite as useful as it appears. In Fig. 12c we show two perturbation contributions to the particle–hole matrix element of a general one-body operator $\mathcal{O}$, where the solid horizontal boundary line separates the W or $|\Psi\rangle$ amplitude contributions from the $W^\dagger$ or $\langle\Psi|$ contributions. These two diagrams combine, via the factorization theorem, to produce what amounts to a factorized (fully "on-shell") version of $W_1^{(3)}$.

The generalized Brillouin condition should therefore be modified to allow for such extra diagrams and the resulting algebraic simplifications that they lead to. Formally, this "extended Brillouin condition" amounts to replacing Eq. (42) by

$$\langle \Psi | a_m^\dagger a_b | \Psi \rangle = 0. \tag{43}$$

This self-consistency requirement clearly helps to optimize the convergence of the perturbation expansion for the expectation value of an *arbitrary* (not necessarily one-body) operator $\mathcal{O}$. It therefore seems fair to say that this requirement is optimizing the convergence of the expansion for the total wavefunction.

The formal distinction between Eqs. (42) and (43) has been discussed by several investigators (Smith and Kutzelnigg, 1967; Kirson, 1968; Kobe, 1969; Schäfer and Weidenmüller, 1971). The advantage of Eq. (43) is one of the reasons why we have long insisted [Brandow, 1966b, 1967, 1970, 1972 (pp. 138–140)] that the $\langle \mathcal{O} \rangle$ expansions provide the most sensitive criterion for the formal definition of the V_{sp} for nuclei, i.e., the nuclear shell model potential. Conversely, this brings into focus a limitation of the Coester–Kümmel–Čižek method. The present consideration is not merely academic, because diagrams of the type shown in Fig. 12c are known to be quantitatively significant in *ab initio* nuclear calculations (Davies *et al.*, 1974). This is not an essential limitation, however, because diagrams such as the second term in Fig. 12c *can* be incorporated within the Coester–Kümmel–Čižek method, once the desirability of doing so is recognized.

Among the other subtleties that must be dealt with in developing a satisfactory theory of the nuclear shell model potential, we shall simply mention (1) the need to also specify the particle–particle and hole–hole elements of V_{sp} (Brandow, 1966b, 1967, 1970, 1972), (2) some practical consequences of the nonlocality of the V_{sp} which results from adopting different recipes for the different types of matrix elements in point 1 (Brandow, 1970), (3) a Hermiticity problem (Brandow, 1970), and (4) the formal connection with the type of variational principle satisfied by the conventional Hartree–Fock theory (Brandow, 1969).

In concluding this survey of the nondegenerate many-body perturbation results, we wish to emphasize that there are completely analogous formal results for Bose fluid systems (Brandow, 1971). On the other hand, the Bose *solid* (solid ^{4}He) is most conveniently dealt with by an analog (Brandow, 1972) of the following open-shell formalism.

F. Outline of the Adiabatic Derivation of Goldstone

We close our discussion of the nondegenerate case with an exposition of the time-dependent adiabatic derivation of Goldstone (1957). Except for diagrams and the use of second quantization, his technique is almost entirely different from the foregoing method. This derivation has been described many times in the literature, but certain key points have not been adequately discussed. We shall focus on the manner in which the various linked-cluster results emerge from the adiabatic approach. What follows is an elaboration of some previous brief remarks (Brandow, 1967, p. 790).

The basic idea is that the perturbation V is switched on gradually over the time interval $-\infty < t < 0$, such that the noninteracting state Φ_0 (the ground eigenstate at $t = -\infty$) should evolve continuously into the fully interacting ground state Ψ at $t = 0$. This is accomplished by solving the time-dependent Schrödinger equation for the Hamiltonian $H_0 + Ve^{\alpha t}$, where α is a small positive constant. The analysis is done within the interaction picture, that is, with the wavefunction

$$\Psi_I(t) = e^{iH_0 t}\Psi_S(t), \tag{44}$$

in which the H_0 part of the time dependence has been removed from the Schrödinger wavefunction Ψ_S. This *interaction wavefunction* evolves according to the differential equation

$$\frac{\partial}{\partial t}\Psi_I(t) = -iH_I(t)\Psi_I(t), \tag{45}$$

where the *interaction Hamiltonian* is

$$H_I(t) = \lambda e^{iH_0 t}Ve^{-iH_0 t}e^{\alpha t}. \tag{46}$$

The coupling constant λ is nominally equal to unity, but it must be carried along explicitly for reasons shown below.

The differential equation (45) is formally solved by iteration to produce an expansion in powers of λ,

$$\Psi_I(0) = \left\{1 + \sum_{n=1}^{\infty}(-i)^n \int_{-\infty}^{0} dt_n \int_{-\infty}^{t_n} dt_{n-1} \cdots \int_{-\infty}^{t_2} dt_1 \right.$$
$$\left. \times\, [H_I(t_n)H_I(t_{n-1}) \cdots H_I(t_1)]\right\}\Phi_0 . \tag{47}$$

The time integrations are all elementary, leading to nth-order terms of the form

$$\Psi_I^{(n)}(0) = \lambda^n \frac{1}{E_0 - H_0 + in\alpha} V$$
$$\times \frac{1}{E_0 - H_0 + i(n-1)\alpha} V \cdots \frac{1}{E_0 - H_0 + i\alpha} V\Phi_0 . \tag{48}$$

This rather abstract result is made more explicit by expressing the various sequences of intermediate states in terms of the now familiar "second quantized" diagrams, where the exclusion principle in intermediate states is ignored. One finds that the most general diagram may contain any number of closed linked pieces, together with any number of open linked pieces. However, the following differences from our previous diagrams should be noted: (1) the mth energy denominator contains the "small quantity" $im\alpha$, in addition to $E_0 - H_0$; (2) the initial "vacuum" state Φ_0 may also occur as an intermediate state, and may do so arbitrarily often.

The relative complexity of this set of diagrams can be greatly reduced by employing Goldstone's version of the factorization theorem. One collects together all nth-order diagrams composed of the same set of linked pieces, such that the only differences among these diagrams are the relative time orders of interactions in *different* linked pieces. (The relative time ordering *within* each linked piece is held fixed.) The sum of this subset of $\Psi_{\mathrm{I}}^{(n)}(0)$ diagrams is generated simply by relaxing the corresponding time-order restrictions within the multiple time integrals of Eq. (47). The result is just the product of the contributions from the various linked parts, each evaluated as if the other linked parts were absent. In other words, each of these linked-part factors is still determined by an expression of the form Eq. (48). Summing now over the particle and hole labels within all of these linked pieces, one finds that weighting factors consisting of inverse factorials are needed whenever two or more of the linked pieces have identical topologies. (The argument is again similar to that of Fig. 9.) Goldstone's factorization therefore leads to the result

$$\Psi_{\mathrm{I}}(0) = \exp(\text{sum of open linked diagrams}) \times \exp(\text{sum of closed linked diagrams}). \tag{49}$$

This type of formal structure was first recognized by Feynman (1949).

We now examine the nature of the adiabatic limit $\alpha \to 0$, corresponding to an infinitely slow switch-on of V. In the open linked diagrams there are (by assumption) no intermediate states degenerate with Φ_0. Thus, all of the energy denominators are well behaved and α can simply be set equal to zero. In this limit, then, the first exponential in Eq. (49) becomes identical to the wavefunction expression (29). This limit is a much more subtle matter for the closed diagrams. Because their final "intermediate state" is always Φ_0, the corresponding final energy denominators are all of the singular form $(in\alpha)^{-1}$. This makes it necessary to Taylor-expand the remaining product of nonsingular denominator factors with respect to α. Referring to Eq. (48), let function $f_n(\alpha)$ be defined by everything except the final (left-most) energy denominator. For small but finite α, we then obtain

$$\frac{1}{in\alpha} f_n(\alpha) = \frac{1}{in\alpha} f_n(0) + \frac{1}{in} f'_n(0) + \mathcal{O}(\alpha). \tag{50}$$

Since the $\mathcal{O}(\alpha)$ remainder becomes negligible as $\alpha \to 0$, it follows that the second factor in Eq. (49) can also be factored into two exponentials,

$$\begin{aligned} &\exp(\text{sum of closed linked diagrams}) \\ &\qquad \to \exp\{\text{sum of } [f_n(0)/in\alpha] \text{ terms}\} \\ &\qquad\quad \times \exp\{\text{sum of } [f'_n(0)/in] \text{ terms}\}. \end{aligned} \tag{51}$$

It turns out that the argument of the second exponential in Eq. (51) is just $+\frac{1}{2}F'_{\mathrm{L}}(0)$, in the notation of Eqs. (31) and (32), whereby this exponential becomes $\langle\Psi|\Psi\rangle^{-1/2}$. This combines with the first exponential in Eq. (49) to produce the desired ground-state wavefunction with unit norm. This result is to be expected because probability is conserved by the differential equation (45), and thus $\langle\Psi_{\mathrm{I}}(t)|\Psi_{\mathrm{I}}(t)\rangle = 1$ for all t, including $t = 0$. To confirm that the f'_n/in terms actually do sum to $+\frac{1}{2}F'_{\mathrm{L}}(0)$, we proceed as follows. Consider the differentiation of the rth energy denominator in Eq. (48), counting from the right. This contributes a term of the form $(-ir/in)(E_0 - H_0)_r^{-2}$. Now observe that for every closed diagram there is also a "mirror inverse" diagram, where the same sequence of intermediate states occurs in the opposite order. Differentiating the corresponding denominator in this inverse diagram produces a term of the form $[-i(n-r)/in](E_0 - H_0)_r^{-2}$. By averaging these two contributions before the general diagram summation, we obtain $-\frac{1}{2}(E_0 - H_0)_r^{-2}$, from which the desired result follows by comparison with Eqs. (31)–(33).

Since the $f_n(\alpha = 0)$ in Eq. (50) is nonsingular and real, the first exponential in Eq. (51) is a pure phase factor $e^{i\phi}$ whose phase $\phi(\alpha)$ diverges as α^{-1}. This too is to be expected on physical grounds. The sum of the $f_n(0)$'s for all nth-order diagrams is simply ΔE_n, the nth-order contribution to ΔE. In the adiabatic formalism this energy shift is accompanied by n switching factors $e^{\alpha t}$, and, thus, the phase effect of this energy shift is the same as if the *full* ΔE_n had acted over a time interval $(n\alpha)^{-1}$. [Consider the final time integration in Eq. (47), $\int dt_n$.] It follows that the net phase must be

$$\phi(\alpha) = -\int_{-\infty}^{0} \Delta E(t)\, dt = -\sum_{n=1}^{\infty} \frac{\Delta E_n}{n\alpha}. \tag{52}$$

Due to its singular nature, it is clearly necessary to eliminate this phase factor before completing the limit $\alpha \to 0$. This can be done by noting that the scalar quantity $\langle\Phi_0|\Psi_{\mathrm{I}}(0)\rangle$ contains the same factor, so that the quotient

$$\Psi_\alpha = \Psi_{\mathrm{I}}(0)/\langle\Phi_0|\Psi_{\mathrm{I}}(0)\rangle \tag{53}$$

remains well-behaved as $\alpha \to 0$. In fact, the foregoing analysis shows that $\Psi_{\alpha=0}$ is just the intermediate-normalized Ψ of Eq. (29). We now see, from the time-integral expression in Eq. (52), together with Eq. (45), that the energy shift can be expressed as

$$\Delta E(t=0) = i\frac{\partial}{\partial t}[\ln\langle\Phi_0|\Psi_{\text{I}}(t)\rangle]\bigg|_{t=0} = \langle\Phi_0|H_{\text{I}}(0)|\Psi_\alpha\rangle. \tag{54}$$

As $\alpha \to 0$ this reduces to the usual expression for ΔE, namely, the $F_{\text{L}}(0)$ defined in Eq. (31). Alternatively, one can use the $\sum_n$ expression in Eq. (52) to obtain the more exotic form

$$\Delta E = \lim_{\alpha\to 0} i\alpha\lambda\frac{\partial}{\partial\lambda}[\ln\langle\Phi_0|\Psi_{\text{I}}(0)\rangle]\bigg|_{\lambda=1}, \tag{55}$$

in which we recognize that the role of the λ differentiation [see Eq. (48)] is to generate the necessary factors of n. Diagrammatic analysis shows, of course, that these two expressions for ΔE are completely equivalent.

We have now described how the various linked-cluster results emerge from the adiabatic formalism. The derivation is incomplete, however, because we have not yet demonstrated *by purely adiabatic means* that these linked-cluster expressions have the physical significance we attributed to them. The necessary confirmation is provided by the adiabatic theorem of Gell-Mann and Low (1951). We shall now present their argument within our present framework (i.e., with Goldstone instead of Feynman–Dyson time ordering).

Starting from Eq. (48), *without* using Goldstone's linked-cluster factorization at this stage, we readily obtain

$$(H_0 - E_0 - in\alpha)\Psi_{\text{I}}^{(n)}(0) = -\lambda V\Psi_{\text{I}}^{(n-1)}(0), \tag{56}$$

and thus, after summing over n,

$$\left(H - E_0 - i\alpha\lambda\frac{\partial}{\partial\lambda}\right)\Psi_{\text{I}}(0) = 0. \tag{57}$$

It then follows from Eq. (53) that

$$\left(H - E_0 - i\alpha\lambda\frac{\partial}{\partial\lambda}\right)\Psi_\alpha = \left\{i\alpha\lambda\frac{\partial}{\partial\lambda}[\ln\langle\Phi_0|\Psi_{\text{I}}(0)\rangle]\right\}\Psi_\alpha. \tag{58}$$

Since the preceding diagrammatic analysis has established that Ψ_α remains well behaved, we can now take the $\alpha \to 0$ limit. The result is just

$$(H - E_0)\Psi_{\alpha=0} = \Delta E\Psi_{\alpha=0}, \tag{59}$$

where this ΔE is defined by Eq. (55). This confirms that $\Psi_{\alpha=0}$ [identical to Eq. (29)] is indeed an eigenstate. Having established this, we can now left-multiply Eq. (59) by Φ_0 to obtain the more familiar expression (54) for ΔE.

Finally, the identification Eq. (32) is made, within the adiabatic framework, by appealing to the norm-preserving (unitary) feature of the time-evolution process in Eqs. (45) and (47).

Compared to our time-independent derivation, the adiabatic approach is clearly more elegant. It has the advantage of producing the Ψ and $\langle\Psi|\Psi\rangle$ results simultaneously with the ΔE result, and, furthermore, it can easily be modified to produce the analogous statistical mechanical results for finite temperatures. On the other hand, this procedure seems more complicated from the pedagogical standpoint, since it requires an understanding of second quantization and the adiabatic theorem, in addition to the factorization theorem. It also seems to us that the desired linked-cluster results are generated "too automatically," thus leaving one unedified about their physical significance. Finally, there is something unsatisfying about relying on a time-dependent technique to obtain ground-state results that must obviously be time-independent. These are the pedagogical reasons why we prefer the more pedestrian time-independent approach.

Throughout our entire discussion of the closed-shell case it has been assumed that the Ψ, which evolves adiabatically from Φ_0, is the actual ground state, and not some excited state. In other words, we assume that there has not been a level crossing (or near crossing) by some other eigenstate. The best-known counterexample is the ground state of a superconducting metal (see, for example, Pines, 1961). Further examples are the ground states of magnetic materials (metals or insulators), if the H_0 is chosen to have a nonmagnetic ground state. All of these "anomalous" cases can be dealt with by the following open-shell formalism, although in many cases it will suffice merely to begin with a more appropriate H_0.

VI. Open-Shell Systems—A General Formalism for Model Hamiltonians and Model Operators

We shall now discuss the extension of the Brueckner–Goldstone formalism to open-shell systems. Besides enabling us to deal with such systems, this provides a very important bonus—a general formalism for deriving effective Hamiltonians and other effective operators. These are operators that, when used within a "model subspace" (degenerate or quasi-degenerate), will correctly reproduce the energies and other operator matrix elements of a corresponding set of exact eigenstates.

In Section II we derived the important result

$$[H_0 + P\mathscr{V}(E) - EI]A = 0, \tag{60}$$

where the $d \times d$ matrix operator $P\mathscr{V}$ was shown to possess a BW type of perturbation expansion. We remarked there that $(H_0 + P\mathscr{V})$ constitutes a type of model Hamiltonian. [Note also that $P\mathscr{V} = P\mathscr{V}P$, according to Eq. (12), thus $P\mathscr{V}$ is actually Hermitian for any fixed E. For simplicity we shall generally not show the left-hand P operator; thus, from now on $\mathscr{V}$ must always be understood to mean $P\mathscr{V}$.]

If our intention had been simply to introduce the concept of an effective or model Hamiltonian, we would have followed the more elementary and direct path of Löwdin (1951, 1962) and Feshbach (1962). They partition the Hilbert space into two subspaces, P and Q, such that the Schrödinger equation becomes a 2×2 block matrix equation:

$$\begin{pmatrix} H_{PP} & H_{PQ} \\ H_{QP} & H_{QQ} \end{pmatrix} \begin{pmatrix} \Psi_P \\ \Psi_Q \end{pmatrix} = E \begin{pmatrix} \Psi_P \\ \Psi_Q \end{pmatrix}. \tag{61}$$

The variable Ψ_Q is then eliminated to produce the "projected" Schrödinger equation

$$[H_{PP} + H_{PQ}(E - H_{QQ})^{-1} H_{QP}]\Psi_P = E\Psi_P. \tag{62}$$

This procedure, often referred to as "partitioning of the Hamiltonian," is obviously independent of perturbation theory. In practice, however, the concept of a model Hamiltonian is usually applied to many-body systems, and therefore we need the perturbative approach to cast Eq. (62) into a suitable linked-cluster form. [Note, by the way, that Eq. (60) follows quite directly from Eq. (62).]

One might suppose that other many-body techniques (such as Green's functions or the Coester–Kümmel–Čižek method) could also accomplish this purpose, but to this date only the perturbative approach has been pursued all the way to a satisfactory general result.[3] We shall now describe a specific perturbation theory, namely, our folded-diagram formalism (Brandow, 1966a, 1967). A number of related derivations have appeared since this work, but none of these later studies are complete and some are quite seriously inadequate. We have recently presented a review of the now-considerable literature on degenerate perturbation theory, many-body and otherwise (Brandow, 1975a). In it we discuss the historical background and also make some critical comparisons with the various alternative

[3] Efforts by several groups to extend the Coester–Kümmel–Čižek formalism to open-shell systems have recently come to our attention, too late for detailed comment. These studies are by Mukherjee *et al.* (1975), by Offermann *et al.* (1976) and Offermann (1976), and yet unpublished works by J. Paldus and M. Saute (Saute, Troisieme Cycle Thesis, Universite de Reims, 1976), and by P. Grange and J. Richert (Laboratoire de Physique Nucleaire Theorique, Strasbourg).

approaches now available.[4] Some connections with other many-body formalisms and results are mentioned at the end of this section, and a specific application is presented in Section VII.

A. Diagrammatic Conventions

As in the closed-shell case, the first step toward a useful linked-cluster result is the introduction of suitable diagrammatic conventions. Our model subspace D [see Eq. (6)] is defined in an obvious manner: the "model" Φ_i's are all those $(N + n)$-body Slater determinants in which a given set of *core orbital* shells are fully occupied by *N core particles*, whereas the remaining *n valence particles* are all distributed among a given set of *valence orbitals*. (Of course there are corresponding results for systems with valence holes instead of, or in addition to, the valence particles, but we shall not consider them here.) For now, we assume exact degeneracy as if all of the valence orbitals belong to a single subshell, but this restriction will be removed later on.

Our goal is to calculate the set of all matrix elements

$$\mathscr{V}_{ij} = \langle \Phi_i | \mathscr{V} | \Phi_j \rangle \tag{63}$$

within the model subspace. To specify one of the model Φ_i's, it is clearly sufficient to know which n of the valence orbitals are occupied, since, by definition, the N core orbitals are always occupied in these model states. This suggests that the various terms in the perturbation expansion of Eq. (63) should be represented by diagrams with n "incoming" valence lines at the bottom, to represent Φ_j, and n "outgoing" valence lines at the top, for Φ_i. The N core particles may be represented by the same "vacuum convention" used previously; that is, we need only indicate the *deviations* from the closed-shell configuration.

With these conventions we now apply the BW expansion (12) to some particular matrix element $\mathscr{V}_{ij}$. The resulting diagrams may be classified into three general categories, examples of which are shown in Fig. 13 for the case $n = 3$. Figure 13a shows a *valence diagram*, in which all of the interactions are connected (directly or indirectly) to one or more of the external valence lines. A *core diagram* is shown in Fig. 13b, in which all of the valence lines are completely passive. Figures 13c and d are "mixed" terms containing both core and valence interaction processes. Note that at each intermediate level (between the first and last interactions V) at least one of the particles must be excited to a higher orbital than in the initial configuration Φ_j,

[4] This subject has also been reviewed by Klein (1974), who presents many interesting observations. Readers should be aware, however, that his proof of the linked-cluster property is based on a false premise (Brandow, 1975a), and his variational analysis of the orthogonal model eigenvectors [see Eq. (80) below] is partially incorrect (Brandow, 1977, Appendix F).

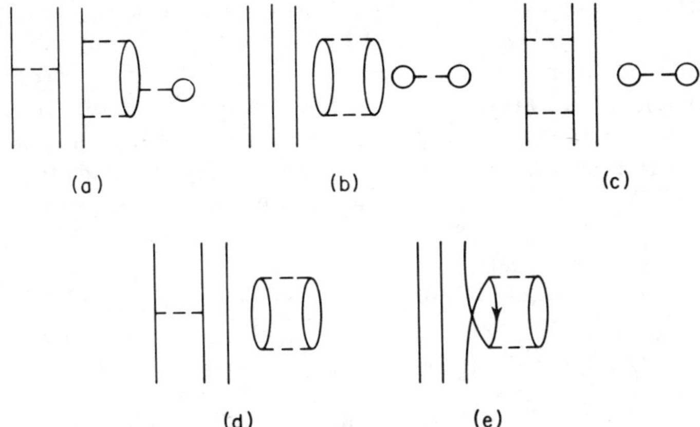

Fig. 13. Diagrams from the BW expansion (12) for $\mathscr{V}_{ij}$: (a) and (e) are valence diagrams, (b) is a core diagram, whereas (c) and (d) are of mixed character.

because none of the allowed intermediate states $\Phi_{i'}$ [from Eq. (7)] are degenerate with the model states.

We continue to ignore exclusion in intermediate states, in order to make the set of core diagrams *precisely the same* as if the valence particles did not exist. This necessitates the introduction of some core-valence interference terms, analogous to Fig. 7b, one of which is shown in Fig. 13(e). This represents a "blocking" of one of the assumed core-correlation processes, due to the presence of a valence particle in one of the orbitals that was presumed to be vacant for the purpose of defining these core processes. All such core–valence interference terms are classified as valence diagrams, similar to example Fig. 13a.

B. Separation of Core and Valence Contributors

We decompose the total energy eigenvalue into core and valence terms, and then further separate each of these into unperturbed and interaction contributions:

$$\begin{aligned} E &= E_c + E_v \\ &= E_{0c} + \Delta E_c + E_{0v} + \Delta E_v \, . \end{aligned} \tag{64}$$

Since in Eq. (60) the H_0 acts only within the model subspace, one sees immediately that

$$H_0 A = (E_{0c} + H_{0v})A = (E_{0c} + E_{0v})A. \tag{65}$$

We now define ΔE_c to be just the interaction energy for the previous non-degenerate case, in which all of the valence particles are physically absent.

This ΔE_c is given by the sum of all of the core diagrams, Fig. 13b, etc., from which it follows that ΔE_v is defined by all of the remaining diagrams. We now pursue the ΔE-expansion argument of Eq. (21), Figs. 4 and 5, but with the following qualification. Although ΔE_c is expanded out of *all* of the present BW energy denominators, ΔE_v is only expanded for *some* of these denominators (see Brandow, 1967, for details). This leads to the following results: (1) ΔE_c reduces to just the sum of linked core diagrams evaluated with RS energy denominators, i.e., the Goldstone result; (2) all diagrams with mixed character, such as c and d of Fig. 13, are canceled identically. We also note that the noninteracting valence lines shown in b of Fig. 13 are equivalent to the unit operator I for the model subspace. These results mean that the model interaction operator simplifies considerably,

$$\mathscr{V}(E) \rightarrow I\ \Delta E_c + \mathscr{V}_v(E_v), \tag{66}$$

where $\mathscr{V}_v$ is given by just the valence diagrams, Figs. 13a and e, etc. The net result is that Eq. (60) now reduces to

$$[H_{0v} + \mathscr{V}_v(E_v) - E_v I]A = 0, \tag{67}$$

an expression in which all explicit reference to "core" quantities has disappeared. The notation indicates that ΔE_v still appears within all of the energy denominators of $\mathscr{V}_v$. This important result was first obtained by Bloch and Horowitz (1958).

We could have used Eq. (65) to reduce this result further to

$$[\mathscr{V}_v(\Delta E_v) - I\ \Delta E_v]A = 0, \tag{68}$$

but we chose not to, for the following reason. For many applications we need results that are not restricted to exact degeneracy. Our strategy is to carry through the entire analysis assuming exact degeneracy, and then extend the results to the quasi-degenerate case by adding a diagonal degeneracy-breaking interaction,

$$\begin{aligned} V_{db} &= \sum_{i \in D} |\Phi_i\rangle\ \Delta E_i \langle\Phi_i| \\ &= \sum_{v} a_v^\dagger a_v\ \Delta\varepsilon_v\,, \end{aligned} \tag{69}$$

to the perturbation V. (The second line expresses the determinantal energy shifts ΔE_i in terms of shifts in the valence orbital eigenvalues.) Thanks to their being diagonal, it is then an easy matter to formally sum out all of the V_{db} interactions. One thereby obtains a result of the same general form as Eq. (67), but with V_{db} now incorporated within H_{0v}.

C. Need for a Fully Linked Valence Expansion

The "valence secular equation" (67) appears at first sight to resemble the conventional shell-model secular equations used to calculate energy spectra in nuclear physics. The Slater–Condon parametrized secular equations of atomic theory also come to mind here. Formally, however, we still have a considerable way to go before we can claim to have a justification for these model Hamiltonians. The main problem is that the valence diagrams in $\mathscr{V}_v$ are not all fully linked. For example, diagram a of Fig. 13 must be regarded as an effective three-body interaction, contrary to the intuitive idea that this represents a two-body interaction together with a "self-energy" or effective one-body potential contribution. The resulting calculational problem may still be manageable, for the case of just three valence particles. The physical interpretation is awkward,[5] however, and this approach rapidly becomes unmanageable as the number of valence particles increases. Another undesirable feature is that, since all of the perturbative terms beyond first order depend on the *total* valence interaction energy ΔE_v, the effective two-body interaction from the present scheme is found to have an unphysical dependence on the total number of valence particles. These difficulties are a reflection of the fact that although the core part is treated by the fully linked RS formalism, the valence part of the system is still being treated by a BW type of expansion. It should now be clear that we need a Rayleigh–Schrödinger expansion for $\mathscr{V}_v$.

D. Degenerate Rayleigh–Schrödinger Perturbation Theory

Let us return to the original degenerate secular equation (60), ignoring all many-body features for now. We can convert the BW expansion for $\mathscr{V}(E)$ to an expansion of RS form by the following iterative procedure, assuming that the degeneracy is exact. We write $E = E_0 + \Delta E$, and expand out the ΔE dependence,

$$\mathscr{V}(E_0 + \Delta E) = \sum_{r=0}^{\infty} \mathscr{V}^{(r)}(-\Delta E)^r, \tag{70}$$

where, according to the Taylor formula,

$$\mathscr{V}^{(r)} = \frac{(-1)^r}{r!} \frac{d^r \mathscr{V}(E_0)}{dE_0^r}. \tag{71}$$

The minus signs are inserted here because

$$\frac{(-1)^r}{r!} \left(\frac{d}{dE_0}\right)^r \left(\frac{1}{E_0 - H_0}\right) = +\left(\frac{1}{E_0 - H_0}\right)^{r+1}. \tag{72}$$

[5] Note that this problem exists even for the case of two valence particles. Without a fully linked formalism, it is impossible to make a clean distinction between the one-body (self-energy or shell-model potential) and the two-body (effective interaction) aspects of $\mathscr{V}_v$.

[The simplicity of this result is to be expected, in view of Eq. (21).] From repeated use of the analog of Eq. (68), we find

$$(-\Delta E_\alpha)^r A_\alpha = [-\mathscr{V}(E_\alpha)]^r A_\alpha\,, \tag{73}$$

where α refers to a particular exact eigenstate. Therefore, *when applied to a particular eigenvector A_α and the corresponding eigenvalue E_α*, $\mathscr{V}(E)$ can be replaced by

$$\mathscr{W}_1(\Delta E_\alpha)A_\alpha = \sum_{r=0}^{\infty} \mathscr{V}^{(r)}[-\mathscr{V}(E_0 + \Delta E_\alpha)]^r A_\alpha\,. \tag{74}$$

Note that the matrix $\mathscr{W}_1(\Delta E_\alpha)$ is generally not identical to $\mathscr{V}(E_\alpha)$. Nevertheless, A_α remains an eigenvector of the secular equation, with the same eigenvalue as before. In $\mathscr{V}$, the ΔE dependence first appears in the perturbative term of order V^2. In $\mathscr{W}_1$, however, this ΔE dependence does not show up explicitly until *fourth* order, namely, in the $\mathscr{V}^{(1)}[-\mathscr{V}^{(0)}(E_0 + \Delta E)]$ term

$$-PV\frac{Q}{(E_0 - H_0)^2}VPV\frac{Q}{E_0 + \Delta E - H_0}VP. \tag{75}$$

The first explicit appearance of ΔE can be pushed out to sixth order by repeating this process, introducing

$$\mathscr{W}_2(\Delta E_\alpha) = \sum_{r=0}^{\infty} \mathscr{V}^{(r)}[-\mathscr{W}_1(\Delta E_\alpha)]^r. \tag{76}$$

This can be repeated *ad infinitum*,

$$\mathscr{W}_n(\Delta E_\alpha) = \sum_{r=0}^{\infty} \mathscr{V}^{(r)}[-\mathscr{W}_{n-1}(\Delta E_\alpha)]^r, \tag{77}$$

whereby the limiting form $\mathscr{W}_\infty$ no longer contains ΔE explicitly in any term of finite order. It is therefore equally valid, formally, for *all* of the eigenstates α of the original secular equation (60). This $\mathscr{W}_\infty \equiv \mathscr{W}$ constitutes the desired RS analog of the matrix operator $\mathscr{V}(E)$.

E. The Folded-Diagram Expansion

We shall now apply this $\mathscr{V} \to \mathscr{W}$ development to the Bloch–Horowitz operator $\mathscr{V}_\text{v}(E_\text{v})$ from Eq. (67), calling the resulting operator $\mathscr{W}_\text{v}$. The matrix multiplications and explicit minus signs in Eqs. (74)–(77) will be dealt with by means of "folded diagrams." Consider a system with just two valence particles. The $\langle\Phi_i|\mathscr{V}_\text{v}^{(1)}[-\mathscr{V}_\text{v}^{(0)}]|\Phi_j\rangle$ terms [similar to (75) but with ΔE deleted] can be represented by diagrams such as Fig. 14a, where labels have been added to show the meaning of the various sections. The

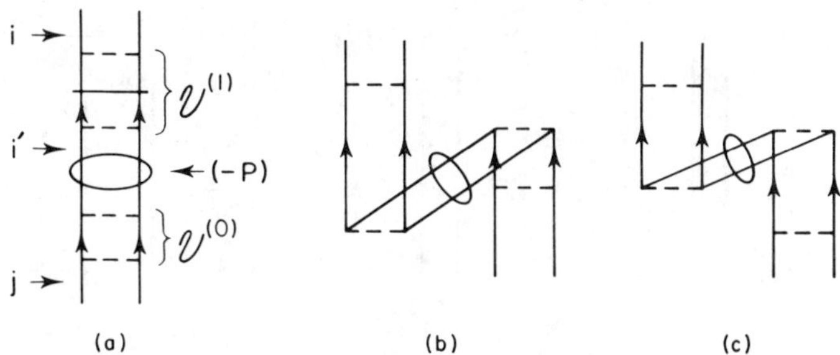

Fig. 14. Examples of folded diagrams: folded diagrams b and c are generated by the "unfolded" diagram a.

solid horizontal line in $\mathscr{V}^{(1)}$ stands for the repeated energy denominator of $\mathscr{V}^{(1)}$, from Eq. (72), exactly as in the case of Figs. 4 and 5. The loop in the middle of this diagram stands for the projection operator P, together with the explicit minus sign from $[-\mathscr{V}_{\mathrm{v}}^{(0)}]$. Note that the associated matrix multiplication involves a sum over all intermediate states $\Phi_{i'}$, with i' in D. This can be accomplished by summing each line segment within the loop over all of the valence orbitals. The loop therefore indicates these orbital summations, together with the explicit minus sign. These valence-orbital summations can be done independently, since the resulting exclusion-violating terms will all cancel in the manner of Fig. 7.

We now "fold" diagram a in a zig-zag fashion, placing the folds at the bottom interaction of $\mathscr{V}_{\mathrm{v}}^{(1)}$ and at the top of $\mathscr{V}_{\mathrm{v}}^{(0)}$, so as to bring the top of $\mathscr{V}_{\mathrm{v}}^{(0)}$ up to the same level as the horizontal bar in diagram a. We also apply the factorization theorem, in the opposite direction to the way it was used in Section III, to obtain finally the two folded diagrams b and c of Fig. 14. These diagrams now have off-shell energy denominators, such that these denominators are now of just the same form as in the first two diagrams of Fig. 6. The matrix multiplications associated with the $r > 1$ terms in Eq. (74), as well as those arising at later stages of the sequence (77), are all to be treated in a similar manner, thus the perturbation series for $\mathscr{W}_{\mathrm{v}}$ includes diagrams with arbitrarily many folds.

The practical value of these folded diagrams becomes apparent when we are faced with unlinked valence diagrams such as Fig. 13a. Consider the unlinked diagrams a and b of Fig. 15, for a system with four valence particles. In the present RS-type expansion, these two diagrams are canceled identically by diagram c. This becomes quite evident when diagram c is manipulated in the manner shown in Fig. 14. This type of cancellation can be shown to be completely general (Brandow, 1967), which means that *only fully linked valence diagrams appear within the expansion for* $\mathscr{W}_{\mathrm{v}}$. To obtain the desired complete cancellation, it turns out to be essential (not

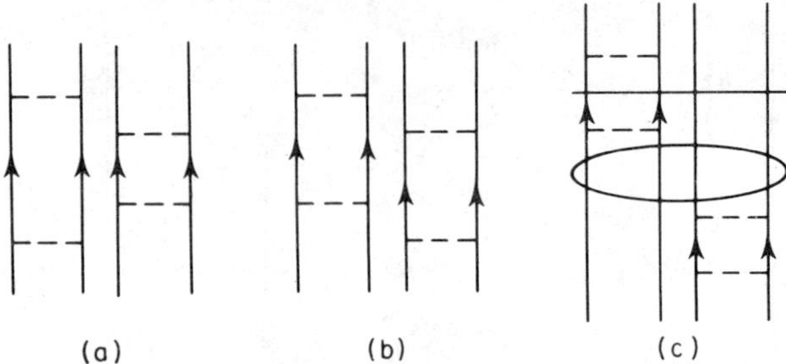

Fig. 15. An example of unlinked diagram cancellation: diagrams a and b are canceled by c.

merely a convenience) to "ignore the exclusion principle" while summing over the sets of valence-orbital lines enclosed by the various $(-P)$ loops. In fact, one must consistently ignore the exclusion principle at *all* intermediate levels between the topmost and bottommost interaction of each diagram, where we are now referring to the "unfolded" forms (Fig. 14a) of the diagrams. In the resulting fully linked valence diagrams, any completely passive (noninteracting) valence lines may now be erased, since these lines do not convey any useful information.

This result does not mean, however, that all folded diagrams disappear from sight. Although the unlinked terms (folded or otherwise) are all canceled, one is still left with an *infinite number* of linked folded diagrams, such as Figs. 14b and c. These leftover folded diagrams are all analogous to the closed-shell diagrams such as Fig. 7b, in which some of the hole lines violate exclusion. In fact, a one-to-one correspondence can be established by applying the present formalism to the case where the n valence particles completely fill the valence shell, so that the system is actually nondegenerate ($d = 1$, $P = |\Phi_0\rangle\langle\Phi_0|$). It should also be noted that these leftover diagrams may contain arbitrarily many folds. [In the alternative treatments which may be more familiar to quantum chemists, the works of Sandars (1969) and Lindgren (1974), diagrams with more than one fold were not considered.] At this stage, where the unlinked diagrams have all been eliminated, one may now simplify these linked folded diagrams by applying the factorization theorem in the "forward direction," i.e. in the manner of Fig. 6. Moreover, the algebraic structure of Eqs. (70)–(77) provides further opportunities for partial summation of these diagrams.

We shall now describe the general time-ordering structure of the $\mathscr{W}_v$ diagrams containing more than one fold. Consider the set of twice-folded diagrams, all of which are derived from the "unfolded" structure shown schematically in Fig. 16a. (Here each heavy line stands for a set of n lines representing valence orbitals, i.e., a state in the model subspace.) There are two corresponding types of folded structures, as shown in Figs. 16b and c.

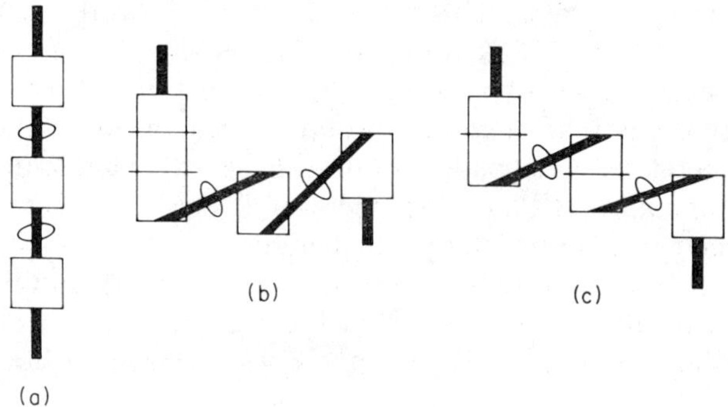

Fig. 16. The two possible types of "time ordering" for twice-folded diagrams. Diagram b represents $\mathscr{V}_v^{(2)}[-\mathscr{V}_v^{(0)}]^2$, whereas diagram c represents $\mathscr{V}_v^{(1)}[-\mathscr{V}_v^{(1)}][-\mathscr{V}_v^{(0)}]$; diagram a shows the corresponding unfolded form.

Structure b represents $\mathscr{V}_v^{(2)}[-\mathscr{V}_v^{(0)}]^2$, whereas structure c represents $\mathscr{V}_v^{(1)}[-\mathscr{V}_v^{(1)}][-\mathscr{V}_v^{(0)}]$. This example illustrates the main features of the general case, since all of the folded structures necessary to represent $\mathscr{W}_v$ can be generated by the following recipe. Let the diagrams containing r folds $(0 \leq r < \infty)$ be drawn, at the outset, in the unfolded form of Fig. 16a, where there are now r intermediate levels within the model subspace P. Now fold this diagram in a zig-zag manner, such that the top of each $\mathscr{V}_v$ block is brought up to a level *higher* than the bottom of the $\mathscr{V}_v$ block to which it is directly connected. That is, one must reverse the relative vertical positions of the ends of each of the looped heavy lines. The expansion for $\mathscr{W}_v$ consists of all possible relative time orderings that can be generated in this manner, subject to the following constraint: the topmost V-interaction of the original unfolded diagram must still be the highest interaction in all of the resulting folded diagrams.

The reader should work through some further examples to convince himself that this recipe is fully equivalent to Eq. (77). What we have just described is the fully off-shell form, which is needed to demonstrate the cancellation of all unlinked terms. The remaining linked diagrams can obviously be simplified by means of the factorization theorem, as already mentioned.

F. Reduction of the Bloch–Horowitz Diagrams

We shall now return to the Bloch–Horowitz result, Eq. (67), to mention a rather subtle problem that requires attention before we carry out the $\mathscr{V}_v \to \mathscr{W}_v$ replacement and introduce the folded diagrams. This seems to be the most obscure aspect of our detailed analysis (Brandow, 1967), since no other papers in this field have succeeded in finding a satisfactory treatment for the present problem.

The basic problem is that the separation of "core" and "valence" aspects, within the result Eqs. (66) and (67), is not as complete as we have a right to expect. The set of $\mathscr{V}_{\mathrm{v}}$ diagrams contains many instances of core-particle excitations that ought, according to general experience, to be factorized and rendered independent of the ΔE_{v} which enters into all of the $\mathscr{V}_{\mathrm{v}}$ energy denominators. In addition, there are "projecting core excitations" that tie together two or more "elementary" or "irreducible" valence-excitation sections, thus forming a composite diagram that fulfills the requirement of having all of its intermediate states outside of the degenerate subspace D. Some examples are shown in Fig. 17, where the horizontal loops

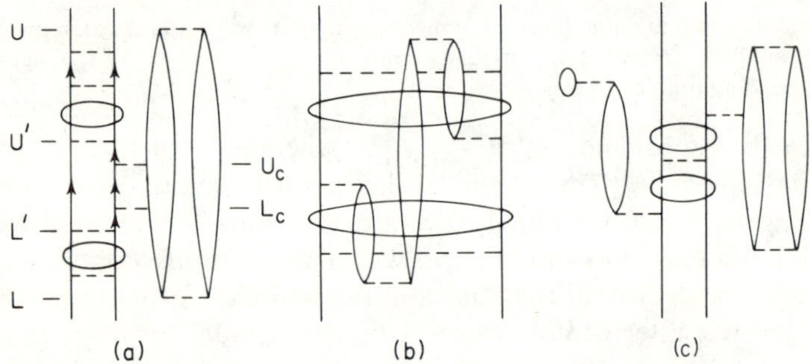

Fig. 17. Examples of reducible Bloch–Horowitz diagrams.

now indicate intermediate levels that would be in D, and would therefore be unallowed, were it not for the presence of projecting core excitations. According to the prescription for $\mathscr{W}_{\mathrm{v}}$ outlined above, the upper and lower folding levels for the $\mathscr{V}_{\mathrm{v}}$ terms shown here would have to be their topmost and bottommost interactions, marked U and L in example a. This would leave some (although not all) of the present complications built into the final result for $\mathscr{W}_{\mathrm{v}}$.

Our procedure is to expand ΔE_{v} formally out of all the energy denominators between the levels marked U and U' in example a, and similarly between L and L', to replace the resulting $(-\Delta E_{\mathrm{v}})$ insertions by diagrams, then use the factorization theorem to take these inserted parts "off shell," so that they cancel the corresponding "reducible" valence diagrams. [This is the basic idea, although diagrams with more complicated overlap structures, such as c, require a slightly more involved treatment.] The immediate result is to strip away the upper and lower sections of valence interactions, such that there are no longer any valence interactions above U' or below L'. A further application of the factorization theorem then simplifies the core-particle excitations that project above U_{c} and below L_{c}. This allows these projecting core excitations to be evaluated "on shell," independent of the valance excitation between U_c and U', and between L_c and L'. As far as the subsequent folded expansion for $\mathscr{W}_v$ is concerned, these projecting core excitations now take place "instantaneously" at the levels U_c, L_c; the appropriate folding levels for this "reduced" $\mathscr{V}_v$ diagram are now located at U' and L'.

It turns out that the *downward* aspect of this reduction (that involving L, L', L_c) could have been accomplished equally well by means of the folded-diagram expansion, simply by folding at the original bottom level L instead of L', and then factorizing the projecting core excitation. This does *not* work for the *upward* (U, U', U_c) aspect, however, or for complex cases such as example c; thus the explicit treatment outlined above seems to be unavoidable. The time-dependent adiabatic approach of Goldstone has been used by several investigators to derive the basic folded-diagram results (Oberlechner *et al.*, 1970; Kuo *et al.*, 1971), but that approach also fails to cope with the problem of upward reducibility.

Incidentally, the surviving portion of Fig. 17b is a core-polarization effect—the exchange of a virtual core phonon. This illustrates the fact that the core-valence separation, Eq. (67), has only eliminated the *static* core effects. All *dynamic* effects of the core (effects related to the valence particles) are still contained within $\mathscr{V}_v$ and $\mathscr{W}_v$.

G. Hermiticity and Effective Operators

It must be recognized that the present effective interaction matrix $\mathscr{W}_v$ is not fully Hermitian. The eigenvalues are all real (by construction they must reproduce certain of the exact eigenvalues), but the *model eigenvectors* A_α are generally not orthogonal. This result can be understood from two different viewpoints. On the one hand, the A_α's represent *projections* of the exact eigenvectors onto the model subspace [see Eq. (6)]; projections of orthogonal vectors are generally not orthogonal unless this feature is enforced by symmetry. On the other hand, different eigenstates α generally have different eigenvalues E_α (apart from symmetry degeneracy), whereas the original matrix operator $\mathscr{V}(E)$ exhibits Hermiticity only when E is held fixed. Diagrammatically, this lack of Hermiticity arises via Eqs. (74)–(77) and the convention of Figs. 14–16, where the "folded-in" insertions are all connected to the *bottoms* of the skeleton parts [the explicit $\mathscr{V}^{(r)}$'s of Eq. (77)]. We have not been concerned about this lack of Hermiticity until now, because the present $\mathscr{W}_v$ is considerably easier to calculate than any of the formally equivalent Hermitian matrices that can be constructed (Brandow, 1975a). Nevertheless, we shall now indicate a relatively simple and convenient way to obtain a Hermitian analog of $\mathscr{W}_v$.

Assuming the intermediate normalization convention,

$$\langle \Psi_{D\alpha} | \Psi_\alpha \rangle = \langle \Psi_{D\alpha} | \Psi_{D\alpha} \rangle \equiv \langle A_\alpha | A_\alpha \rangle = 1, \tag{78}$$

the orthogonality condition for the *exact* eigenstates becomes

$$\begin{aligned}\langle \Psi_\alpha | \Psi_\beta \rangle &= \langle \Psi_{D\alpha} | \Omega_\alpha^\dagger \Omega_\beta | \Psi_{D\beta} \rangle \\ &= \langle \Psi_{D\alpha} | \Omega_\alpha^\dagger (P + Q) \Omega_\beta | \Psi_{D\beta} \rangle \\ &\equiv \langle A_\alpha | (I + \Theta) | A_\beta \rangle = N_\alpha \delta_{\alpha\beta} . \end{aligned} \tag{79}$$

The present Ω's carry subscripts because of their dependence on the eigenvalues E_α, E_β via Eqs. (7) and (12). In obtaining the last line of Eq. (79), we have (1) used the RS technique of Eqs. (70)–(77), together with folded diagrams, to eliminate the explicit eigenvalue dependence of these Ω's, and (2) expressed $\Omega^\dagger \Omega$ as a matrix $(I + \Theta)$ acting within the model subspace, using the fact that $\Omega^\dagger P \Omega = P = I$ here, whereby $\Theta = \Omega^\dagger Q \Omega$. The last line of Eq. (79) suggests the introduction of a new set of model eigenvectors,

$$\hat{A}_\alpha = (I + \Theta)^{1/2} A_\alpha N_\alpha^{-1/2}, \tag{80}$$

which are now orthonormal by construction.[6] The eigenvalue equation

$$(H_{0\text{v}} + \mathscr{W}_\text{v}) A_\alpha = E_\text{v} A_\alpha , \tag{81}$$

obtained from Eq. (67) by the replacement $\mathscr{V}_\text{v} \to \mathscr{W}_\text{v}$, can now be expressed in terms of the $\hat{A}$ vectors,

$$(H_{0\text{v}} + \mathscr{W}_\text{v})(I + \Theta)^{-1/2} \hat{A}_\alpha = E_\text{v} (I + \Theta)^{-1/2} \hat{A}_\alpha , \tag{82}$$

and thus

$$(I + \Theta)^{1/2} (H_{0\text{v}} + \mathscr{W}_\text{v})(I + \Theta)^{-1/2} \hat{A}_\alpha \equiv (H_{0\text{v}} + \mathscr{K}_\text{v}) \hat{A}_\alpha = E_\text{v} \hat{A}_\alpha . \tag{83}$$

The new interaction matrix $\mathscr{K}_\text{v}$ is Hermitian by construction (it has real eigenvalues and orthogonal eigenvectors), but it is far from being *manifestly* Hermitian. (Its perturbation expansion confirms, for the low orders that can be worked out explicitly, that $\mathscr{K}_\text{v}$ is indeed Hermitian.) In practice, this defect can be remedied by explicit Hermitization,

$$\mathscr{K}_\text{v} \to \tfrac{1}{2} [(I + \Theta)^{1/2} (H_{0\text{v}} + \mathscr{W}_\text{v})(I + \Theta)^{-1/2} + \text{h.c.}] - H_{0\text{v}} . \tag{84}$$

This recipe can be used, for example, to study the errors incurred by the obvious approximation

$$\mathscr{K}_\text{v} \approx \tfrac{1}{2} (\mathscr{W}_\text{v} + \mathscr{W}_\text{v}^\dagger), \tag{85}$$

which is often employed in nuclear physics. For many-body systems one must deal with Eq. (83) or (84) by expanding the square-root factors into binomial series in Θ. [Note that truncation of Eq. (84) at any finite order in

[6] Formal aspects of this orthogonalization technique are discussed in Brandow (1970, Appendix A; 1975a; 1977, Appendix F).

Θ will preserve Hermiticity.] We have shown that the resulting series for $\mathscr{K}_v$ is fully linked, although this is not true for Θ by itself (Brandow, 1967).

An added benefit from the use of $\mathscr{K}_v$ and the resulting $\hat{A}$ model vectors is that this leads to a convenient expression for the "model operators" $\mathscr{M}(\mathscr{O})$, whose effect within the model subspace is the same as that of $\mathscr{O}$ acting on the corresponding exact eigenstates:

$$\langle \hat{A}_\alpha | \mathscr{M}(\mathscr{O}) | \hat{A}_\beta \rangle \equiv \frac{\langle \Psi_\alpha | \mathscr{O} | \Psi_\beta \rangle}{(\langle \Psi_\alpha | \Psi_\alpha \rangle \langle \Psi_\beta | \Psi_\beta \rangle)^{1/2}}. \tag{86}$$

The diagonal ($\beta = \alpha$) terms of this transition matrix are, of course, the ordinary expectation values $\langle \mathscr{O} \rangle_\alpha$. By using Eq. (80), one finds that

$$\mathscr{M}(\mathscr{O}) = (I + \Theta)^{-1/2}(\Omega^\dagger \mathscr{O} \Omega)(I + \Theta)^{-1/2}. \tag{87}$$

In an open-shell system, the various matrix operators in Eq. (87) should all have valence subscripts v. The complete $\mathscr{M}(\mathscr{O})$ operator then contains an additional diagonal term $\langle \mathscr{O} \rangle_c \delta_{\alpha\beta}$, where $\langle \mathscr{O} \rangle_c$ is just the contribution from the core particles described in Section V,A, calculated as if all of the valence particles had been physically removed. The folded-diagram expansion for $\mathscr{M}_v(\mathscr{O})$ has also been shown to be fully linked (Brandow, 1967). An example of where this expansion should be useful is in the calculation of electromagnetic transition amplitudes.

In concluding this discussion, we should mention that there is a formal (and sometimes practical) problem in ascertaining just which d of the infinite number of exact eigenstates should be reproduced when using various approximation techniques to evaluate $\mathscr{W}_v$ or $\mathscr{K}_v$. This question has been the object of much research in nuclear physics, where it is known as the problem of "intruder states." The reader will find this discussed in several of the articles in a recent conference proceedings (Barrett, 1975) and in a review by Ellis and Osnes (1977). We have also commented about this problem in earlier reports (Brandow, 1967, pp. 801–803 and Appendix A; 1969, pp. 68–69).

H. Relation to Other Many-Body Theories and Results

From the standpoint of practical applications, it is very helpful to understand the connections with other many-body formalisms and their principal results. The extent of the connections established to date will now be summarized.

For the case of a *single* valence particle (or hole) beyond the closed shells, the diagrammatic expansion for $\mathscr{W}_v$ should be formally equivalent to the usual mass-operator (self-energy operator) expansion from the "field-theoretic" form of many-body perturbation theory (Pines, 1961). This has been confirmed in detail (Brandow, 1971, Appendix C). The correspondence is nontrivial, since the present treatment of self-energy effects is "reducible" (in the usual field-theoretic sense), and hence some further cancellation and partial summation is needed to obtain the more familiar "irreducible" form. The latter is certainly the most useful form. Two other connections with the theory of causal one-body Green's functions have also been examined (Brandow, 1969)—these deal with the strength "Z" of the quasi-particle pole, and with the Ward identity of quantum electrodynamics. Unfortunately, the Z aspect has not yet been fully explored; we hope to present a thorough discussion at some future date.

The standard form of the random-phase approximation (RPA) for collective oscillations has been derived from the present formalism (Brandow, 1967). For this purpose one must take the model subspace D to consist of all determinants with one particle and one hole, as compared to the ground state of a closed-shell system. The corresponding ground-state correlation energy and wavefunction (the contributions due to zero-point motion of the collective modes) have been thoroughly discussed by Ellis (1970).

The Beliaev–Hugenholtz–Pines expression for the elementary excitation spectrum of a Boson system (phonon–roton spectrum) is closely related to the RPA formalism. This has been demonstrated in the course of a detailed derivation of the phonon–roton energy expression by the present techniques (Brandow, 1971). This study may also be of interest for Fermi systems, since it indicates that by successively refining the various "vertex functions" (the usual A and B matrices), the RPA can be developed into a formally exact theory of excitations in closed-shell systems.

Connections with the Landau theory of Fermi liquids have also been worked out in detail (see Brandow, 1967, 1969, 1971; Brown, 1971, 1972; Babu and Brown, 1973). In this case certain of the folded diagrams can be neglected for a macroscopic quantum fluid (liquid or gas), because each of the effective two-body matrix elements (from linked $\mathscr{V}_v$ pieces) is only of order N^{-1}, and thus the folded $(r > 0)$ $\mathscr{W}_v$ diagrams involving repeated effective two-body interactions become negligible compared to the corresponding $\mathscr{V}^{(0)}$ contribution. However, the folded two-body terms do become significant in finite systems and in macroscopic systems with localized orbitals (quantum crystals; see also the following section).

At this point we should mention the connections between folded diagrams and the usual applications of quantum electrodynamics. A physical electron is in constant interaction with the physical vacuum, thus its self-energy δm is nonzero (actually divergent). In terms of the present formalism, this δm becomes ΔE, and thus the diagrams with folds are quantitatively significant. They play an important role in the process of passing from the reducible to the irreducible form of the self-energy operator. Other effects which are algebraically equivalent to our folded diagrams can be identified within the Feynman–Dyson formalism, for example wavefunction renormalization. Although we treat this by means of folds (Brandow, 1969, Sec. XI E), in the field-theoretic formalism this arises from diagrams having the "unfolded" topology of Fig. 16a. (Note that "propagation backwards in time" would now imply a virtual antiparticle, i.e., a physically different process.) The wavefunction renormalization now arises via details of the adiabatic switching process (Schweber, 1961, pp. 539–543), in a manner analogous to our discussion of the $f_n'(0)$ terms from Eq. (50). This effect is quantitatively significant for the same reason as δm, namely, the fact that a physical electron is formally equivalent to a *bound state* of the bare electron and the physical vacuum. Another example more obviously related to Eq. (70) *et seq.* is the case of bound states treated via the Bethe–Salpeter equation, where ΔE is now an ordinary binding energy. On the other hand, the corresponding ΔE is infinitesimal for physical scattering, due to the finite time duration of the actual scattering process. Thus, the effects that we would represent by diagrams with folds are, in this case, strictly negligible. The point of these comparisons is to emphasize that the effects corresponding to our folded diagrams are intrinsically bound-state ($\Delta E \neq 0$) aspects, as opposed to physical scattering ($\Delta E = 0$) aspects.

For most of the connections listed above, the basic formal results are obtained considerably more easily by the frequency-dependent field-theoretic approach (Pines, 1961) than by our present time-independent methods. On the other hand, our approach has the virtue of being more closely related to the method of configuration mixing for few-body systems. Furthermore, this time-independent approach is much better suited for systems with strong short-range ("hard-core") repulsions.

The reason for the latter is as follows. The field-theoretic formulation (via Green's functions) has the virtue of treating particles and holes in a highly symmetrical manner. This is physically appropriate for problems such as superconductivity, or for the thermal behavior of liquid ^{3}He at very low temperatures. In these cases the "active" orbitals are all very close to the Fermi surface, so that the finiteness of k_F or ε_F, or alternatively the curvature of the Fermi surface, plays essentially no role. By contrast, consider the case of nuclear matter. There the hard-core interaction punches holes in the total wavefunction, giving it somewhat the character of Swiss cheese. By Fourier-transforming the "hole" or "defect function" of the two-body correlations, one finds that the typical intermediate-state momenta ("particle" momenta) are of order $(\pi/2c)$, where c is the hard-core radius. For ordinary nuclear matter this quantity is quite large—about 2.5 k_F. This strongly indicates a lack of physical symmetry between the virtual particles and holes of which the ground-state correlations are composed. The formal consequences of this asymmetry are analyzed in

Brandow (1966b, 1969), where the advantages of the Brueckner–Goldstone approach are described in detail. Nevertheless, after the hard-core aspect has been taken care of by Brueckner–theoretic techniques, it may well be appropriate to use field-theoretic techniques to deal with the correlations near the Fermi surface (see the preceding references concerning the Landau theory).

I. Potential Applications

Most applications to date for this open-shell formalism have been in the field of nuclear physics. Excellent surveys may be found in a recent conference proceedings (Barrett, 1975) and in a review by Ellis and Osnes (1977). Quantitative applications in quantum chemistry are still few and mostly of recent origin (see Kaldor, 1973, 1975, 1976; Garpman *et al.*, 1975; Stern and Kaldor, 1976), although some earlier calculations of electric dipole transitions (Kelly, 1969) are closely related. In a phenomenological sense, however, model Hamiltonians are routinely used for many types of systems, and also model operators to a lesser extent. The following are some well-known examples: (1) the nuclear shell model; (2) the Slater–Condon parametrization in atomic spectroscopy; (3) the Landau description of liquid ^{3}He at low temperatures, (4) the BCS Hamiltonian in superconductivity theory; (5) use of the known phonon–roton spectrum in calculating other properties of liquid ^{4}He, (6) effective π-electron Hamiltonians for planar unsaturated hydrocarbon molecules, and (7) for magnetic ions in insulating materials—crystal-field theory, effective g-factors in magnetic resonance experiments, and effective spin Hamiltonians of the Heisenberg type. At least in principle, the present formalism should be applicable to all of these types of model Hamiltonians.

In Section IV we emphasized that although the linked-cluster perturbation formalism can be regarded as the most complete and general expression of the linked-cluster viewpoint, it may not be the most useful channel for feeding in physical intuition in order to devise a good first approximation for a particular type of system. For the open-shell case, an example of a good approximation devised by alternative methods is the work of Westhaus and Sinanoğlu (1969) on atomic dipole transition strengths. As in Section IV, we should emphasize that the chief virtue of the present formalism may in some cases be its rigor and generality, as needed for calculations where the goal is to proceed *beyond* the appropriate first approximation. Another important virtue is the capability of dealing with excited states (Kaldor, 1973, 1975; Stern and Kaldor, 1976) and their various transition amplitudes. These states are generally of open-shell form, even for systems with closed-shell ground states. Finally, it is significant that this formalism can cope with a large and even macroscopic number of valence particles, as illustrated in the following section.

VII. Formal Derivation of Effective Spin Hamiltonians

As a specific application of the open-shell formalism, we shall now show how this can be used to resolve a very old many-body problem in the theory of magnetic insulator materials. Moreover, it appears to us that the following approach should offer some insight into the empirical success of π-electron Hamiltonians for unsaturated hydrocarbon molecules.

Soon after the classic work of Heitler and London (1927) on the singlet–triplet energy difference in the H_2 molecule, attempts were made to apply their technique to a macroscopic system of hydrogen atoms arranged on a simple lattice—this being the simplest prototype for a magnetic insulator material. It was soon noted, however (Slater, 1930; Inglis, 1934), that this approach led to a mathematical divergence problem, the so-called nonorthogonality catastrophe. Much effort has been devoted to this problem (see the review of Herring, 1966), but prior to our study of the Mott insulator problem (Brandow, 1976, 1977), for which the following constitutes an important ingredient, no fully satisfactory formal solution had been found.

A. Two Nonorthogonality Catastrophes

We consider a system of N hydrogen atoms arranged on a lattice. The Heitler–London (HL) approach is to construct an antisymmetrized product wavefunction Ψ_{HL} out of atomic hydrogen orbitals, one of these orbitals being centered on each of the N sites. Since there are two possible spin assignments for each site (spin up or down), there are 2^N different Heitler–London wavefunctions for the system. These Ψ_{HL}'s have a one-to-one correspondence with the 2^N Ising configurations of an N-spin system. Ideally, one would then approximate the exact spin eigenstates as various linear combinations of these 2^N Ψ_{HL}'s. The object, of course, is to establish that the resulting Ritz variational problem (for this set of 2^N basis functions) is equivalent to the use of some effective spin Hamiltonian. A more modest program is to focus on individual Ψ_{HL}'s, calculating the expectation value of H for each one of these determinantal functions, and then to demonstrate a close correspondence between (1) the small energy differences among these expectation values, and (2) the corresponding energy differences predicted by a simple Heisenberg spin Hamiltonian,

$$\mathscr{H}_{\text{Heis}} = -\sum_{ij} J_{ij}\, \mathbf{S}_i \cdot \mathbf{S}_j\,. \tag{88}$$

The problem encountered in this latter approach is that every one of these N-electron expectation values has the form of a quotient of two series involving very high powers of N,

$$\frac{\langle \Psi_{HL} | H | \Psi_{HL} \rangle}{\langle \Psi_{HL} | \Psi_{HL} \rangle} = \frac{a_1 N + a_2 N^2 + a_3 N^3 + \cdots}{1 + b_1 N + b_2 N^2 + \cdots}. \tag{89}$$

One expects on physical grounds (Section IV) that the result must be essentially just $a_1 N$, together perhaps with some small correction terms (surface effects) which are of order N^0, N^{-1}, N^{-2}, etc. In other words, all of the higher powers of N ought to cancel each other identically, except perhaps for the small corrections just mentioned. It is important, however, to have a formal proof of this cancellation, because, although the higher coefficients a_2, a_3, etc. and b_1, b_2, etc., are certainly small compared to unity, they are vastly overwhelmed by the macroscopic size of $N(\sim 10^{22}$, say) for typical experimental samples of magnetic insulator materials. This may be called the *first* nonorthogonality catastrophe. Clearly, this is just the problem of properly normalizing the various Ψ_{HL}'s. It arises because the N atomic orbitals for a given Ψ_{HL} are not all mutually orthogonal; all orbitals with the same spin must have small but nonvanishing overlaps.

It has been recognized for many years that this particular problem can easily be avoided by suitably applying the symmetrical orthogonalization technique (Löwdin, 1950, 1956) for converting nonorthogonal orbitals into an orthogonal orbital basis. The trick is to apply this technique *separately* to the sets of up-spin and down-spin orbitals in each given Ψ_{HL}, so that a pair of orbitals with *different* spins (one up and one down) need not be space-orthogonal (Carr, 1953). [A *fully* spaceorthogonal orbital basis would necessarily produce ferromagnetic spin coupling (Slater, 1951), in contrast to the fact that nearly all magnetic insulator materials are antiferromagnetic, thus one should not orthogonalize any more than the minimum required here. The remaining lack of space orthogonality is now recognized to be the main source of the antiferromagnetic spin coupling (Anderson, 1963).]

Unfortunately, however, there is a *second* and far more intractable nonorthogonality catastrophe. This arises from the fact that, even when properly normalized, the various Ψ_{HL}'s are not all mutually orthogonal. For each value of S_z, all Ψ_{HL}'s with this conserved quantum number will have small but nonvanishing overlaps. This means that in the Ritz problem described above, the many-body spin eigenstates are determined by a secular equation of the form

$$|\mathscr{H}_{\text{HL}} - EM| = 0, \tag{90}$$

where the metric M is nondiagonal. ($\mathscr{H}_{\text{HL}}$ is the matrix of H in the Ψ_{HL} basis.) This contrasts with the usual idea of an effective spin Hamiltonian $\mathscr{H}_{\text{spin}}$, which is intended for use with a simple diagonal metric,

$$|\mathscr{H}_{\text{spin}} - EI| = 0. \tag{91}$$

There is an obvious way to convert Eq. (90) into the latter form, namely, by employing the matrix transformation

$$\mathscr{H}_{\text{spin}} = M^{-1/2}\mathscr{H}_{\text{HL}}M^{-1/2}. \tag{92}$$

The problem here is that the dimensionality of 2^N is so astronomically large that, in practice, one cannot hope to carry out this transformation explicitly. It is not much help that the various matrices can all be block-diagonalized according to S_z, since the dimension of a typical block is still of order $2^N/N$. The detailed relation between $\mathscr{H}_{\mathrm{HL}}$ and $\mathscr{H}_{\mathrm{spin}}$ is, therefore, subject to a large and potentially catastrophic uncertainty.

Arai (1964, 1966, 1968) has made a very determined attack on this problem, by developing special linked-cluster techniques to deal with the multitude of orbital-overlap factors arising in the Heitler–London expressions. He eventually succeeded in obtaining linked-cluster expressions for *all* of the matrix elements of an $\mathscr{H}_{\mathrm{spin}}$, thus demonstrating that these elements are all well-behaved with respect to N (Arai, 1968). This certainly constitutes a resolution of the nonorthogonality catastrophe, in the sense of showing that none of the matrix elements behave catastrophically. However, the matrix transformation used in this latter work was far less symmetrical than that of Eq. (92). As a consequence, Arai was unable to demonstrate that the effective spin couplings are rotationally invariant ($\sim \mathbf{S}_i \cdot \mathbf{S}_j$), a feature one expects for systems where spin-orbit coupling can be ignored. A number of earlier and less successful efforts have been reviewed by Herring (1966).

B. Choice of H_0

Rather than pursue the Heitler–London approach, which is inherently a rather crude approximation anyway (Herring, 1966), we shall apply our present degenerate perturbation theory. This will lead to a many-body generalization of the exact perturbative form of the superexchange theory (Anderson, 1963). (Anderson's treatment is limited to an isolated pair of magnetic cations, thus avoiding the problems of large N. He has also presented an inexact but more intuitive version, based on Hartree–Fock approximations.) It turns out that the main problems we are now faced with are those concerning the selection of an appropriate form of H_0.

For simplicity, most of our discussion will refer to the same sort of hydrogenic lattice systems considered in the previous Heitler–London studies. The method can easily be generalized to deal with realistic systems (Brandow, 1977).

1. *Localized Orbitals*

The first problem is that we need a basis in which at least some of the orbitals are localized. These orbitals should be as nearly "self-consistent" as possible, in the sense of Hartree–Fock theory for example, in order to obtain a rapidly converging expansion as well as to have reasonably simple physical interpretations for the various terms of the perturbation series. The difficulty here is that the ordinary HF orbitals are necessarily nonlocalized. That is, the *canonical* HF eigenfunctions, which satisfy equations of the usual eigenvalue form,

$$\mathscr{F}\psi_k = \varepsilon_k \psi_k , \tag{93}$$

are necessarily of the Bloch instead of the Wannier form (Brandow, 1975b). Here $\mathscr{F}$ is the usual Fock operator, including the kinetic, ionic, direct, and exchange terms. We shall resolve this problem by means of the "ferromagnetic" prescription of Anderson (1963).

To complete the definition of $\mathscr{F}$, we assume (for the moment) a complete ferromagnetic spin alignment, where all of the electrons have spin up, for example, so that the spin-up 1s band is full. Assuming that the resulting Hartree–Fock problem has been solved self-consistently, we then make a unitary transformation to the corresponding Wannier basis φ_n, where n is a site label. In terms of these localized orbitals, Eq. (93) becomes

$$\mathscr{F}\varphi_n = \sum_{n'} \lambda_{n'n} \varphi_{n'} , \tag{94}$$

$$\lambda_{n'n} = \langle \varphi_{n'} | \mathscr{F} | \varphi_n \rangle. \tag{95}$$

We then convert Eq. (94) to a "pseudocanonical" form by transferring all of the off-diagonal λ's to the left-hand side, together with suitable orbital transfer operators:

$$H_{01}\varphi_{n\nu} \equiv \left[\mathscr{F} - \sum_{ll'\nu'}' | \varphi_{l'\nu'} \rangle \lambda_{l'l\nu'} \langle \varphi_{l\nu'} | \right] \varphi_{n\nu} = \lambda_{nn\nu} \varphi_{n\nu} , \tag{96}$$

where the primed summation indicates omission of the $l' = l$ terms. A band index ν' has been added so that the Wannier functions from certain of the higher Bloch bands, as defined by the original Hartree–Fock equations (93), can also be generated by this one-body H_{01} term. [It turns out that even Eq. (96) may be regarded as a canonical form for the HF equations, from a novel point of view (Davidson, 1972; Domany *et al.*, 1974).]

Observe now that the *spin-up* orbitals generated by the $\mathscr{F}$ of Eq. (93) constitute a complete set of spatial functions, when all of the higher Bloch bands are considered. We are only interested in these spin-up orbitals; the corresponding spin-down orbitals are to be disregarded. (The latter have somewhat different spatial forms, since they do not experience an exchange potential.) These spin-up spatial functions are then to be used to define a complete spin-orbital basis; these same functions are to be used for the spin-down orbitals $\varphi_{n\nu\beta}$ as for the spin-up orbitals $\varphi_{n\nu\alpha}$. The resulting basis is said to be "spin-independent," since the corresponding up- and down-spin orbitals now have precisely the same spatial forms. This feature introduces some departure from full Hartree–Fock self-consistency, but that can be taken care of later by the resulting perturbation series. So far, we have simply followed the prescription of Anderson (1963), although Eq. (96) does not appear explicitly in his work.

At this point we must recognize the difficulty of defining Wannier functions for the higher Bloch bands, especially for a three-dimensional system. Beyond some minimum Bloch eigenvalue, *all* of the higher Bloch bands will be overlapping in energy. That is, there will no longer be any true band gaps. However, studies by Kohn (1959, 1973) and des Cloizeaux (1964) have revealed that the problem of obtaining reasonably well-localized Wannier functions is fraught with grave difficulties unless the corresponding Bloch band (or set of several bands) is bounded above, as well as below, by a true band gap. This means that the v' index in Eq. (96) can, in practice, only be summed over a few (perhaps only one) of the low-energy Bloch bands. This is not a serious obstacle, however, since the lack of $\lambda_{l'l v'}$ terms for the higher bands simply means that H_{01} generates Bloch instead of Wannier basis functions for these bands. The resulting basis is still complete. The maximum number of bands for which the Wannier representation is practical, via Eq. (96), will naturally depend on the lattice spacing of the system. All that is truly essential for the present scheme is that the *lowest* band (the 1s band) must be treatable in this manner. In other words, there *must* be a gap above the 1s band. But we are assured that this is actually the case here, since, by assumption, we are dealing with a magnetic insulator. In the case of actual magnetic insulator materials, the insulating gap guarantees that all of the occupied bands can be represented by Wannier functions. This is more than sufficient for our purposes, since the only essential requirement is that the "magnetic" electrons (those responsible for the local moments) can be so represented. For the remaining bands (lower, if any, as well as higher), the use of Wannier functions is merely a convenience.

2. *Elimination of Unphysical Degeneracy*

At this stage, there is a complete degeneracy among all possible ways of distributing the N electrons among the $2N$ Wannier orbitals of the 1s band. We must now remove some of this degeneracy by adding a two-body term to H_0, to raise the energy by an amount U_{nn} for each doubly occupied site in the N-body configuration Φ_i. This U_{nn} is the single-site member of the set of Coulomb repulsion elements

$$U_{nn'} = \left\langle \phi_{ns}\phi_{n's} \left| \frac{e^2}{r_{12}} \right| \phi_{ns}\phi_{n's} \right\rangle. \tag{97}$$

The desired degeneracy-breaking term can be written in second-quantized notation as

$$\begin{aligned} H_{02} &= \sum_n a^\dagger_{ns\alpha} a^\dagger_{ns\beta} U_{nn} a_{ns\beta} a_{ns\alpha} \\ &= \sum_n U_{nn} \hat{n}_{ns\alpha} \hat{n}_{ns\beta}, \end{aligned} \tag{98}$$

where $\hat{n} = a^\dagger a$ is the number operator. Finally, we take

$$H_0 = H_{01} + H_{02} \tag{99}$$

and

$$V = H - H_0 . \tag{100}$$

C. Results and Discussion

The model subspace D is now defined by the condition that all of the electrons are in 1s orbitals, and none of the sites are doubly occupied. The dimension of this D is 2^N, in agreement with the Heitler–London model. Unlike the Ψ_{HL}'s, however, the present model Φ_i's are completely orthogonal, thus avoiding the problem of Eqs. (90)–(92). Since there are no "core orbitals" for the present system, all N of the electrons must be treated as valence electrons. The diagrams for $\mathscr{V}_{\mathrm{v}}$ should, therefore, all involve N external lines, a feature which is utterly impractical for the present case. Fortunately, however, in the linked expansions for $\mathscr{W}_{\mathrm{v}}$ and $\mathscr{K}_{\mathrm{v}}$ it is possible to ignore all of the noninteracting valence lines.

The leading diagrams that contribute to the effective two-site spin interaction are shown in Fig. 18. Diagrams a and b + c represent, respectively,

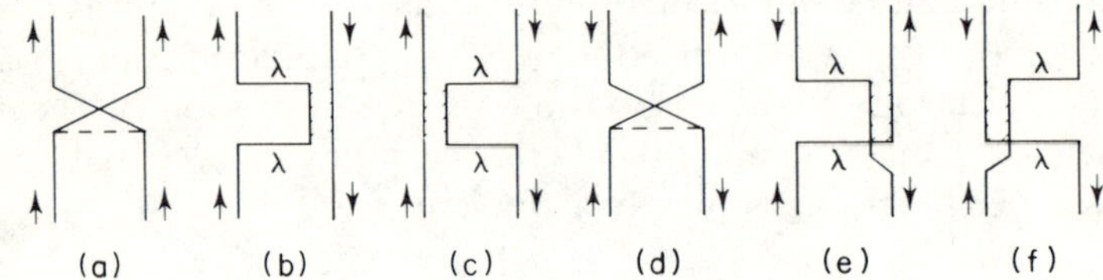

Fig. 18. Leading diagrams for the two-site term $J_{ij}\mathbf{S}_i \cdot \mathbf{S}_j$ of the effective spin Hamiltonian.

the *potential exchange* and *kinetic exchange* mechanisms of Anderson (1963). These diagrams contribute to the difference between the diagonal two-body matrix elements,

$$J_{ij} = \langle i\uparrow j\uparrow | \mathscr{H}_{\mathrm{spin}} | i\uparrow j\uparrow \rangle - \langle i\uparrow j\downarrow | \mathscr{H}_{\mathrm{spin}} | i\uparrow j\downarrow \rangle, \tag{101}$$

for a given pair of sites ij. Diagrams d and e + f show the corresponding contributions to the off-diagonal matrix element

$$J'_{ij} = \langle i\downarrow j\uparrow | \mathscr{H}_{\mathrm{spin}} | i\uparrow j\downarrow \rangle. \tag{102}$$

The diagrammatic rules (Brandow, 1967) show immediately that $J'_{ij} \equiv J_{ij}$, which confirms the rotational invariance ($\mathbf{S}_i \cdot \mathbf{S}_j$ character) of $\mathscr{H}_{\mathrm{spin}}$, at least to the order shown. One can easily see, however, from the rotational invariance of operator H_0 (and thus of $V = H - H_0$), that this identity must hold for *all* orders in V. (In fact this identity applies, separately, to the set of terms for each order in V.) This is a significant result. However, the most important result is simply that the N-site effective spin Hamiltonian has been formally decomposed into a sum of two-site terms, each of which is identical (to the extent that H_0 and V are unchanged) to the effective spin interaction for an *isolated* pair of spins, i.e., for an $N = 2$ system. Although this result is intuitively obvious, a proper formal justification has not been found previously. [Herring (1966) had established this result only for the asymptotic case of widely separated atoms.] In higher orders of the perturbation expansion, one will find terms describing corrections to the basic Heisenberg form, Eq. (88), namely, effective interactions between four spins, six spins, etc. This is to be expected on physical grounds (Herring, 1966).

The conventions in Fig. 18 require some explanation. By construction, the $(-H_0)$ term in $V = H - H_0$ does not produce any transition from $\phi_{is\alpha}$ to $\phi_{js\alpha}$, but the *full* H does have this effect, with a strength just equal to Eq. (95). This explains the λ's in the second-order diagrams of Fig. 18. Quantity U_{nn} of Eqs. (97) and (98) enters into the energy denominators of the expansion, but for the purpose of distinguishing between linked and unlinked diagrams this must be represented as if it were a two-body interaction. We have represented these "H_{02} interactions" by the shaded regions seen in Fig. 18. On physical grounds one expects the denominators for the kinetic exchange process to be of the form $U_{ij} - U_{ii}$, instead of just the $(-U_{ii})$ denominators indicated in Fig. 18. The expected correction is generated by a partial summation of higher order perturbation terms, namely, ladders of "particle–hole" interactions $(-U_{ij})$. Moreover, one knows empirically that the appropriate value of U_{nn} for the kinetic exchange expression is significantly smaller than the Hartree–Fock value given by Eq. (97). Terms describing correlation and screening-polarization corrections to U_{nn} can be identified among the higher order diagrams. It appears to us that the task of developing quantitatively reliable calculations for the overall effective reduction of U_{nn} is the foremost basic challenge remaining in this field (Brandow 1976, 1977).

Beginning in third order, one will also encounter the folded diagrams generated by Eqs. (74)–(77). We have in mind the $\mathscr{H}_v$ form of the effective interaction, for which these folded terms occur in a manner consistent with overall Hermiticity. It seems unlikely, however, that these folded diagrams could be of any quantitative significance, at least not in the near future, in view of the present state of the art (Freeman, 1973) for *ab initio* calculations of J_{ij}. [On the other hand, the folded diagrams may well be of significance for the *cohesive* energy, i.e., for the part of the correlation energy which does not contribute to the magnetic energy differences, Eq. (101). Anderson (1963) has provided a general diagrammatic recipe for distinguishing between the cohesive and the magnetic perturbation terms, the trick being to focus attention on the off-diagonal elements, Eq. (102).]

Rather similar formal many-body treatments of $\mathscr{H}_{\text{spin}}$ have been presented earlier by Bulaevskii (1967), and Klein and Seitz (1973). Bulaevskii was unable to obtain a fully linked result, apparently because of some confusion concerning the proper diagrammatic treatment of the "H_{02} interactions" mentioned above. Klein and Seitz started with a Hubbard (1963) model Hamiltonian, instead of the full electronic Hamiltonian, and employed a less flexible diagrammatic convention. They calculated through sixth order in the nearest-neighbor transfer integral (our $\lambda_{n'n}$) and explicitly demonstrated the existence of effective four-site spin interactions in $\mathscr{H}_{\text{spin}}$. Neither of these investigations considered the problem of defining the basis orbital functions $\phi_{n\nu}$ in a reasonably optimum manner.

A very important virtue of the present formalism is that it is sufficiently flexible and general to easily handle the added complications of actual magnetic insulator materials. These complicating features include the following: orbital degeneracy of d or f electrons; partial lifting of this degeneracy by the crystal field; effects of anions and of polarizations of the closed inner cation shells; presence of more than one magnetic electron per site [implying $(2S + 1) > 2$ model states per cation]; and effects of spin-orbit coupling and crystal-field degeneracy, as encountered for example with Co^{2+}, Fe^{2+}, and V^{3+} ions, together with the rare earths and actinides. Moreover, for orbitally degenerate systems, the one-site terms of the present formalism automatically incorporate the crystal-field (ligand-field) aspects, including the exciton aspects (see, for example, Hubbard *et al.*, 1966). The necessary formal generalizations are discussed elsewhere (Brandow, 1977).

D. Effective π-Electron Hamiltonians

Quantum chemists should recognize that the present spin Hamiltonian problem is closely related to the problem of deriving effective π-electron Hamiltonians. [For reviews of the latter problem, the reader may consult previous articles in the present series by Lykos (1964), I'Haya (1964), and Ohno (1967), as well as the book by Parr (1964).] The main formal difference between these problems lies in the choice of model subspace D. We can use the present system to illustrate this point: if the H_{02} term, Eq. (98), is transferred from H_0 to V, the dimension of D is enlarged from 2^N to $(2N)!/(N!)^2$, the number of ways of distributing N electrons among $2N$ orbitals. For an appropriate (small) value of N, the resulting model system is then an s-electron analog of a typical π-electron model system. This s-electron system differs from the model of Hubbard (1963) chiefly through the presence of Coulomb repulsion terms, Eq. (97), for $n' \neq n$. [For a qualitative discussion of the conditions for which the reduced dimensionality of 2^N is relevant for the low excited states, see Brandow (1976).]

The present formalism appears very well suited to explain one of the most striking features of π-electron phenomenology—the empirical fact that the various Hamiltonian parameters depend mainly on just the local environment within the molecule, such that each parameter is transferrable from one molecule to another containing a similar local environment. The present localized-orbital basis is very similar to the use of Löwdin-orthogonalized atomic orbitals (see the discussion of Ohno, 1967). This means that the localization feature becomes apparent in the lowest order of the perturbation series, where the basic structure of the π Hamiltonian is already manifest. In the higher orders, one will automatically encounter the effects due to polarization of the σ electrons, as well as correlations between π electrons that are not mediated by the σ's. All of the parameters will be affected by these higher order terms, the transfer integrals (our off-diagonal λ's) as well as the one- and two-site Coulomb repulsion terms. The influence of the σ electrons is, therefore, included properly, even though the *apparent* π-σ separability is built into the structure of this scheme. The numbers of π and σ electrons need not be conserved, of course, in the various intermediate states that enter into $\mathscr{H}_{\mathrm{v}}$.

Other virtues of the present approach are (*1*) the clean separation of the solution of the model Hamiltonian (the usual configuration-interaction problem) from the calculation of the model parameters themselves, and (*2*) the linked-cluster feature, which makes it possible to focus attention on any single one of the model parameters, and thus to calculate each one independently of the others. We therefore suggest to investigators in this field that the present approach deserves serious consideration. An entry to the literature of previous *ab initio* studies may be found in a recent paper by Westhaus *et al.* (1975). Most of these efforts are based on other types of formalism, especially the canonical transformation version of degenerate perturbation theory. This is formally equivalent to the present approach, but for high-order perturbation terms (which may be needed for partial summations), the transformation approach turns out to be less convenient (Brandow, 1975a). One previous study (Kvasnicka *et al.*, 1974, 1975) has, however, advocated essentially the present approach, although the authors were unaware of the formally exact Hermitian version $\mathscr{H}_{\mathrm{v}}$ of the linked-cluster perturbation expansion. Another recent work (Iwata and Freed, 1976a) involves an algebraic instead of diagrammatic development of Rayleigh–Schrödinger perturbation theory. This also obscures the linked-cluster feature, as in the canonical transformation approach; the authors have apparently verified this feature only in the lowest nontrivial order (third order in energy). Another feature of their work is that no valence-particle terms appear in the energy denominators; all terms involving valence particles are treated as part of the basic perturbation V. We believe that this must significantly reduce the rate of convergence.

A fundamental problem for the *ab initio* determination of *any* effective many-body Hamiltonian is that any reasonably systematic formalism will generate a number of additional terms, such as effective three-body interactions, which have no counterpart in the types of effective Hamiltonians used to date. Much further study will be needed to elucidate the effects of these "nonclassical" terms and the extent to which their effects can or cannot be incorporated by adjustments of the usual parameters. An interesting discussion of this problem has been given by Iwata and Freed (1976b), for π-electron Hamiltonians. This problem has been discussed previously (Brandow, 1967, p. 799) in connection with the Landau theory.

Acknowledgments

I am grateful to the organizers of the 1976 Sanibel Symposia for the opportunity to lecture on the present subject matter during the "workshop" session. The present review originated from the notes for these lectures. I wish to thank several of the participants, especially Yngve Öhrn, Janos Ladik and Paul Westhaus, for helpful discussions about π-electron Hamiltonians, and also to thank Uzi Kaldor, Paul Ellis, and John Zabolitsky for reviewing the manuscript.

References

Ambegaokar, V. (1969). *In* "Superconductivity" (R. D. Parks, ed.), Vol. 2, p. 1359. Dekker, New York.
Anderson, P. W. (1963). *Solid State Phys.* **14**, 99.
Arai, T. (1964). *Phys. Rev.* **134A**, 824.
Arai. T. (1966). *Prog. Theoret. Phys.* **36**, 473.
Arai, T. (1968). *Phys. Rev.* **165**, 706.
Babu, S., and Brown, G. E. (1973). *Ann. Phys. (N.Y.)* **78**, 1.
Baker, G. A. (1971). *Rev. Mod. Phys.* **43**, 479.
Barr, T. L., and Davidson, E. R. (1970). *Phys. Rev. A* **1**, 644.
Barrett, B. R., ed. (1975). "Effective Interactions and Operators in Nuclei." Springer-Verlag, Berlin and New York.
Bethe, H. A. (1965). *Phys. Rev.* **138**, B804.
Bloch, C. (1958). *Nucl. Phys.* **7**, 451.
Bloch, C., and Horowitz, J. (1958). *Nucl. Phys.* **8**, 91.
Brandow, B. H. (1966a). *Proc. Int. Sch. Phys. "Enrico Fermi"* **36**, 496.
Brandow, B. H. (1966b). *Phys. Rev.* **152**, 863.
Brandow, B. H. (1967). *Rev. Mod. Phys.* **39**, 771.
Brandow, B. H. (1969). *Lect. Theor. Phys.* **11b**, 55.

Brandow, B. H. (1970). *Ann. Phys.* (*N.Y.*) **57**, 214.
Brandow, B. H. (1971). *Ann. Phys.* (*N.Y.*) **64**, 21.
Brandow, B. H. (1972). *Ann. Phys.* (*N.Y.*) **74**, 112.
Brandow, B. H. (1975a). *In* "Effective Interactions and Operators in Nuclei" (B. R. Barrett, ed.), p. 1. Springer-Verlag, Berlin and New York.
Brandow, B. H. (1975b). *Phys. Rev. B* **12**, 3464.
Brandow, B. H. (1976). *Int. J. Quantum Chem., Symp.* **10**, 417.
Brandow, B. H. (1977). *Adv. Phys.* (to be published).
Brown, G. E. (1971). *Rev. Mod. Phys.* **43**, 1.
Brown, G. E. (1972). "Many-Body Problems," p. 5. North-Holland Publ., Amsterdam.
Brueckner, K. A. (1955). *Phys. Rev.* **100**, 36.
Brueckner, K. A. (1959). *In* "The Many-Body Problem" (C. DeWitt, ed.), Dunod, Paris.
Bulaevskii, L. N. (1967). *Sov. Phys.—JETP* (*Engl. Transl.*) **24**, 154.
Carr, W. J. (1953). *Phys. Rev.* **92,** 28.
Čižek, J. (1966). *J. Chem. Phys.* **45**, 4256.
Coester, F. (1958). *Nucl. Phys.* **7**, 421.
Coester, F. (1969). *Lect. Theor. Phys.* **11b**, 157.
Coester, F., and Kümmel, H. (1960). *Nucl. Phys.* **17**, 477.
da Providencia, J. (1963). *Nucl. Phys.* **46**, 401.
Davidson, E. R. (1972). *J. Chem. Phys.* **57**, 1999 and 2005.
Davies, K. T. R., McCarthy, R. J., Negele, J. W., and Sauer, P. U. (1974). *Phys. Rev. C* **10**, 2607.
Day, B. D. (1969). *Phys. Rev.* **187**, 1269.
des Cloizeaux, J. (1964). *Phys. Rev.* **135**, A685 and A698.
Domany, E., Katriel, J., and Kirson, M. W. (1974). *Phys. Lett. A* **48**, 477.
Ellis, P. J. (1970). *Nucl. Phys. A* **155**, 625.
Ellis, P. J., and Osnes, E. (1977). *Rev. Mod. Phys.* (in press).
Feshbach, H. (1962). *Ann. Phys.* (*N.Y.*) **19**, 287.
Feynman, R. P. (1949). *Phys. Rev.* **76**, 749.
Frantz, L. M., and Mills, R. L. (1960). *Nucl. Phys.* **15**, 16.
Freed, K. F. (1971). *Annu. Rev. Phys. Chem.* **22**, 313.
Freeman, S. (1973). *Phys. Rev. B* **7**, 3960.
Garpman, S., Lindgren, I., Lindgren, J., and Morrison, J. (1975). *Phys. Rev. A* **11**, 758.
Gell-Mann, M., and Low, F. (1951). *Phys. Rev.* **84,** 350 (Appendix).
Glassgold, A. E., Heckrotte, W., and Watson, K. M. (1959). *Phys. Rev.* **115**, 1374.
Goldstone, J. (1957). *Proc. Roy. Soc., London, Ser. A* **239**, 267.
Heitler, W., and London, F. (1927). *Z. Phys.* **44**, 455.
Herring, C. (1966). *Magnetism* **2B**, 1.
Hubbard, J. (1957). *Proc. R. Soc. London, Ser. A* **240**, 539.
Hubbard, J. (1963). *Proc. R. Soc. London, Ser. A* **276**, 238.
Hubbard, J., Rimmer, D. E., and Hopgood, F. R. A. (1966). *Proc. Phys. Soc.* **88**, 13.
Hugenholtz, N. M. (1957). *Physica* (*Utrecht*) **23**, 481.
Huzinaga, S., and Arnau, C. (1970). *Phys. Rev. A* **1**, 1285.
I'Haya, Y. (1964). *Adv. Quantum Chem.* **1**, 203.
Inglis, D. R. (1934). *Phys. Rev.* **46**, 135.
Iwata, S., and Freed, K. F. (1976a). *J. Chem. Phys.* **65**, 1071.
Iwata, S. and Freed, K. F. (1976b). *Chem. Phys. Lett.* **38**, 425.
Kaldor, U. (1973). *Phys. Rev. Lett.* **31**, 1338.
Kaldor, U. (1975). *J. Chem. Phys.* **63**, 2199.
Kaldor, U. (1976). *J. Comput. Phys.* **20**, 432.
Kelly, H. P. (1963). *Phys. Rev.* **131**, 684.

Kelly, H. P. (1964a). *Phys. Rev.* **134**, A1450.
Kelly, H. P. (1964b). *Phys. Rev.* **136**, B896.
Kelly, H. P. (1969). *Adv. Chem. Phys.* **14**, 129.
Kirson, M. W. (1968). *Nucl. Phys. A* **115**, 49.
Klein, D. J. (1974). *J. Chem. Phys.* **61**, 786.
Klein, D. J., and Seitz, W. A. (1973). *Phys. Rev. B* **8**, 2236.
Kobe, D. H. (1969). *Phys. Rev.* **188**, 1583.
Kohn, W. (1959). *Phys. Rev.* **115**, 809.
Kohn, W. (1973). *Phys. Rev. B* **7**, 4388.
Kümmel, H. (1962). *In* "Lectures on the Many-Body Problem" (E. R. Caianiello, ed.), p. 265. Academic Press, New York.
Kümmel, H., Lührmann, K. H., and Zabolitsky, J. G. (1977). *Phys. Rep.* (in press).
Kuo, T. T. S., Lee, S. Y., and Ratcliff, K. F. (1971). *Nucl. Phys. A* **176**, 65.
Kvasnicka, V. (1975). *Phys. Rev. A* **12**, 1159.
Kvasnicka, V., Holubec, A., and Hubac, I. (1974). *Chem. Phys. Lett.* **24**, 361.
Langhoff, S. R., and Davidson, E. R. (1974). *Int. J. Quantum Chem.* **8**, 61.
Lee, T., Dutta, N. C., and Das, T. P. (1971). *Phys. Rev. A* **4**, 1410.
Lefebvre, R., and Moser, C. (1969). *Adv. Chem. Phys.* **14** (entire volume).
Lindgren, I. (1974). *J. Phys. B* **7**, 2441.
Löwdin, P.-O. (1950). *J. Chem. Phys.* **18**, 365.
Löwdin, P.-O. (1951). *J. Chem. Phys.* **19**, 1396.
Löwdin, P.-O. (1956). *Adv. Phys.* **5**, 1.
Löwdin, P.-O. (1962). *J. Math. Phys.* **3**, 969.
Lührmann, K. H. (1977). *Ann. Phys. (N.Y.)* **103**, 253.
Lykos, P. G. (1964). *Adv. Quantum Chem.* **1**, 171.
Manne, R. (1977). *Int. J. Quantum Chem., Symp.* (in press).
Miller, J. H., and Kelly, H. P. (1971). *Phys. Rev. A* **3**, 578, and *A* **4**, 480.
Morita, T. (1963). *Prog. Theoret. Phys.* **29**, 351.
Mukherjee, D., Moitra, R. K., and Mukhopadhyay, A. (1975). *Mol. Phys.* **30**, 1861.
Nesbet, R. K. (1958). *Phys. Rev.* **109**, 1632.
Nesbet, R. K. (1965). *Adv. Chem. Phys.* **9**, 321.
Nesbet, R. K. (1967). *Phys. Rev.* **155**, 51 and 56.
Nesbet, R. K., Barr, T. L., and Davidson, E. R. (1969). *Chem. Phys. Lett.* **4**, 203.
Oberlechner, G., Owono-N'-Guema, F., and Richert, J. (1970). *Nuovo Cimento B* **68**, 23.
Offermann, R., Ey, W., and Kümmel, H. (1976). *Nucl. Phys.* **A273**, 349.
Offermann, R. (1976). *Nucl. Phys.* **A273**, 368.
Ohno, K. (1967). *Adv. Quantum Chem.* **3**, 240.
Paldus, J., and Čižek, J. (1975). *Adv. Quantum Chem.* **9**, 105.
Paldus, J., Čižek, J., and Shavitt, I. (1972). *Phys. Rev. A* **5**, 50.
Parr, R. G. (1964). "The Quantum Theory of Molecular Electronic Structure." Benjamin, New York.
Pines, D. (1961). "The Many-Body Problem." Benjamin, New York.
Rajaraman, R., and Bethe, H. A. (1967). *Rev. Mod. Phys.* **39**, 745.
Sandars, P. G. H. (1969). *Adv. Chem. Phys.* **14**, 365.
Schaefer, L., and Weidenmuller, H. A. (1971). *Nucl. Phys. A* **174**, 1.
Schweber, S. (1961). "An Introduction to Relativistic Quantum Field Theory." Harper, New York.
Silverstone, H. J., and Yin, M.-L. (1968). *J. Chem. Phys.* **49**, 2026.
Sinanoğlu, O. (1962). *J. Chem. Phys.* **36**, 706 and 3198.
Sinanoğlu, O. (1964). *Adv. Chem. Phys.* **6**, 315.

Slater, J. C. (1930). *Phys. Rev.* **35**, 509.
Slater, J. C. (1951). *J. Chem. Phys.* **19,** 220.
Smith, V. H., and Kutzelnigg, W. (1967). *Arkiv Fysik* **38,** 309.
Stern, P. S., and Kaldor, U. (1976). *J. Chem. Phys.* **64**, 2002.
Thouless, D. J. (1958). *Phys. Rev.* **112**, 906.
Thouless, D. J. (1961). "The Quantum Mechanics of Many-Body Systems," 1st ed., p. 47. Academic Press, New York.
Westhaus, P., and Sinanoğlu, O. (1969). *Phys. Rev.* **183**, 56.
Westhaus, P., Bradford, E. G., and Hall, D. (1975). *J. Chem. Phys.* **62**, 1607.
Zabolitsky, J. G. (1976). *Phys. Rev. C* **14**, 1207.

Quantum-Mechanical Approach to the Conformational Basis of Molecular Pharmacology

BERNARD PULLMAN

Institut de Biologie Physico-Chimique
Fondation Edmond de Rothschild
Paris, France

I. Introduction 251
II. Histamine 259
III. Antihistamines 268
IV. Serotonin and Related Indolealkylamines 275
V. Phenethylamines and Phenethanolamines 281
VI. Acetylcholine 289
VII. Conformation of Acetylcholine in Solution 297
VIII. Acetylcholine Derivatives and Analogs 302
A. Acetylthiocholine (XXVI) 302
B. Derivatives Modified at the Cationic Head 303
C. Local Anesthetics Derived by a Modification of the Ester Terminal . . . 308
D. Anticholinergic Compounds 311
IX. Conclusions 314
Addendum 323
References 323

I. Introduction

Among the many factors considered as conditioning the pharmacological activity of drugs, a particularly prominent role is attributed to their conformation. In recent years the conformations of pharmacological molecules have been the subject of extensive studies both experimental and theoretical, of which quantum-mechanical molecular orbital investigations are the most numerous. Their advent became possible due to the elaboration of molecular orbital methods dealing simultaneously with all valence, σ and π, or even all (including inner-shell) electrons. With these methods, one is able to evaluate the total molecular energy corresponding to any given configuration of the constituent atoms and thus to choose the preferred ones. The most prominent among these methods are the extended Hückel theory (EHT) (Hoffmann, 1963, 1964), the iterative extended Hückel theory (IEHT) (Carroll *et al.*, 1966; Rein *et al.*, 1966), the complete neglect of

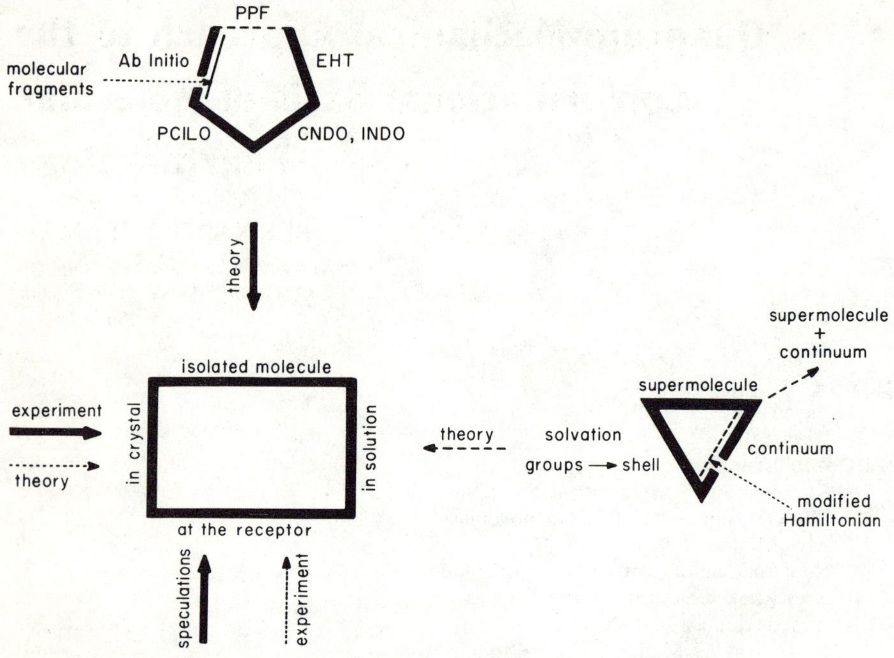

Fig. 1. The four-faced problem in conformational studies of pharmacological molecules and theoretical exploration of two phases. (PPF = partitioned potential functions.)

differential overlap (CNDO/2) method (Pople and Segal, 1965, 1966), the intermediate neglect of differential overlap (INDO) method (Pople *et al.*, 1967), the perturbative configuration interaction using localized orbitals (PCILO) method (Diner *et al.*, 1969), and the *ab initio* or nonempirical SCF procedure (Roothaan, 1951). Because these methods are all well known to the readers of these volumes, they will not be described again here. A general presentation of them has been given in recent reviews on their utilization for the study of the conformational properties of proteins and their constituents (B. Pullman and Pullman, 1974) and of nucleic acids and their constituents (B. Pullman and Saran, 1976).

The conformational properties of pharmacological compounds present us with a four-faced problem (Fig. 1), corresponding to the four principal conditions under which they are studied: the isolated molecule, the molecule in the crystal, the molecule in solution, and the molecule at the receptor. In the present stage of our knowledge there is a very uneven utilization of different methodologies in order to obtain the relevant information. The bulk of the computations refer, of course, to the isolated molecules, but recently important progress, of great potential significance for correlations with pharmacological activity, is taking place through the inclusion of the solvation effect. These investigations are being carried out by different groups of authors using different approaches: the supermolecular-microscopic model developed essentially by A. Pullman and co-workers (Alagona *et al.*, 1973; Port and Pullman, 1973a,b, 1974a; Pullman and Port, 1973a; Pullman and Armbruster, 1974; Pullman *et al.*, 1974; Pullman and Pullman, 1975; for a related "empirical" study, see Krimm and Venkatachalam, 1971; Venkatachalam and Krimm, 1973); the continuum model used essentially by Beveridge and co-workers (1974a,b); the solvation groups (shell) model developed by Hopfinger (1973; Weintraub and Hopfinger, 1974); and the modified Hamiltonian procedure of Hylton *et al.* (1974).

Possibilities also appear for a theoretical treatment in the near future of the crystal-packing effect. Different groups of authors are developing actively the general methodology in this field, and some applications are even available for biological molecules although not for pharmacological ones (e.g., Coiro *et al.*, 1971, 1972, 1973; McGuire *et al.*, 1971; Ferro and Hermans, 1972; Motherwell and Isaacs, 1972; Motherwell *et al.*, 1973; Kitaigorodsky, 1973; Momany *et al.*, 1974a,b; Gavuzzo *et al.*, 1974; Hagler and Lifson, 1974; Caillet and Claverie, 1974). On the contrary, very little is known about the conformations adopted by drugs on receptors, but from this point of view the theoretical methods are in a no worse (although no better either) state than the experimental techniques and will remain so until more is learned about the nature of the receptors. Here, too, however, if judged from recent progress in the study of substrate–enzyme interactions, developments may be expected in the not too remote future.

At present one of the most significant advances occurs, as mentioned above, with the inclusion of the solvent effect. Although biophase is not just water and although the microenvironment of a drug at the receptor may produce strong local effects, it seems probable that the conformations occurring in water may be considered as closer *precursors* to conformations linked with activity than are the conformations of the free molecule or of the molecule in the crystal. The knowledge of conformations in solution represents thus an important step forward toward the elucidation of the conformational dynamics of drugs and, possibly, the understanding of the role of conformations in pharmacological activity.

It may, therefore, be interesting to summarize here the essentials of some of the aforementioned techniques that are being used in the study of the conformation of pharmacological molecules in solutions and that are less well-known than the standard quantum-mechanical methods for studying the conformations of the free molecules.

The *supermolecular approach* to the solvation problem consists of fixing water molecules in the most favorable hydration sites of the solute and of evaluating the influence of this first hydration shell on the properties of the solute—in the present case, its conformation. The most favorable hydration sites are determined in general by *ab initio* studies on model compounds, and a large number of such determinations have been carried out by A. Pullman and co-workers (for general review, see Pullman and Pullman, 1975). The model tends to individualize a certain number of water molecules and hydration sites, and it is, therefore, particularly well adapted to describe the situation when strong interactions take place between specific groupings on the solute and the solvent ("bound" water). It implicitly assumes that it is this preferential hydration that has a dominant effect on the molecular conformation. Even its protagonists do not claim that this approach gives a complete account of the solution behavior of the solute. Rather the aim of such a treatment is to obtain a reasonable indication of the direction and magnitude of changes in conformational preferences of the isolated molecule when it enters aqueous solution, and it is expected then that the inclusion in the computations of the essential water molecules of the first hydration shell should be particularly significant.

In the *continuum* model, the total energy of a molecule in a given geometry in the presence of the solvent is to be considered in the form

$$E_{\text{total}}^{\text{eff}} = E_{\text{solute}} + E_{\text{solvation}}^{\text{eff}},$$

where E_{solute} is the total quantum-mechanical energy of a solute molecule in the free space approximation, and $E_{\text{solvation}}^{\text{eff}}$ is the effective solute–solvent interaction energy. The solvation energy is partitioned, empirically, into three distinct contributions:

$$E_{\text{solvation}}^{\text{eff}} = E_{\text{es}} + E_{\text{dis}} + E_{\text{cav}},$$

where E_{es} represents the electrostatic solute–solvent binding energy, E_{dis} is the solute–solvent binding energy due to dispersion forces, and E_{cav} is the energy required for formation of a cavity in the solvent to accommodate a solute molecule. Terms E_{es} and E_{dis} are negative, but E_{cav} is positive. Each of the terms depends parametrically upon quantities that vary as functions of molecular geometry.

The electrostatic term arises from the interactions of permanent and induced electric moments of the solute with the solvent. For the calculation of this term, the solute is treated as a point dipolar ion of charge Q and total dipole moment m at the center of a sphere of effective radius a imbedded in the solvent. The solvent is represented as a polarizable dielectric continuum of dielectric ε. The solute induces a reaction field E_{R} in the solvent, which acts back on the solute system. Generally, differential effects in the monopolar term are considered negligible and the relevant expression for E_{es} reduces to

$$E_{es} = -mE_R$$

$$E_R = \frac{2(\varepsilon - 1)m}{(2\varepsilon + 1)a^3}.$$

Both m and a depend on molecular geometry and are evaluated as a function of conformation from the molecular wavefunction and the molecular volume, respectively.

The solute–solvent interaction energy due to dispersion forces is estimated in the continuum model as

$$E_{dis} = \frac{Q}{2}\int_0^\infty V^{eff}(r)g^{(2)}(r)4\pi r^2\, dr,$$

where Q is the number density of the solvent, $V^{eff}(r)$ is an effective pairwise potential function for solute–solvent interactions, and $g^{(2)}(r)$ is a radial distribution function. The cavity term is estimated from the cavity area a and solvent surface tension γ as

$$E_{cav} = f4\pi a^2\gamma,$$

where f is a factor designed to relate the macroscopic to microscopic dimensions. Both E_{dis} and E_{cav} vary parametrically with geometry through the cavity size a. In calculating the various contributions to $E^{eff}_{solvation}$, macroscopic values of ε and γ are used.

More details about the procedure may be found in Beveridge *et al.* (1974a).

In *the solvation groups (shell) model*, as used by Hopfinger (1973), it is assumed that all solvent molecules distribute themselves about a solvation group or atom of the solute molecule so that they may be encapsulated in a hydration shell of radius R_v. Each atom or group will allow a maximum of n solvent molecules in the hydration shell. Of course, only a limited hydration shell volume will be available to each of the n solvent molecules. This available volume is $(V_s + V_f)$, where V_s is the steric volume of the solvent molecule, and V_f is the free volume of packing associated with one solvent molecule in the hydration shell. The van der Waals spheres of other atoms and/or groups in the solute molecule will, as a function of conformation, intersect the hydration shell and sterically force the solvent molecules from the hydration shell. If Δf is the change in free energy associated with the removal of one solvent molecule from the hydration shell for the solute molecule in conformation k, then the total change in free energy, $F(k)$, associated with the hydration shell is

$$F(k) = -\Delta f\left[n - \left(\frac{V(k)}{V_s + V_f}\right)\right], \qquad \text{when } V(k) < n(V_s + V_f),$$

$$= 0, \qquad \text{when } V(k) \geq n(V_s + V_f),$$

where $V(k)$ is the total volume of the intersection of the van der Waals spheres with the hydration shell.

The total molecule–solvent interaction free energy is obtained by simply taking the sum of the $F(k)$ for all hydration shell in the solute molecule.

Tables are available that give the values of the molecule–solvent interaction parameters n, Δ_f, R_v, and V_f for a number of groups or atoms in aqueous solution.

The *modified Hamiltonian* method has not as yet been applied to any molecular system of appreciable dimensions and will, therefore, not be described here.

It is evident that these different procedures for the inclusion of the solvent effect on molecular conformations represent different approximations for the treatment of a rather complex and difficult problem. We may consider them, in their present stage as theoretical experiments. They all suffer from obvious although different limitations. Which of them is the best adapted to the problem and gives the most promising results can be judged reliably by comparing the predictions with experimental data. Although only a few such confrontations are available, they are sufficient, as will become evident later in this paper, for at least preliminary conclusions in this respect.

We shall illustrate the precise nature of the problem and the principal features of the results that may be obtained by examples taken in the field of biogenic amines and related compounds. These substances have been the subject of research by a number of quantum chemists and constitute the *most fundamental* compounds in the field of molecular pharmacology. These amines may be described by the general formula of Fig. 2, where R_0 represents a conjugated (hetero)aromatic ring, e.g., R_0 = phenyl in phenethylamines, indole in indolalkylamines, and imidazole in histamines. (These different rings may, of course, carry different substituents.) The R_0 may also represent a conjugated group, e.g., the ester linkage in acetylcholine; R_1, R_2, and R_3 represent in a similar manner the H atoms or CH_3 (or higher) aliphatic groups. The H atoms on C_α and C_β may also be replaced by different groups. In spite of these structural variations, all these molecules correspond to the same basic pattern of Fig. 2. Some of them may, of course, also be considered in their neutral form with a terminal NH_2 or $N(CH_3)_2$

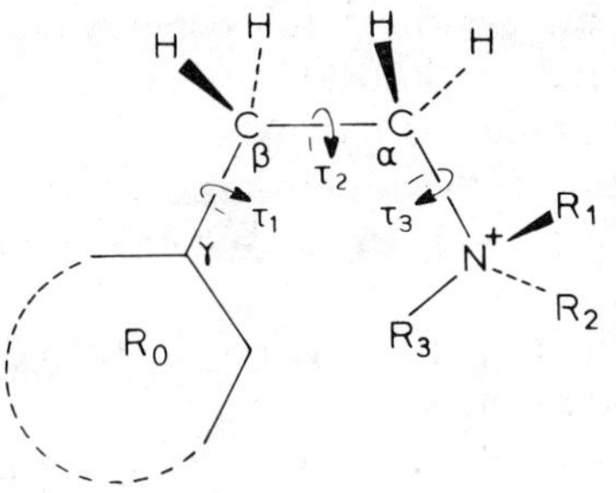

Fig. 2. General representation of biogenic amines. The principal torsion angles ($\tau_1 = \tau_2 = \tau_3 = 0°$).

group, and the conformational properties of such neutral forms have also been investigated in some cases. Although there is no reason to eliminate *a priori* their possible involvement in the microenvironment related to their interaction with a biological receptor, attention has, nevertheless, centered primarily on the cationic forms.

The essential conformational problem in this type of molecule, although not the only one (see below), concerns the overall orientation of the side chain with respect to the conjugated ring or fragment and is described by the two principal torsion angles τ_1 and τ_2 (Fig. 2). The first of these torsion angles defines the position of the plane of the side chain with respect to the plane of the ring, e.g., coplanar or perpendicular, and the second, the orientation of the cationic head with respect to the ring, e.g., trans or gauche. The torsion angles in the principal types of compound that we shall discuss will be defined more precisely later in this paper. Here we recall only that the torsion angle τ about the bond B—C in the sequence of atoms A—B—C—D is the angle through which the far bond C—D is rotated relative to the near bond A—B, the cis-planar position of bonds A—B and C—D, representing $\tau = 0°$ (Fig. 3). The torsion angles are considered positive for a right-handed rotation: when looking along the bond B—C, the far bond C—D rotates clockwise relative to the near bond A—B. Alternatively, the positive angles are defined as 0° to 180°, measured for a clockwise rotation, and negative angles as 0° to $-180°$, measured for a counterclockwise rotation. The theoretical results are generally presented in the form of

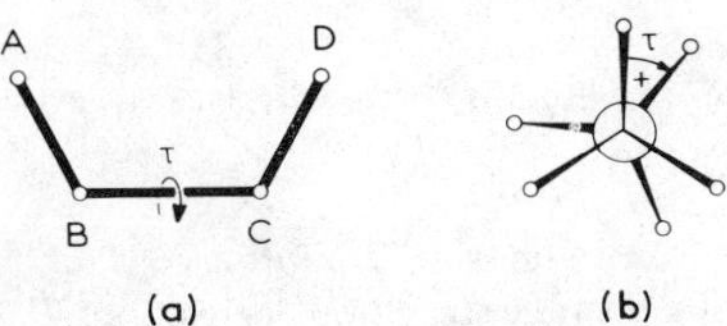

Fig. 3. The definition of a torsion angle: (a) choice of zero angle and (b) sense of rotation.

conformational energy maps indicating isoenergy curves as a function of the torsion angles considered (in kilocalories per mole) with respect to the global energy minimum taken as energy zero.

Besides the elucidation of the conformational properties of the individual compounds or groups of compounds studied, we aim also at some more general findings as follows.

1. Common conformational properties likely to occur in all or at least a number of molecules of this category.

2. A critical examination of certain proposals concerning *apparent* conformational analogies among different molecules of this type or between some representatives of these molecules and related biological or pharmacological molecules. An example of a particularly interesting problem in this respect concerns a series of divergent proposals that have been put forward in recent years in order to find structural analogies between the hallucinogenic drug lysergic acid diethylamide (LSD; I) and mescaline (II), a phenethylamine, or/and psilocin (III), an indolealkylamine. Some of the proposals stress the significance of the folded (Snyder and Richelson, 1968) but others that of the extended structures (Baker *et al.*, 1973a and b; Chothia and Pauling, 1969; Kang and Green, 1970; Kang *et al.*, 1973a) of II and III in this respect (Fig. 4). These proposals are, however, based on two-

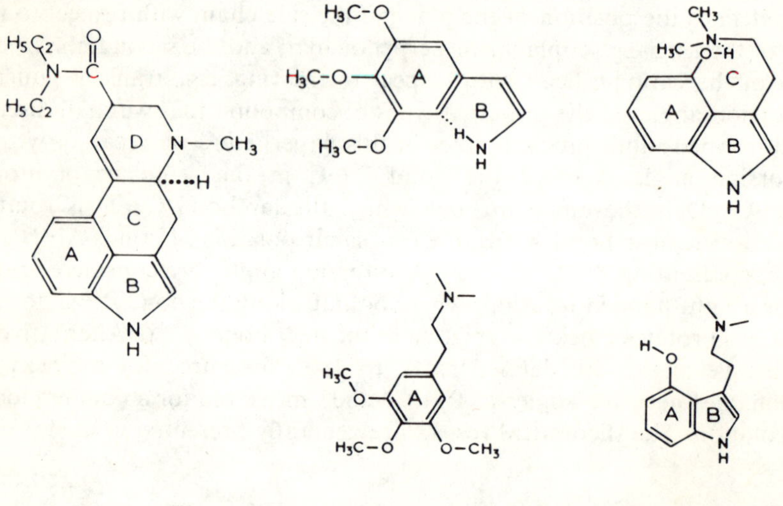

(I) D-Lysergic acid diethylamide (LSD)

(II) 3,4,5-Trimethoxyphenethylamine (mescaline)

(III) 4-Hydroxy-N,N-dimethyltryptamine (psilocin)

Fig. 4. Diagrammatic representations of conformations of lysergic acid diethylamide, mescaline, and psilocin, showing possible structural correlations. (Upper part following Snyder and Richelson, 1968; lower part following Baker *et al.*, 1973a and b, and Kang *et al.*, 1973a.)

dimensional drawings that may obviously be far removed from the three-dimensional reality.

3. A critical evaluation of the validity of the results obtained by the various theoretical methods. In this regard, the biogenic amines and related pharmacological drugs offer an almost unique opportunity because of the very large number of computations performed using nearly all the available quantum-mechanical procedures. Parallel, a critical comparison from the same point of view of the quantum-mechanical results with those obtained by the so-called empirical computations (based on partitioned potential functions) may also be carried out at least in some cases.

4. A critical appraisal of the significance of the theoretical results with respect to the available experimental, physicochemical data. In particular, this involves evaluation of the significance, from that point of view, of the theoretical results obtained for the isolated molecule with respect to the experimental reality. Concomitantly, it is necessary to appraise the proposed methodologies for the introduction of the solvent effect. Again, for this purpose, some of the biogenic amines represent practically the only examples for which results obtained by different procedures are available.

5. An appraisal, if possible, of the significance of the results for molecular pharmacology.

It is expected that such a multidirectional analysis will lead a step forward both in the appreciation of the value of the theoretical computational procedures and in the understanding of the overall conformational possibilities and preferences of the compounds studied. In this way, it will contribute to the elucidation of the conformational basis of molecular pharmacology. In addition, it should offer useful indications to the growing number of experimenters who try to employ the available quantum-mechanical procedures but who, not being specialists in the field, have difficulty in evaluating their reliability.

II. Histamine

We shall start this review by examining the case of histamine (Fig. 5) which encompasses the majority of typical problems encountered in biogenic amines. In the first place, the computation of the conformational properties of this compound raises the question of the validity of the various procedures used because different theoretical methods have led to strikingly different results.

The first molecular orbital calculation performed by Kier (1968a) using the extended Hückel theory for the monocation (Fig. 5a), which is the prevailing tautomeric (N_3—H also called N^{τ}—H) and ionic form of histamine at physiological pH (Ganellin, 1973a, 1974), indicated a slight preference for a trans conformation ($\tau_1 = 60°$, $\tau_2 = 180°$) with two secondary

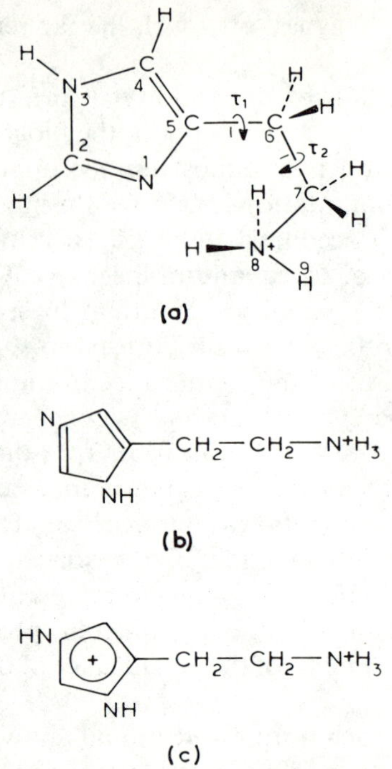

Fig. 5. Cations of histamine. (a) The predominant monocation at physiological pH showing torsion angles $\tau_1 = \tau\ (N_1{-}C_5{-}C_\beta{-}C_\alpha)$ and $\tau_2 = \tau\ (C_5{-}C_\beta{-}C_\alpha{-}N^+)$; (b) the rare monocation; (c) the dication.

gauche energy minima: one at $\tau_1 = 60°$, $\tau_2 = 60°$ at 0.5 kcal/mole above the global minimum and the second at $\tau_1 = 90°$, $\tau_2 = -60°$ at 1 kcal/mole above the global minimum, with the plane of the side chain always largely inclined with respect to that of the ring. These EHT results have been confirmed by Ganellin and collaborators (1973a,b,c) who evaluated at 55–60% the predicted population of the trans conformer in the equilibrium mixture. This last group of authors have also extended the EHT computations to the tautomeric monocationic form (Fig. 5b) ($N_1{-}H$, also called $N^\pi{-}H$) which they estimate to be present in the proportion of 20% at physiological pH (Ganellin, 1973a) and to the diprotonated form (Fig. 5c) which should prevail at low pH and for which they predict a slightly greater preference ($\approx 65\%$) for the trans form: global energy minimum at $\tau_1 = 70°$, $\tau_2 = 180°$, with again two gauche minima, one at $\tau_1 = 70°$, $\tau_2 = 60°$, at 0.5 kcal/mole above the global one and another at $\tau_1 = 110°$, $\tau_2 = -60°$ at 1.3 kcal above the global one.

Strikingly different results are obtained with the PCILO method (Coubeils *et al.*, 1971). The PCILO conformational energy map for the principal tautomeric form of the monocation, Fig. 5a, is reproduced in Fig. 6 (on which are also shown, for easy comparison, the above specified minima of the EHT computations). The results indicate a very neat global energy minimum for a gauche form with the plane of the side chain only moderately inclined with respect to that of the ring at $\tau_1 = 30°$, $\tau_2 = -60°$, which is strongly stabilized by a close approach between the negatively charged N_1 atom of the ring and the ammonium group. The trans form is calculated to

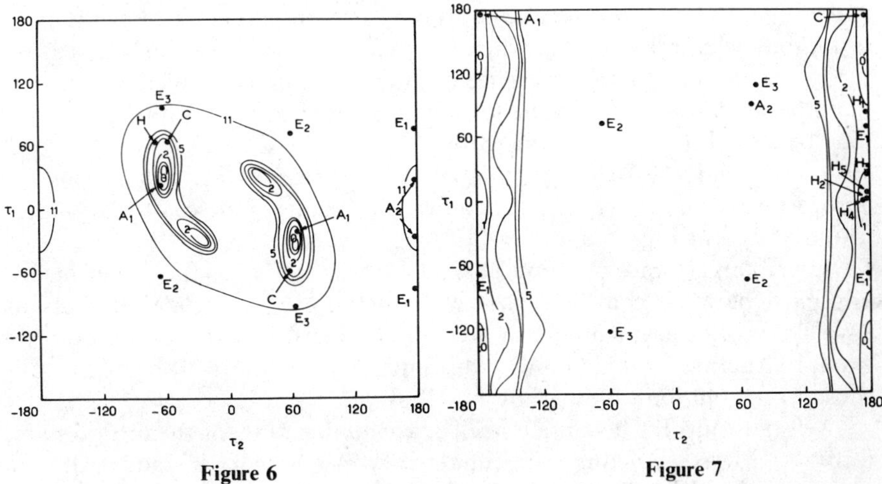

Fig. 6. The PCILO conformational energy map of the monocation (Fig. 5a) of histamine. Isoenergy curves in kilocalories per mole with respect to the global energy minimum considered as energy zero. E_1, E_2, E_3: energy minima of the EHT calculations in order of increasing energy. C = the global minimum of CNDO computations. A_1 = the global *ab initio* energy minimum of the gauche forms; A_2 = the local *ab initio* energy minimum of the trans forms, 21 kcal/mole above A_1. H = crystal conformation of histidine.

Fig. 7. The PCILO conformational energy map of the histamine dication (Fig. 5c). Isoenergy curves in kilocalories per mole with respect to the global energy minimum considered as energy zero. E_1, E_2, E_3: energy minima of the EHT calculations in order of increasing energy. C = the global minimum of the CNDO computations. A_1 = the global *ab initio* energy minimum of the trans forms, A_2 = the local *ab initio* energy minimum of the gauche forms, 8 kcal/mole above A_1. H_1–H_5: crystal conformations of the dictation of histamine.

be about 11 kcal/mole above this global minimum, and, therefore, the compound is expected to exist almost completely in the gauche form in its isolated state. The PCILO results for the diprotonated form (Fig. 5c), presented in Fig. 7, indicate just the opposite. They predict a very strong preference of the compound for a trans conformation with the global energy minimum at $\tau_1 = 120°$, $\tau_2 = 180°$ and a close local energy minimum at $\tau_1 = 0° \pm 30°$, $\tau_2 = 180°$ at 1 kcal/mole above the global one. In fact, when

$\tau_2 = 180°$, the whole range of the trans forms at all the values of τ_1 is enclosed by the 2 kcal/mole contour line, the stability of these forms being due to the minimization in this conformation of the repulsion between the two positive centers. The gauche forms are predicted to be about 8 kcal/mole higher in energy than the trans form, which may therefore be expected to represent essentially the isolated state of the compound.

It is thus evident that the two procedures, EHT and PCILO, lead to substantially different results. The first predicts the coexistence of gauche and trans forms for both cations, with a slight predominance of the trans form, somewhat greater in the dication than in the monocation, whereas the second predicts a strong preference of the two cations for different forms: of the monocation for the gauche form and of the dication for a trans form. It may be added that calculations carried out by the CNDO (Ganellin *et al.*, 1973a,b) and INDO (Margolis *et al.*, 1971) methods are *grosso modo* in agreement with the PCILO predictions. The results of these calculations are also depicted in Figs. 6 and 7.

Interestingly, and perhaps at first sight surprising, the experimental approach parallels the discrepancies found in the theoretical predictions. Two types of experimental data are available in this field: results from the crystal structure obtained by X-ray and neutron diffraction studies and those from solution obtained by NMR studies.

X-Ray results for histamine itself are available only for the diprotonated form which was studied in four different crystals, namely, histamine diphosphate monohydrate (Veidis *et al.*, 1969), tetrachlorocobaltate of histamine (Bonnet and Jeannin, 1972), histamine bromide (Decon, 1964), and histamine sulfate monohydrate (Yamane *et al.*, 1973) which contains two crystallographically independent molecules. The four crystals show a trans arrangement of the ammonium group relative to the ring ($\tau_2 = 180°$), but three different values for τ_1, which is approximately equal to 90°, 0°, and 30°, thus corresponding closely to the general predictions of the PCILO map (Fig. 7). There are no X-ray data available for the monocation of histamine.[1] However, the closely related histidine monocation has been abundantly investigated by several groups (Donohue and Caron, 1964; Bennett *et al.*, 1970; Candlin and Harding, 1970; Madden *et al.*, 1972a,b), and in this case the NH_3^+ group appears always, in both orthorhombic and monoclinic varieties of the crystal, to be gauche with respect to the ring (and the COO^-

[1] Recently (Prout *et al.*, 1974), X-ray results have been obtained for the crystal of histamine monohydrobromide, where it exists as a monocation. The molecule is found in this crystal in the extended form with $\tau_1 = 89.7°$ and $\tau_2 = 177.3°$. It seems highly probable that this is due to the packing forces. Molecules are linked together in pairs by two hydrogen bonds between the cationic head of each and the ring N_1 atom of its neighbor and, moreover, are subject to the perturbing influence of Br^- ions.

group trans), with $\tau_1 \approx 57°$ and $\tau_2 \approx -59°$, very close to the values for the global energy minimum of Fig. 6. This gauche arrangement corresponds to an intramolecular interaction between the N^+H_3 group and the lone-pair-bearing atom N_1 of the ring. Indeed, a recent neutron diffraction study, which indicates the positions of the hydrogen atoms (Lehmann *et al.*, 1972), explicitly shows this interaction as a bent hydrogen bond $N_1 \cdots H—N_8^+$.

The second group of experimental data refer to aqueous solution as indicated by measurements of the NMR coupling constants for the ethanelike protons on atoms C_6 and C_7 of the side chain of histamine. On the basis of the assumption that trans and gauche forms are the only ones present, it is possible to estimate from these measurements their relative proportions. Two such studies have been performed for histamine, one by Ganellin and co-workers (1973a,b) and the other by Ham *et al.* (1973; Ison, 1974), and both estimate essentially equal proportions of the trans and gauche forms in monoprotonated histamine with a small but significant favoring of the trans form in the diprotonated species. Thus, the second protonation increases the proportion of the trans conformer from 47 to 67%, according to Ham *et al.* (1973), and from 45 to 55%, according to Ganellin *et al.* (1973a). The energy differences between the trans and gauche conformers in solution are thus predicted to be extremely small (0.5 kcal/mole). No information is provided by these data on the torsion angle τ_1.

On the whole, the X-ray crystallographic results seem to underline the significance of the PCILO computations, whereas those of the NMR solution studies plead in favor of the plausibility of the EHT computations; this impression is further emphasized by the fact that this apparent agreement extends to studies on different methylated histamines (Ganellin *et al.*, 1973b; Ganellin, 1973b).

Faced with these problems, a reasonable first step is to look for an arbitrage by a more refined theoretical procedure. The problem has thus been studied by the *ab initio* nonempirical SCF method (B. Pullman and Port, 1974) using the program Gaussian 70 in an STO-3G basis (Hehre *et al.*, 1969). Rather than reevaluate the whole conformational map, which is expensive in the *ab initio* treatment, the authors simply calculated the energies of all the trans conformers and of all the gauche conformers by rotating the ring (varying τ_1) as the side chain is fixed with $\tau_2 = 180°$ or $60°$, respectively.

The variation of the *ab initio* energy of the trans and gauche forms of the histamine monoprotonated form as the ring is rotated is shown in Fig. 8. The energy of the trans form varies little as the ring is rotated, not surprisingly since in this form the ring side-chain interactions are minimized. The gauche form, on the other hand, shows a pronounced minimum in the region of $\tau_1 = -30°$ corresponding clearly to the interaction of the N^+H_3

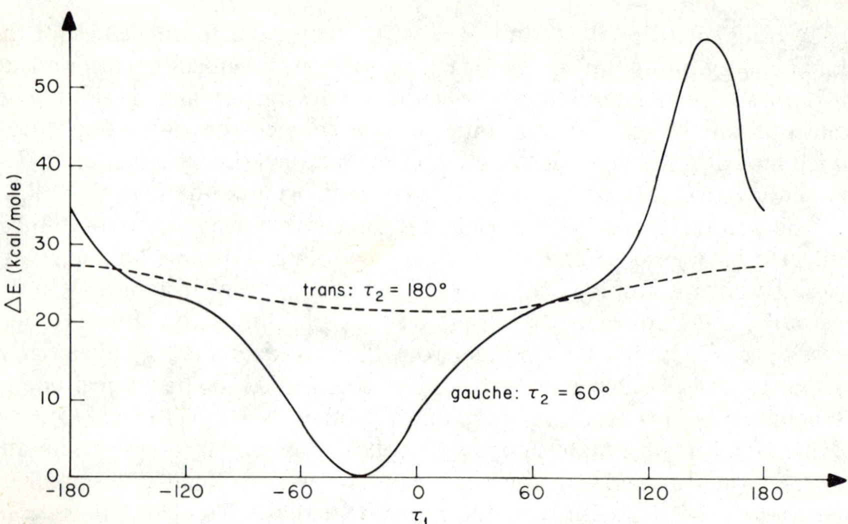

Fig. 8. *Ab initio* computations for monoprotonated histamine (Fig. 5a), showing the energies of the gauche ($\tau_2 = 60°$) and trans ($\tau_2 = 180°$) forms as a function of τ_1 (in kilocalories per mole with respect to the global energy minimum taken as zero energy).

group with the lone-pair of atom N_1 of the ring. This interaction stabilizes this gauche form over the most stable trans form ($\tau_1 = -30°$ to $30°$, $\tau_2 = 180°$) to an extent of 21 kcal/mole. These two minima lie close to the corresponding PCILO minima of Fig. 6.

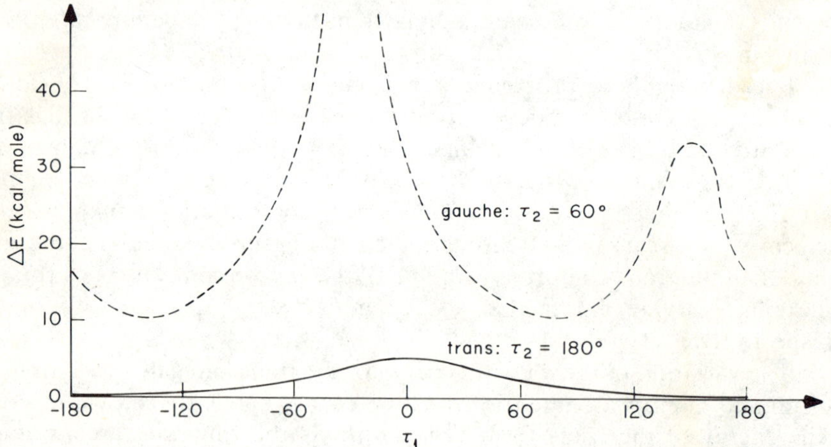

Fig. 9. *Ab initio* computations for diprotonated histamine (Fig. 5c), showing the energies of the gauche ($\tau_2 = 60°$) and trans ($\tau_2 = 180°$) forms as a function of τ_1 (in kilocalories per mole with respect to the global energy minimum taken as zero energy).

The variation of the *ab initio* energy of the trans and gauche forms of diprotonated histamine is shown in Fig. 9. Here again the energy of the trans form varies little as the ring rotates but has a small minimum toward $\tau_2 = 180°$. The gauche form is always higher in energy than the trans one, with a minimum in the region of $\tau_1 = \pm 90°$. This minimum is, however, about 11 kcal/mole higher in energy than the lowest trans form. Again this situation agrees with the PCILO results of Fig. 7.

The *ab initio* results, therefore, support the PCILO predictions of extremely strong preference for the gauche form by the histamine monocation and for the trans form by the dication. The energy differences are as large as in the PCILO predictions, and these findings force us to conclude that, so far as the theoretical methods are concerned, they do indeed apply to the situation in the free molecule, which seems to be preserved in the crystal.

Under these conditions the greatly reduced energy differences observed by NMR experiments in solution, as indicated by the coexistence of different forms, must presumably be the result of strong solvent effects. In order to verify this assumptions, this effect has to be introduced into the computations. This has been done (B. Pullman and Port, 1974) within the "microscopic supermolecular" approach. The most favorable hydration sites of histamine were determined by *ab initio* studies on model compounds, following the procedure presented by A. Pullman and Pullman (1975). The construction of the conformational map of the new "supermolecule" representing the hydrated histamine was carried out by the PCILO method because the compound is then too large for an *ab initio* computation. As the PCILO results have been shown above to reflect correctly the *ab initio* results the procedure is certainly acceptable.

Ab initio calculations of the ethylammonium cation and of histamine suggest four principal hydration sites, which are shown in Fig. 10, together with the energies of interaction calculated on an STO-3G basis. In both mono- and diprotonated histamine the ammonium group can form three very energetically favorable hydrogen bonds to water (28 kcal/mole each). In the monoprotonated molecule a fourth water molecule can act as proton donor to the lone pair on N_1 (energy, 5 kcal/mole), whereas in the diprotonated species, it can act as a proton acceptor from the N_1^+H bond (energy, 19 kcal/mole). A notable feature is the high hydrogen bond energy when the nitrogen is quaternized. Its value is probably overevaluated in this treatment due to the smallness of the basis set adopted. An energy of 20 kcal/mole would seem a more appropriate value, but it is still a high value and justifies the individualization of these water molecules and the hypothesis that they are following the cationic head during conformational changes.

Based on this water-fixation scheme, the PCILO conformational energy maps for hydrated cations are presented in Figs. 11 and 12. Figure 11 gives

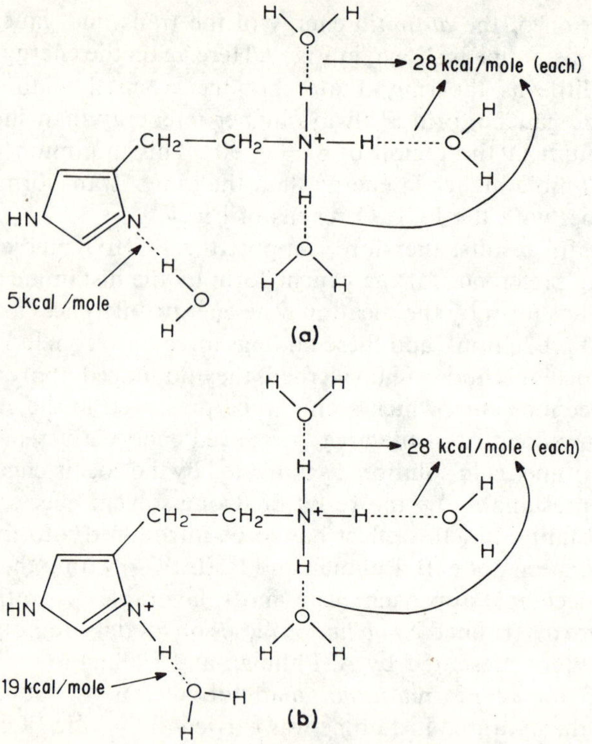

Fig. 10. Hydration sites in (a) mono- and (b) diprotonated histamine. Energies in kilocalories per mole calculated *ab initio* (STO-3G).

the results for the hydrated monocation and indicates a profound change with respect to the results for the isolated molecule depicted in Fig. 6. In place of one deep minimum for the gauche form in Fig. 6, we now observe gauche ($\tau_1 = \pm 60°, \tau_2 = \pm 60°$) and trans ($\tau_1 = \pm 30°, \tau_2 = 180°$) minima of equal energy. The trans minima are more extended and will thus be statistically favored. The results lead one to expect a mixture of conformers in solution with a predominance of the trans one. No hydrogen bond is now expected between the cationic head and the imidazole ring, in agreement with recent experimental studies on the ionization constants of histamine and derivatives in aqueous solution (Paiva *et al.*, 1970).

Figure 12 presents the results for the hydrated dication, and in this case also substantial changes are observed with respect to the results for the isolated molecule depicted in Fig. 7. Although the global energy minimum still corresponds to a trans form ($\tau_1 = 180°$, $\tau_2 = 180°$), it is now limited to narrow values of τ_1 associated with the antiplanar arrangement of the side chain with respect to the ring. Moreover, a local energy minimum appears now at 3 kcal/mole above the global one, for a gauche conformer at $\tau_1 = 180°$, $\tau_2 = 90°$, and another one at 5 kcal/mole above the global one at $\tau_1 = -90°$, $\tau_2 = -60°$ suggesting the possible occurrence of such conformers in equilibrium with the predominant trans one but in a proportion smaller than that expected for the monocation.

Altogether, the results for the hydrated cations indicate thus a striking influence of the solvation effect on molecular conformation. Furthermore, this effect, which tends to moderate the extreme situation prevalent for the isolated molecules (a strong predominance of one of the conformers, trans or gauche), brings the computed results in qualitative agreement with the NMR data for molecules in solution (mixture of tautomers with a predomin-

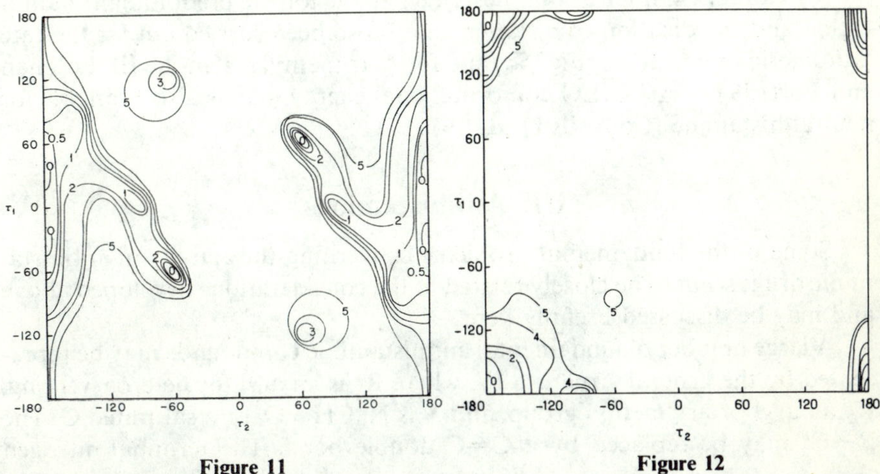

Figure 11 **Figure 12**

Fig. 11. Conformational energy map of the hydrated monocation of histamine. Isoenergy curves in kilocalories per mole with respect to the global energy minimum taken as zero energy.

Fig. 12. Conformational energy map of the hydrated dication of histamine. Isoenergy curves in kilocalories per mole with respect to the global energy minimum taken as zero energy.

ance of the trans one, this predominance being greater in the dication than in the monocation). The trend predicted for the solution effect by the super molecule model seems thus certainly significant. It has a buffering effect in that, on the one hand, it diminishes the intramolecular attractive interactions prominent in the isolated monocation and, on the other hand, reduces the intramolecular repulsive interactions prominent in the isolated dication, thus bringing the two species closer together from the viewpoint of conformational possibilities. No other treatments are available on solvated histamine.

The results enable also a rational evaluation of the significance of the EHT computations and of their *apparent* agreement with findings in solution. It is manifest that the agreement is accidental or, more precisely, that by underevaluating the intramolecular interactions, attractive and repulsive, in the isolated molecules to the point of finding no significant differences between the structure of the free mono- and dications, the extended Hückel method corresponds to some degree to the condition produced by the intervention of the solvent molecules and, thus, describes accidentally the state in aqueous solution. On the other hand, this accidental coincidence, once established for the basic skeleton, is systematically carried over to its derivatives and, thus, accounts for the satisfactory correlation found by Ganellin *et al.* (1973b) between the EHT predictions and the NMR results for a series of methyl histamines.

A treatment similar to that described above for the predominant monocation and the dication of histamine have also been carried out for the rate tautomeric monocation (Fig. 5b) and for *N*-trimethyl histamine (B. Pullman and Port, 1974). A PCILO conformational energy map is also available for neutral histamine (Coubeils *et al.*, 1971).

III. Antihistamines

Some of the fundamental problems concerning the activity of antihistaminic drugs seem to be closely related to the considerations developed above and may be discussed usefully here.

A large number of fundamental antihistaminic compounds may be represented by the general structure IV, where R_1 is an aryl (or heteroaryl) ring, R_2 an aryl or arylmethyl group, and X is N, CH—O, or a saturated C. The X—C^{β} may be replaced by a C=C double bond. The terminal nitrogen

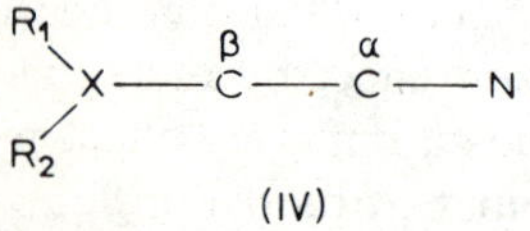

(IV)

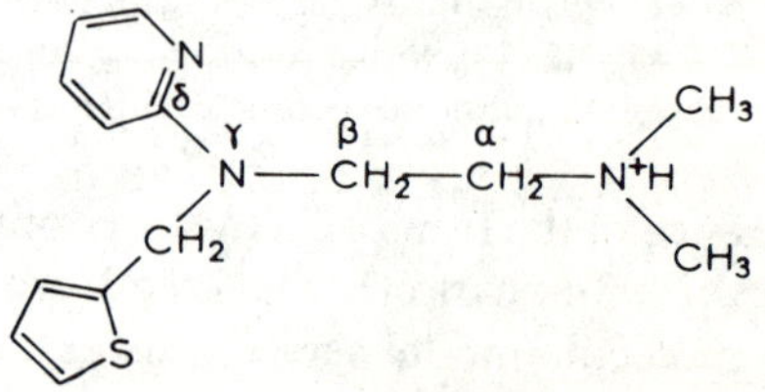

(V) HISTADYL

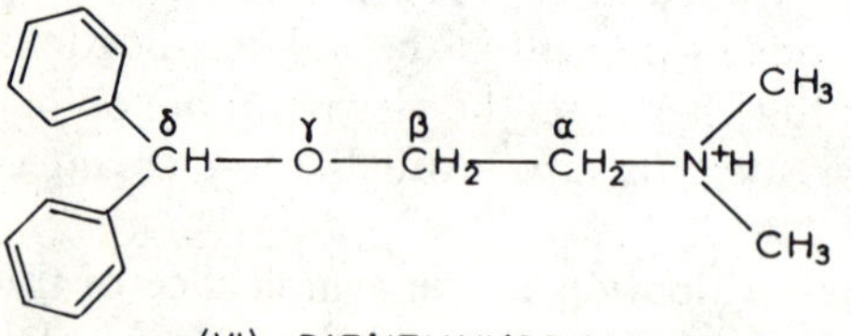

(VI) DIPHENHYDRAMINE

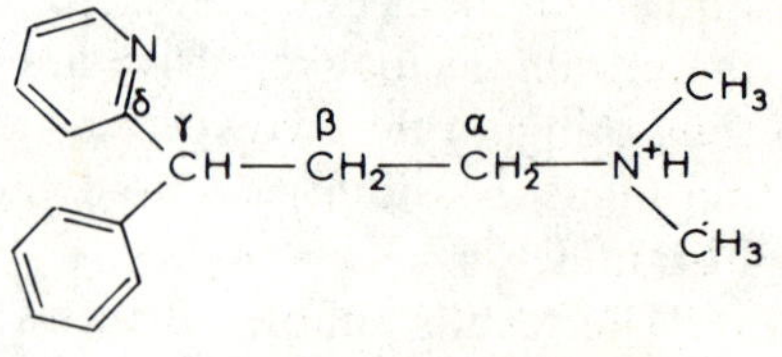

(VII) PHENIRAMINE

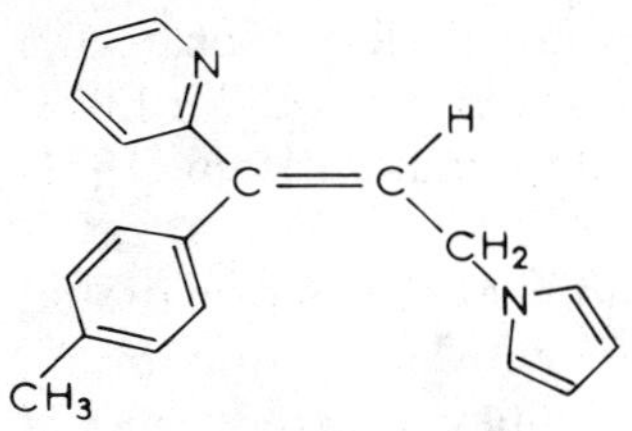

(VIII) TRIPROLIDINE 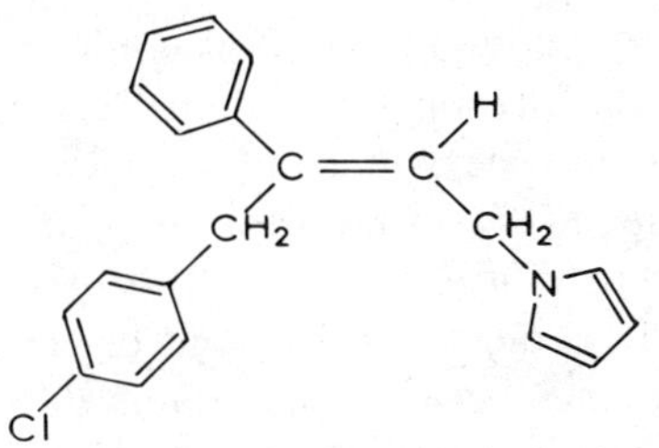 (IX) PYRROBUTAMINE

atom is part of a tertiary acyclic or alicyclic basic grouping. Typical representatives of these different classes of antihistaminic drugs are ethylenediamine derivatives [such as histadyl (V)], diphenhydramine (VI), pheniramine (VII), triprolidine (VIII), and pyrrobutamine (IX).

The discussion on the structure–activity relationship in this series of molecules has been dominated by the search for conformational analogies among these compounds and histamine. It has been postulated on the basis of similarities found in the crystal structure of histamine, histadyl hydrochloride (Clark and Palenik, 1972), and brompheniramine or chlorpheniramine maleates (James and Williams, 1971, 1974a) that a fully extended trans conformation about the $C^{\alpha}—C^{\beta}$ bond is essential for antihistaminic activity. The demonstration that, in order to have a high antihistaminic activity in the conformationally restricted compounds VIII and IX, it is necessary to have a *trans*-Ar—C=C—CH_2—N arrangement with the aromatic nucleus (α-pyridyl or phenyl) coplanar with the double bond (Casy and Ison, 1970; Ison *et al.*, 1973a) seemed to corroborate this point of view, the more so as a recent X-ray determination of the crystal structure of triprolidine hydrochloride monohydrate (James and Williams, 1974b) indicated an essentially trans arrangement of the two N-containing rings.

The belief that the crystallographically observed conformations are significant for antihistaminic activity is, however, subject to caution, if not objections. In the first place, the crystallographic results for histamine refer mostly to a diprotonated form and may be different for the physiologically probably more significant monoprotonated one. Second, studies with other conformationally restricted potent antagonists, *trans*- and *cis*-1,5-diphenyl-3-dimethylaminopyrrolidines (Hanna and Ahmed, 1973), which cannot attain a fully extended *trans*-N—C—C—N conformation, show that a range of values for the relevant torsional angle is acceptable for effective drug–receptor interactions. Similarly, although the crystallographic conformation of diphenhydramine is unknown, results for related systems containing the O—C—C—N^+ chain, e.g., acetylcholine and derivatives (see in following) or some phenylcholine ethers (Kneale *et al.*, 1974a,b; Hamodrakas *et al.*,

1974), show a variety of conformations with, moreover, a preference for a gauche arrangement. Last but not least, studies in solution indicate the frequent coexistence of a number of gauche and trans forms both for histamine (as described above) and for different antihistamines (Ham, 1971; Testa, 1974).

In order to understand better the overall possibilities of conformational analogies between histamine and the antihistamines, it is interesting to investigate what are the *intrinsic* conformational possibilities and preferences of the " flexible " antihistaminic drugs and to what extent are they influenced or modified by environmental factors and in particular by solvent water. Of interest from this point of view are the nature of both the most stable and, if available, secondary stable conformations and also the energy barrier between them. This has been done recently with respect to the two essential torsion angles $\tau_1 = \tau$ (C^{δ}—X^{γ}—C^{β}—C^{α}) and $\tau_2 = r$ (X^{γ}—C^{β}—C^{α}—N^+) that define the degree of extension or folding of the backbone. Within the usual nomenclature relevent to histamine receptors, the compounds discussed here may be considered essentially as H_1-receptor antagonists (see, e.g., Black *et al.*, 1972).

Figures 13 to 15 present PCILO conformational energy maps of histadyl (V), diphenhydramine (VI), and pheniramine (VII) as a function of the torsion angles τ_1 and τ_2 (B. Pullman *et al.*, 1975). These three maps have in common the presentation of a global energy minimum corresponding to a gauche conformation at $\tau_1 = 60°$, $\tau_2 = -90°$ for histadyl (Fig. 13) and pheniramine (Fig. 15) and, at $\tau_1 = 180°$ or $-120°$, $\tau_2 = 60°$, for diphenhydramine (Fig. 14). There are secondary energy minima on these maps associated with extended forms, e.g., at $\tau_1 = -60°$, $\tau_2 = 180°$ in Fig. 13 (2 kcal/mole above the global minimum), at $\tau_1 = -90°$, $\tau_2 = 180°$ in Fig. 14 (1 kcal/mole above the global minimum), and at $\tau_1 = 60°$, $\tau_2 = 180°$ in Fig. 15 (2 kcal/mole above the global minimum).

In fact, the computations for these molecules have been pushed a step forward toward the evaluation of the *populations* of the different gauche and trans rotamers through the computation of the partition function of the compounds and of the statistical weight and probabilities of the different conformations (B. Pullman *et al.*, 1975). These results indicate a very strong predominance, of over 90%, of the gauche tautomers in the three cases. The examination of the models corresponding to the three gauche conformers shows, however, striking differences in their structure pointing to differences in the origin of their stability. The stable gauche conformer of diphenhydramine (Fig. 14) is essentially due to a strong electrostatic interaction between the cationic head and the esteric oxygen, typical (see in following) of the O—C—C—N^+ interaction as established in particular in the studies on acetylcholine and related compounds. Quite differently, the gauche confor-

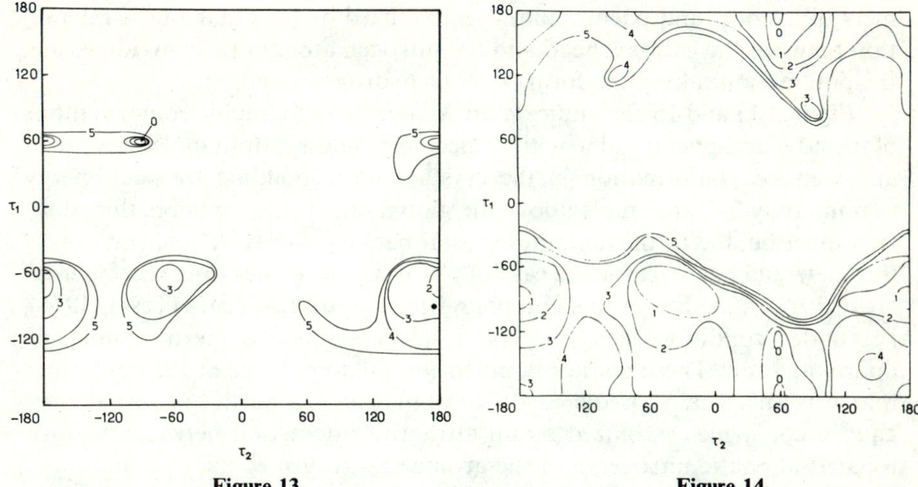

Fig. 13. Conformational energy map for histadyl with $\tau_3 = 41.4°$ (crystallographic value). Isoenergy curves in kilocalories per mole with respect to the global energy minimum taken as zero energy. (●) X-Ray crystallographic conformation (Clark and Palenik, 1972).

Fig. 14. Conformational energy map for diphenhydramine with $\tau_3 = 51.8°$. Isoenergy curves in kilocalories per mole with respect to the global energy minimum taken as zero energy.

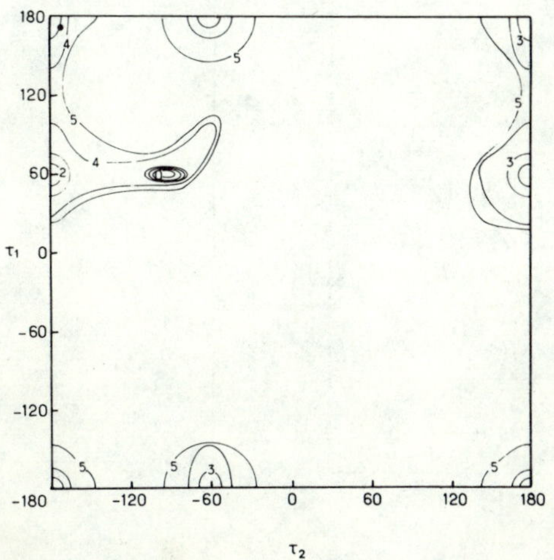

Fig. 15. Conformational energy map of pheniramine with $\tau_3 = 51.8°$ (crystallographic value). Isoenergy curves in kilocalories per mole with respect to the global energy minimum taken as zero energy. (●) X-Ray crystallographic conformation of brompheniramine maleate (James and Williams, 1971).

mers of histadyl and pheniramine are stabilized by the electrostatic interaction between the cationic head and the nitrogen atom of their pyridine ring, this interaction taking the form of weak hydrogen bonding.

Figures 13 and 15 also indicate the X-ray crystallographic conformations of histadyl and pheniramine with respect to τ_1 and τ_2. Both molecules are in an extended conformation in the crystals, corresponding to local energy minima only 2–3 kcal/mole above the global one. In our opinion, this situation must be due to the action of crystal packing forces. A confirmation of this viewpoint, at least for the case of pheniramine, comes from recent circular dichroism studies of this compound in nonpolar solvents (Testa, 1974), i.e., under conditions closer to those corresponding to the free molecule approximation. These studies point to the predominance of the conformer that was specifically predicted by computations to be the preferred one: gauche conformer stabilized by an attractive interaction between the protonated aliphatic nitrogen and the aromatic nitrogen of the pyridine ring.

The results of Figs. 13–15 have been obtained with $\tau_3 = \tau(C^\beta—C^\alpha—N^+—H) \approx 60°$, which allows the N^+H bond of the cationic head to be oriented toward the backbone of these molecules. Computations were also performed with $\tau_3 = 180°$. In this orientation of the cationic head, its proton is continuously directed toward the outside and cannot interact with the esteric oxygen or the pyridine nitrogen. The results are extremely interesting. Those for histadyl (Fig. 16) and pheniramine (Fig. 17) show a

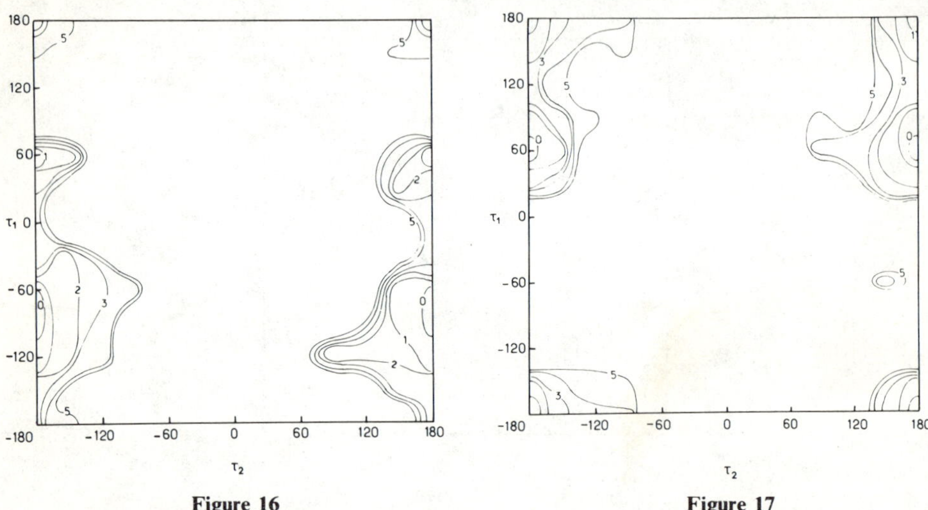

Fig. 16. Conformational energy map for histadyl with $\tau_3 = 180°$. Isoenergy curves in kilocalories per mole with respect to the global energy minimum taken as zero energy.

Fig. 17. Conformational energy map for pheniramine with $\tau_3 = 180°$. Isoenergy curves in kilocalories per mole with respect to the global energy minimum taken as zero energy.

profound modification of the conformational energy map. The attractive interactions between the cationic head and the pyridine nitrogen are no longer visible and the most stable conformer should now be an extended one with $\tau_1 = -90°$, $\tau_2 = 180°$ in histadyl and $\tau_1 = 60°$, $\tau_2 = 180°$ in pheniramine. On the contrary, in diphenhydramine (Fig. 18), the global energy minimum continues to be associated with a gauche form, at $\tau_1 = 180°$, $\tau_2 = 60°$ analogous to the one predicted and observed, as we shall see later, in acetylcholine. The strength of the attractive electrostatic association between the cationic head and the esteric oxygen atom of diphenhydramine, thus, exceeds significantly the attraction between the cationic head and the pyridine nitrogens of histadyl and pheniramine and does not depend on the possibility of H bonding.

The computations indicate that the global energy minimum of Fig. 13 is about 3 kcal/mole and that of Fig. 14, about 2 kcal/mole lower than that of Fig. 16 and of Fig. 18, respectively. On the other hand, the global energy minima of Figs. 15 and 17 are practically degenerated.

We may now turn to the study of the effect of water on the conformational properties of the antihistamines. This was studied with the "supermolecule" approach, similar to the corresponding study of histamine. The problem is in this case simplified by the existence of one particularly important hydration site, represented in all the molecules considered here by the N^+—H bond of the cationic head.

The extent of the perturbation is illustrated in Figs. 19–21 which repre-

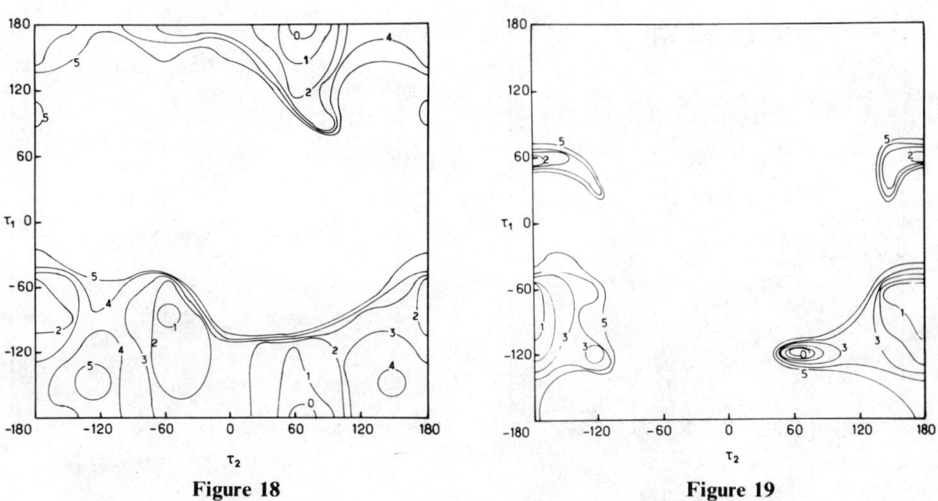

Figure 18

Figure 19

Fig. 18. Conformational energy map for diphenhydramine with $\tau_3 = 180°$. Isoenergy curves in kilocalories per mole with respect to the global energy minimum taken as zero energy.

Fig. 19. Conformational energy map for hydrated histadyl. Isoenergy curves in kilocalories per mole with respect to the global energy minimum taken as zero energy.

sent the conformational energy maps of hydrated histadyl, diphenhydramine, and pheniramine, respectively. The effect of hydration on the conformational characteristics of the compounds is obviously appreciable. Thus, although the global energy minimum for hydrated histadyl (Fig. 19) still corresponds to a gauche form at $\tau_1 = -120°$, $\tau_2 = 60°$, there appears now a broad local energy minimum for the trans conformer at $\tau_1 = -60°$ to $-100°$, $\tau_2 = 150°$ to $180°$ only 1 kcal/mole above the global one. A similar situation is observed for hydrated diphenhydramine (Fig. 20). In hydrated pheniramine (Fig. 21) the global energy minimum corresponds to a trans form at $\tau_1 = 60°$, $\tau_2 = -90°$. Considered in terms of populations the gauche/trans ratio is 72/28 in diphenhydramine, 57/43 in histadyl, and 1/99 in pheniramine. Thus the hydration of the cationic head has a moderate effect on the percentage of the gauche form in diphenhydramine, a strong effect in histadyl, and a very strong effect in pheniramine. The attachment of a water molecule has as consequence to disrupt the interaction between the cationic head and the pyridine nitrogens in histadyl and pheniramine. The extent of the gauche form in the equilibrium mixture seems now to be governed by the electronegativity of the main chain X atom with which the hydrated cationic head may interact: it decreases when X changes from O to N to C.

The practically complete disappearance of the gauche form in the case of hydrated pheniramine represents such an extreme effect that it seemed suitable to elaborate on the computations in this particular case. This has been

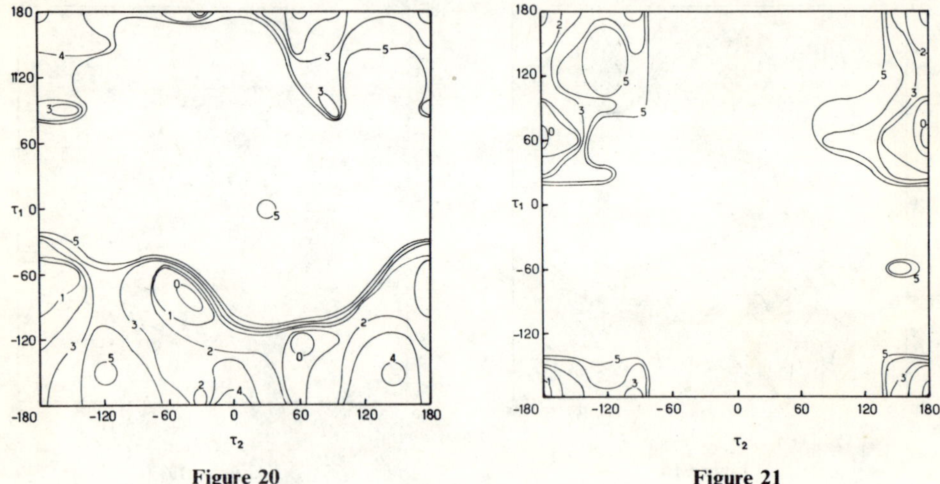

Figure 20 **Figure 21**

Fig. 20. Conformational energy map for hydrated diphenhydramine. Isoenergy curves in kilocalories per mole with respect to the global energy minimum taken as zero energy.

Fig. 21. Conformational energy map for hydrated pheniramine. Isoenergy curves in kcal/mole with respect to the global energy minimum taken as energy zero.

done by allowing a large flexibility for a number of variables in the input data. The result (B. Pullman *et al.*, 1975) is to increase the proportion of the gauche rotamer up to 20%. These refined computations may thus be considered as modifying somewhat but in no way drastically the results obtained within the rigid frame approximation. Possibly a similar increase in the proportion of the gauche form may be estimated for a similar refinement for the other flexible antihistaminics.

We may now confront the theoretical results with whatever available experimental data there are about these or related compounds. Although rather scarce, these data seem, nevertheless, in general agreement with the computations. As concerns the antihistamines themselves, two studies as mentioned above are available. One by Ham (1971) who uses NMR spectroscopy, shows that, although diphenhydramine exists predominantly in water in the gauche form, cationic ethylenediamine derivatives (pyrilamine, tripelennamine, and methapyrilene) exist rather as approximately equivlent mixtures of gauche and trans conformers. The second study, by Testa (1974), who uses circular dichroism measurements, refers to pheniramine and confirms the predominance of the gauche form in inert solvents (see above) but of the trans form in water. A related NMR study by Testa (1974) on norpheniramine indicates a nearly equivalent mixture of trans and gauche conformers in solution. As the methyl groups are expected to increase the proportion of the trans form, this result permits also to estimate that the trans form would dominate in solution for pheniramine. Altogether, this group of data confirms our finding that in solution the proportion of the gauche form should increase with the electronegativity of the X atom of the X—C—C—N^+ chain of antihistamines.

Thus the theoretical and experimental data described in this section converge toward a clear-cut indication that in distinction to X-ray crystallographic data the situation of antihistamines in solution, as that of histamine itself, corresponds to the presence of a mixture of gauche and trans conformers. From the point of view of the relative proportion of the two conformers, it is the antihistaminic ethylenediamine derivatives that resemble most closely histamine itself (slight predominance of the trans conformer). Whatever it be, it is obvious that the availability in solution and possibly in the biophase of a multiplicity of conformers must be borne in mind in the elaboration of theories on antihistaminic activity of drugs.

IV. Serotonin and Related Indolealkylamines

The computations developed in the study of histamine set up a pattern that was followed in the study of other biogenic amines. It was first applied to indolealkylamines. Again the primitive theoretical computations were inconclusive.

The first quantum-mechanical computation on serotonin (X), the principal representative compound of this series, carried out by Kier (1968b) using the EHT method predicted only one stable conformation corresponding to $\tau_1 = 90°$, $\tau_2 = 180°$. A more precise EHT treatment (Kang *et al.*, 1973b) indicated, however, the existence of low local energy minima at $\tau_1 = 90°$,

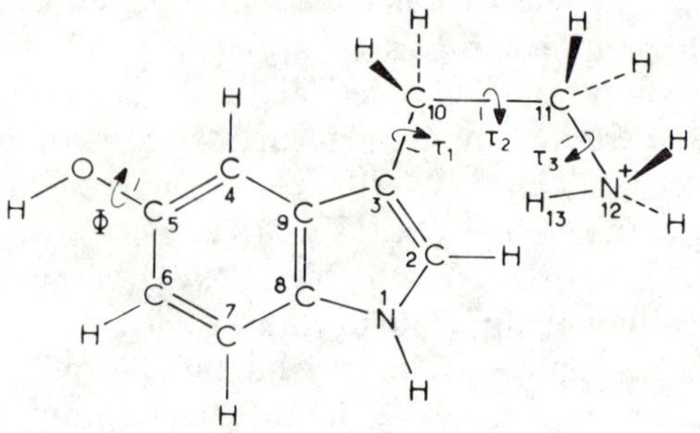

(X) SEROTONIN

$\tau_1 = \tau(C_2 - C_3 - C_{10} - C_{11})$

$\tau_2 = \tau(C_3 - C_{10} - C_{11} - N^+)$

$\tau_2 = \pm 60°$, 1 kcal/mole above the global one. An "empirical" computation performed by the same authors (Johnson *et al.*, 1973) reversed the relative stabilities of these forms indicating the gauche forms to be 1–3 kcal/mole more stable than the trans one. Computations using more refined theoretical methods yielded quite different results. With the PCILO treatment (Courrière *et al.*, 1971; B. Pullman *et al.*, 1974b), the most stable conformation was predicted as highly folded, corresponding to $\tau_1 = 140°$, $\tau_2 = -20°$ (and $\tau_1 = -140°$, $\tau_2 = 20°$ for the enantiomorph) with a secondary local energy minimum at $\tau_1 = 100°$, $\tau_2 = 60°$ ($\tau_1 = -100°$, $\tau_2 = -60°$) 2 kcal/mole above the global one. The INDO calculations (Kang and Cho, 1971) predicted the most stable form at $\tau_1 = 60°$, $\tau_2 = 0°$ with the conformation at $\tau_1 = 90°$, $\tau_2 = 180°$ as the least probable.

This confused theoretical situation may be compared with the available experimental data which come essentially from X-ray crystallographic studies. Two results are available for serotonin itself, corresponding to different surroundings: one for serotonin–creatinine sulfate complex (Karle *et al.*, 1965) with $\tau_1 = 13.3°$, $\tau_2 = 172.6°$ and one for serotonin–picrate monohydrate (Thewalt and Bugg, 1972a) with $\tau_1 = 112.5°$, $\tau_2 = -66.6°$. For reasons which will become clear later in this section, we may consider here also the results on tryptamine hydrochloride (Wakahara *et al.*, 1973) which indicate $\tau_1 = 110.8°$, $\tau_2 = -60.5°$.

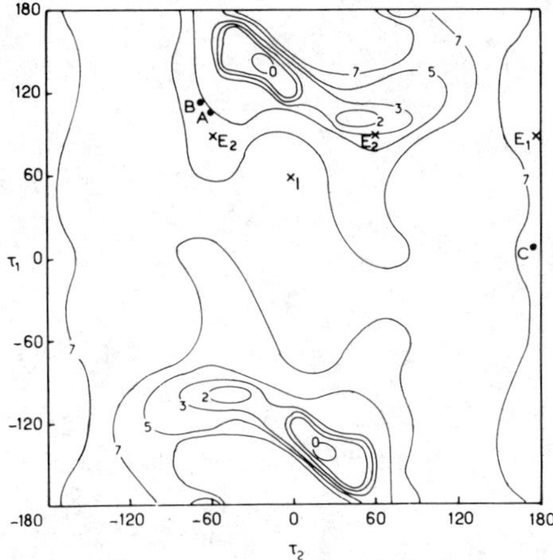

Fig. 22. The PCILO conformational energy map for cationic serotonin. Isoenergy curves in kilocalories per mole with respect to the global energy minimum taken as zero energy. Also indicated are E_1, global minimum of EHT computation (Kier, 1968) and E_2, secondary minimum of EHT computations. In empirical computations (Kang *et al.*, 1973), these minima are reversed.

The theoretical and experimental data are plotted on Fig. 22 from which it is, in fact, obvious that the agreement between theory and experiment is a limited one. Although two of the three experimental conformations are close to the predicted PCILO global energy minimum, the remaining one (representing serotonin in the creatinine sulfate complex) does not even correspond to a local energy minimum and falls in a region of relatively high-conformational energy: 7 kcal/mole above the global minimum. It represents an extended, trans form with the N^+ atom nearly coplanar with the ring.

Under these circumstances, first an essay was made to ascertain the validity of the various semiempirical theoretical computations by carrying out *ab initio* ones. This has been done using an extension, performed in our laboratory, of the program Gaussian 70 to 105 orbitals in an STO-3G basis.

The results for serotonin are indicated in Fig. 23 (Port and Pullman, 1974b). The global energy minimum ($\tau_1 = 90°$, $\tau_2 = 60°$) and the next local energy minimum ($\tau_1 = 120°$, $\tau_2 = -60°$), 1.4 kcal/mole above the global one, correspond to gauche forms with the side chain perpendicular or nearly so to the ring. There is a local energy minimum at $\tau_1 = 110°$, $\tau_2 = 180°$, 2.6 kcal/mole above the global one, corresponding to an extended form. The *ab initio* computation confirms thus the PCILO results indicating a preference by cationic serotonin for a gauche form. There is, however, an inversion of the energies of the two gauche forms with respect to the indications of

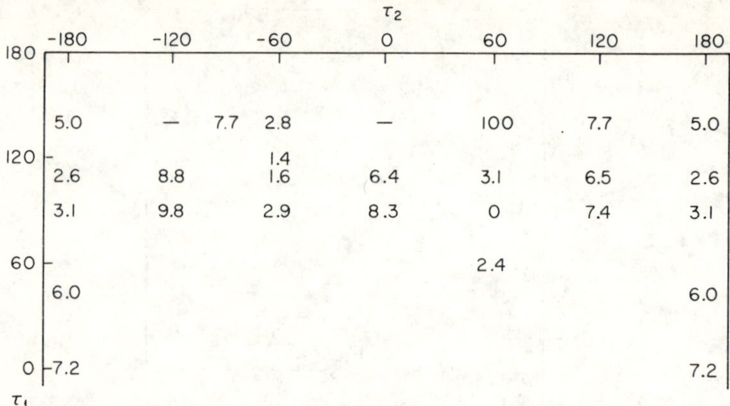

Fig. 23. *Ab initio* conformational energy map for cationic serotonin. Energies in kilocalories per mole with respect to the global energy minimum taken as zero energy.

PCILO. The planar extended conformation ($\tau_1 = 0°$, $\tau_2 = 180°$) is 7.2 kcal/mole above the global one and does not correspond to an energy minimum. The result is again very similar to that obtained by PCILO. It seems, therefore, necessary to admit that the nearly planar-extended conformation ($\tau_1 = 13°$, $\tau_2 = 173°$) assumed by serotonin complexed with creatinine sulfate must be due to crystal packing forces.

As a next step the conformational energy map was constructed for the solvated serotonin (B. Pullman *et al.*, 1974b). The principal hydration sites were again determined by the *ab initio* supermolecular approach and they correspond to the fixation of 3 water molecules at the cationic head. Using this water fixation scheme, the PCILO conformational energy map for the hydrated cation of serotonin is given in Fig. 24. It is profoundly different from that of the nonhydrated species. The high selectivity for a gauche form, which was one of the outstanding features of the conformational energy map for isolated serotonin, disappears for the hydrated form leading to a near equivalence of the energy minima related to the gauche and trans forms. These may, therefore, be expected to coexist in solution in comparable amounts.

A recent experimental result (Ison *et al.*, 1972), based on NMR spectroscopy, is a particularly satisfying confirmation of this theoretical evaluation. The measurement of vicinal coupling constants for the protons of the $C_\alpha H_2$—$C_\beta H_2$ bond of cationic serotonin leads to two possible proportions of the gauche and trans forms: one corresponding to a slight preponderance of the trans rotamer ($n_t = 0.45$, $n_g = 0.55$), and the other to a slight preponderance of the gauche rotamers ($n_t = 0.28$, $n_g = 0.72$) (when the three rotamers are of equal energy, namely, $n_t = 0.33$, $n_g = 0.67$). The experiment does not permit one to decide between these two possibilities, but it does indicate that the energy difference between the trans and gauche rotamers in solution is very small (for $n_t = 0.45$, $n_g = 0.55$, the $E_g - E_t = 0.3$ kcal/mole) which is in agreement with the results of our computations.[2] The effect of the solvent is, therefore, as it is also in the case of the histamine cations, to reduce the energy gap between the possible rotamers, and, thus, to assure their coexistence in solution.

It is useful to add that some of the theoretical studies, in particular the PCILO ones, involved not only serotonin but a large number of indolealkylamines. One of the important results of the PCILO computations was that the understanding of the conformational properties of these molecules and of the relationship between the theoretical computations and the experimental crystallographic or solution findings necessitates a subdivision of these molecules into subgroups corresponding to their possible occurrence in the neutral or cationic species with an amino or dimethylamine terminal grouping. Each subgroup has its conformational preferences with respect to the two principal torsion angles, τ_1 and τ_2, and has to be studied separately.

Thus when this subdivision is omitted and when all the compounds of the series studied crystallographically are plotted on the conformational map of serotonin considered as representative of the whole series, the deceptive results of Fig. 25 are obtained. The situation improves appreciably when the above-mentioned division is introduced. Thus, without going into details about the conformational preferences of each subgroup, it may be noticed, for example, that the two molecules (A and B) of N_1 N-dimethyltryptamine (XI) and of bufotenine (XII) not only have a dimethylamine terminal group

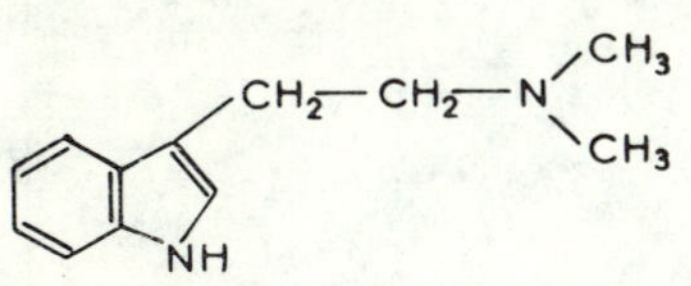

(XI) N,N-DIMETHYLTRYPTAMINE

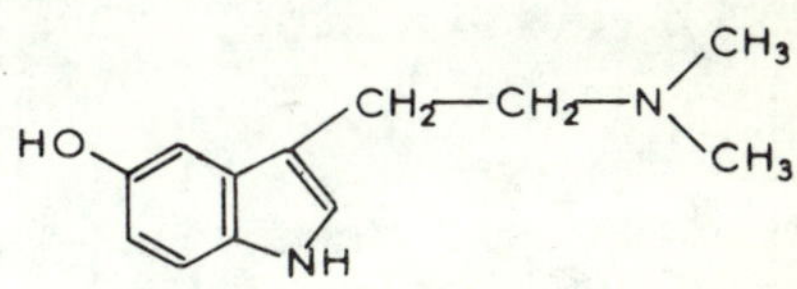

(XII) BUFOTENINE

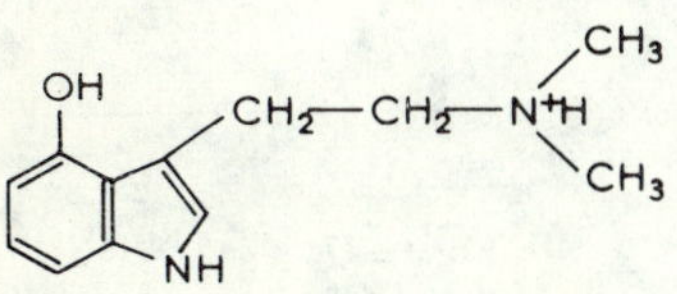

(XIII) PSILOCIN

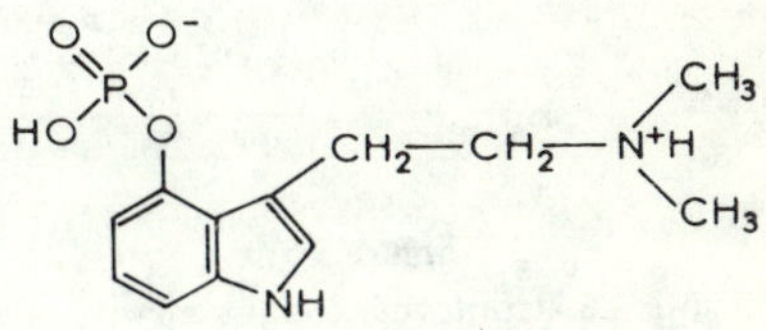

(XIV) PSILOCYBIN

[2] Very recently (Culvenor and Ham, 1974) the problem was solved in favor of the trans rotamer.

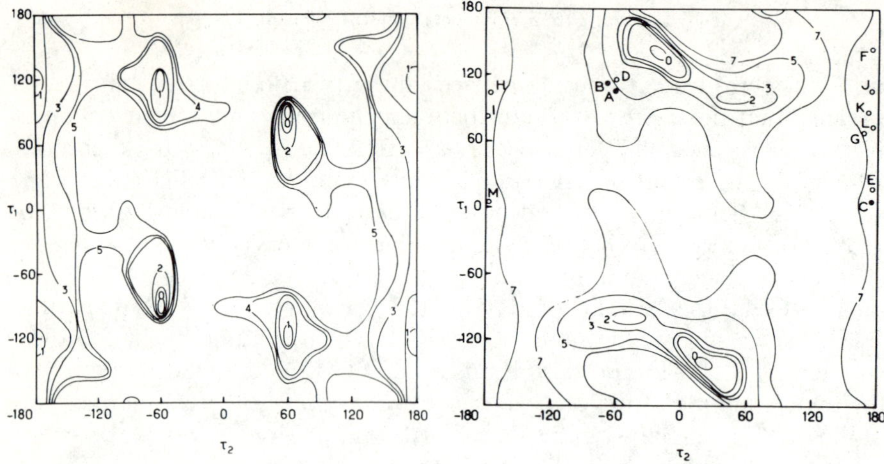

Figure 24 **Figure 25**

Fig. 24. Conformational energy map of hydrated cationic serotonin. Isoenergy curves in kilocalories per mole with respect to the global energy minimum taken as zero energy.

Fig. 25. The conformational energy map of serotonin considered as representative for the whole series of indolealkylamines. X-Ray crystallographic conformations (Carlström *et al.*, 1973) of (A) tryptamine HCl; (B) serotonin–picrate monohydrate; (C) serotonin–creatine sulfate; (D) 5-methoxytryptamine; (E) 5-methoxy-*N*,*N*-dimethyltryptamine; (F) LSD; (G) psilocybin A; (H) psilocybin B; (I) *N*,*N*-dimethyltryptamine A; (J) *N*,*N*-dimethyltryptamine B; (K) bufotenine A; (L) bufotenine B; (M) melatonine.

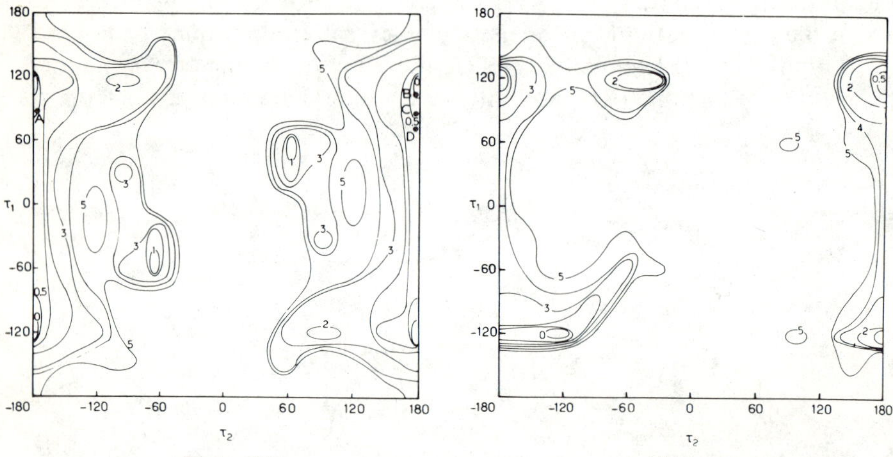

Figure 26 **Figure 27**

Fig. 26. Conformational energy map of neutral bufotenine. Isoenergy curves in kilocalories per mole with respect to the global energy minimum taken as zero energy. X-Ray crystal conformations of (A) *N*,*N*-dimethyltryptamine molecule A; (*B*) *N*,*N*-dimethyltryptamine molecule B (Falkenberg, 1972a); (C) bufotenine molecule A; (D) bufotenine molecule B (Falkenberg, 1972b).

Fig. 27. Conformational energy map of cationic psilocin (for $\tau_3 = 60°$). Isoenergy curves (kilocalories per mole) with respect to the global energy minimum taken as zero energy.

but, moreover, are neutral in the crystal. The conformational energy map representative of this subgroup of indolealkylamines is given in Fig. 26. It is seen that it is substantially different from that of cationic serotonin and that it shows a global energy minimum for an extended conformation $\tau_1 = \pm 100$, $\tau_2 = 180°$ corresponding closely to the observed crystallographic forms.

Because of its important position in a number of works looking for structural correlations among different types of halucinogenic compounds, we are reproducing here in Fig. 27 a PCILO conformational energy map for psilocin (XIII) in its cationic form. This molecule differs from serotonin in that it has the OH group attached to C_4 of the indole ring (and a dimethylated amino group). The predicted most stable conformation is a gauche one. A similar prediction is made by the INDO method (Kang *et al.*, 1973a). A recent X-ray study of psilocin (Petcher and Weber, 1974) indicates that it has the OH group attached to C_4 of the indole ring (and a mation ($\tau_1 = -20.3°$, $\tau_2 = 172°$). The ionization state of the molecule is, however, not certain. On the other hand, psilocybin (XIV), the phosphorylated metabolic precursor of psilocin, exists in the crystal (Weber and Petcher, 1974) in two forms (both zwitterions): molecule A with $\tau_1 = -72.4°$, $\tau_2 = -166°$ and molecule B with $\tau_1 = -107.6°$ and $\tau_2 = 174.7°$; thus, it is appreciably different from psilocin.

V. Phenethylamines and Phenethanolamines

A number of molecules of this group play a fundamental role in molecular pharmacology. Their structures are indicated in Fig. 28. The line of study described above for histamine and indolealkylamines was also applied to the investigation of the conformational properties of the essential phenethylamines and phenethanolamines. The overall results are summed up in Table I.

Early theoretical studies left the problem somewhat unresolved. Computations carried out by the extended Hückel theory for norepinephrine (Kier, 1969), ephedrine (Kier, 1968c), isoproterenol (George *et al.*, 1971), and dopamine (Kier, 1973a) predict systematically a preference for a trans conformation. An extensive PCILO computation for a large series of phenethylamines found practically degenerate energy minima for the gauche and trans forms in nearly all of them (B. Pullman *et al.*, 1972, 1974a). Norepinephrine has also been investigated by the CNDO (Katz *et al.*, 1973, 1974) and INDO (Pedersen *et al.*, 1973) methods. The first predicts a preference for the gauche form, the second the near equivalence of the gauche and trans forms. A preference for the gauche form is predicted by the CNDO/2 method for

TABLE I

CONFORMATIONAL PROPERTIES OF PHENETHYLAMINES[a]

Compound	Theory								Experiment	
	Empirical PPF[b]	EHT	CNDO	INDO	PCILO	STO-3G	4-31G	7s, 3p, 3s contracted 2s, 1s, 3s	Crystal X-ray	% of t in solution
Phenethylamine (XV)	g	—	—	—	g, t	g	g	g	t	56
Amphetamine (XXII)	g	—	—	—	g, t	g	—	—	t	50
Norepinephrine (XVI)	—	t	g	g, t	g, t	g	—	—	t	76
Dopamine (XVIII)	—	g, t	—	—	g, t	g	—	—	t	43
Epinephrine (XXI)	—	—	—	—	g	—	—	—	t planar	77
Ephedrine (XVII)	—	t	—	—	{g / g, t}	—	—	—	t	—
Isoproterenol (XXIII)	—	t	—	—	g, t	g	—	—	t	83

[a] Orientation: g = gauche; t = trans.

[b] PPF = partitioned potential functions.

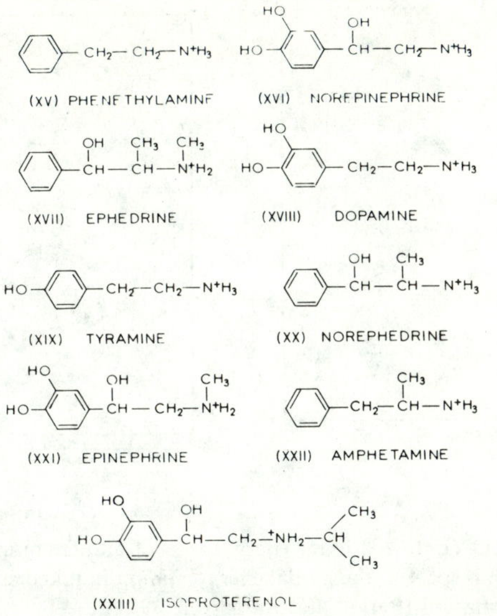

(XV) PHENETHYLAMINE (XVI) NOREPINEPHRINE

(XVII) EPHEDRINE (XVIII) DOPAMINE

(XIX) TYRAMINE (XX) NOREPHEDRINE

(XXI) EPINEPHRINE (XXII) AMPHETAMINE

(XXIII) ISOPROTERENOL

Fig. 28. The principal phenethylamines and phenethanolamines.

isoproterenol (Petrongolo *et al.*, 1974). Finally, empirical (partitioned potential functions) computations for isolated cationic phenethylamine and amphetamine predict a preference for a gauche form (Weintraub and Hopfinger, 1973).

As an example of the general aspect of the PCILO conformational energy maps of this type of molecule, we reproduce in Figs. 29 and 30 the maps for amphetamine and norepinephrine, respectively. (Others may be found in B. Pullman *et al.*, 1972, 1974a.) These figures indicate also the experimental crystallographic conformation of the molecules studied. Both of them, as in fact all phenethylamines studied till now with the exception of epinephrine, exist in the crystal in a trans conformation ($\tau_2 \approx 180°$) with $\tau_1 \approx 90°$ corresponding to one of the degenerate global energy minima (for a general review, see Carlström *et al.*, 1973). Epinephrine although trans in the crystal has the ethylamine chain nearly coplanar to the benzene ring (Carlström, 1973). This conformation is about 3 kcal/mole above the computed global energy minimum and does not correspond to any local energy minimum. However, these compounds exist in water as a mixture of trans and gauche forms. The trans form is generally predominating but in a variable amount (Ison *et al.*, 1973b; Roberts, 1974): of the order of 50% in phenethylamine or amphetamine, but it amounts to about 80% in norepinephrine or isoproterenol (last column of Table I).

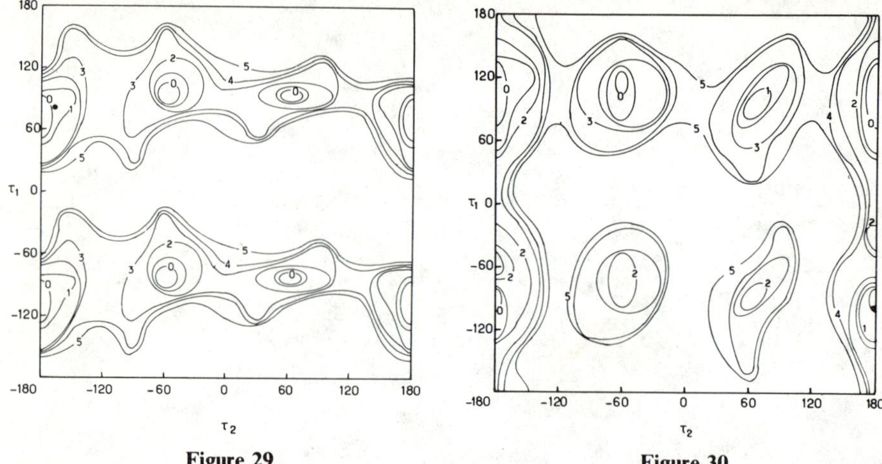

Fig. 29. The PCILO conformational energy map of amphetamine. Isoenergy curves in kilocaries per mole with respect to the global energy minimum taken as zero energy. (●) X-Ray crystallographic conformation (Bergin and Carlström, 1971).

Fig. 30. The PCILO conformational energy map of norepinephrine. Isoenergy curves in kilocalories per mole with respect to the global energy minimum taken as zero energy. (●) X-Ray crystallographic conformation (Carlström and Bergin, 1967).

Ab initio self-consistent field (SCF) computations have been carried for selected representative phenethylamines including amphetamine and norepinephrine (B. Pullman *et al.*, 1974b). For reasons of economy, we did not recalculate the whole conformational energy maps but only the essential parts as suggested by previous PCILO and other results. Fundamentally, the most striking result is that in all cases the method leads to a preference for the gauche conformation, which appears systematically 1–2 kcal/mole more stable than the trans one. There is a barrier between the two of the order of 5–6 kcal/mole with the exception of norepinephrine for which the barrier is higher, of the order of 8 kcal/mole, and of isoproterenol, for which the barrier is negligible. As an example, the *ab initio* results for amphetamine and norepinephrine are reproduced in Figs. 31 and 32.

It may be interesting to mention that in this particular case, an attempt was made to verify the foregoing results by using more refined basis sets. Thus an *ab initio* SCF computation with the more extended 4-31G basis set (Ditchfield *et al.*, 1971) was carried out for the two essential conformations of phenethylamine: gauche at $\tau_1 = -90°$, $\tau_2 = -60°$ and trans at $\tau_1 = -90°$, $\tau_2 = 180°$. The result confirms the general findings obtained with the STO-3G basis set: the gauche conformation is found to be the most stable and its 3 kcal/mole below the trans one. Because of the high cost of

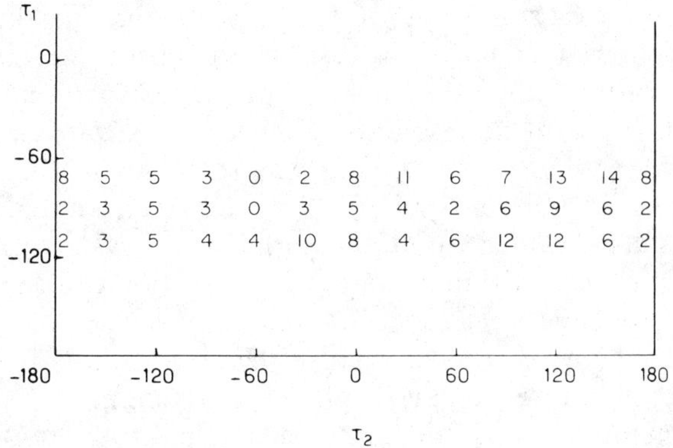

Fig. 31. *Ab initio* STO-3G conformational energy map of amphetamine. Energies in kilocalories per mole with respect to the global energy minimum taken as zero energy.

the 4-31G computations, no calculations have been performed for other compounds of the series. It is, nevertheless, tempting to consider the result found for phenethylamine as possibly representative for the whole series. On the other hand, another *ab initio* SCF computation has been carried out for the same two gauche and trans conformations of phenethylamine using Clementis 7s, 3p/3s basis contracted to a 2s 1p/1s set (Clementi *et al.*, 1967). The result again confirms a preference for the gauche conformation which, in this computation, is 1.8 kcal/mole below the trans one. We feel again inclined to consider it as probably representative for the series.

It must therefore be considered that the *ab initio* computations resolve the ambiguity of the PCILO results and point to a small but constant preeminence of the gauche conformation for the isolated molecule. Under

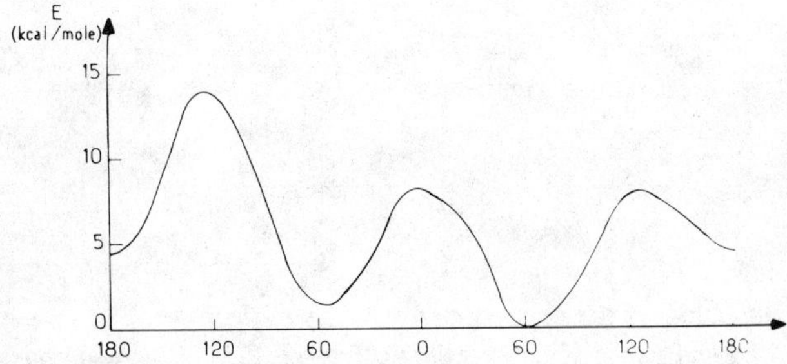

Fig. 32. *Ab initio* STO-3G conformational energy curve for norepinephrine for $\tau_1 = -90°$. Energies in kilocalories per mole with respect to the global energy minimum taken as zero energy.

these circumstances it seems appropriate to admit that it is the environmental forces which are responsible for the occurrence of the trans form exclusively in the crystal and predominantly in aqueous solution.

The situation prevalent in solution has been investigated in this case with the help of two methodologies: the supermolecule model (B. Pullman *et al.*, 1974a) and the hydration groups (shell) model (Weintraub and Hopfinger, 1973). The results obtained with the supermolecule model for amphetamine and norepinephrine (hydration limited to the cationic head) are shown in Figs. 33 and 34. It is seen that the conformationally allowed space for hydrated amphetamine is substantially reduced with respect to the free molecule and that, moreover, although the trans and gauche energy minima are still degenerate, there is a definite advantage for the trans form on the probability scale, as judged from the increase of the trans area included within the 0 kcal/mole isoenergy curve and the decrease of the similar gauche area. In fact, the explicit evaluation of the rotamer populations indicates 60% of the trans form and 40% of the gauche form which is in satisfactory agreement with experiment.

For hydrated norepinephrine the conformationally allowed space is still further reduced with respect to that of the isolated molecule and, moreover, the hydrated species has a global energy minimum for the trans conformation, with only a local energy minimum for a gauche form 1 kcal/mole above the global minimum. The explicit evaluation of the rotamer populations

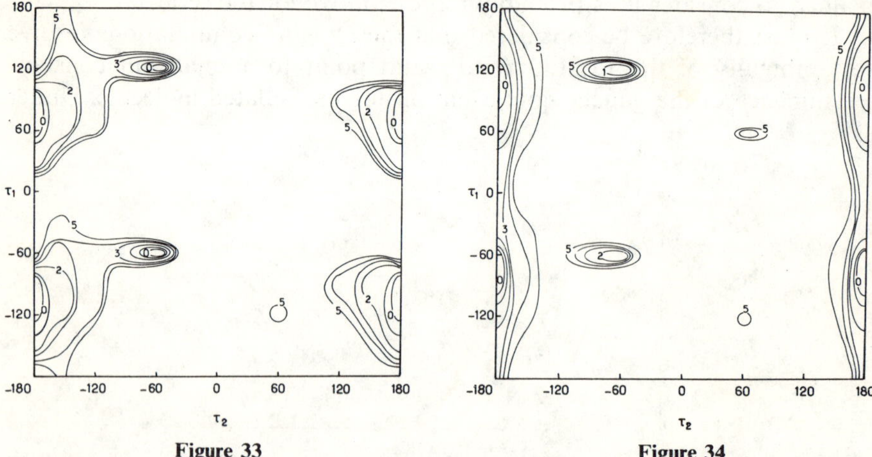

Fig. 33. The PCILO conformational energy map of hydrated amphetamine. Isoenergy curves (kilocalories per mole) with respect to the global energy minimum taken as zero energy.

Fig. 34. The PCILO conformational energy map of hydrated norepinephrine. Isoenergy curves (kilocalories per mole) with respect to the global energy minimum taken as zero energy.

indicates a trans/gauche ratio of 99/11%, again in satisfactory agreement with experiment.

In all, hydration seems to increase the probability of the extended form. These results account for the general characteristics of the behavior of phenethylamines in solution: the possibility for the coexistence of the trans and gauche forms and an increase in the proportion of the trans form when going from amphetamine to norepinephrine. No computations have been carried out for the hydrated forms of the other phenethylamines, the preceding results being considered as representative of the overall phenomenon. It may be remarked that the effect of hydration on the conformational energy map is altogether smaller in this mode of approach in the series of phenethylamines than in the histamines or indolalkylamines and also that the differences in energy between the trans and gauche conformers remain very small. This last result is in agreement with the evaluation of the energy difference between conformers from the experimentally observed populations in solutions, as obtained, for example, for dopamine by Bustard and Egan (1971).

On the other hand, the results obtained by Hopfinger's procedure with hydration groups (shell) are substantially different. In the series of phenethylamines considered here, data are available only for phenethylamine and amphetamine. For the cationic form of both molecules, these findings indicate that the effect of water produces a complete change in the conformational energy map: the energy minimum related to the gauche form moves to very high energies in their conformational energy map, thus leading to the prediction that these molecules should exist 98–99% in the trans form in solution. This is in striking disagreement with the NMR experimental results.

Recently (Germer, 1974), a CNDO/2 treatment has been carried out to ascertain the influence of hydration on the conformation of noradrenaline. Two classical models of the continuum type were used to describe the solvent effect. In the first model the molecule is pictured as a charge distribution within an infinite dielectric continuum of dielectric constant ε. The total energy of the system is assumed to be the sum of the molecular energy and the reversible work required to charge the dielectric medium:

$$U = -\left(\frac{1}{8}\pi\right)\left(1 - \frac{1}{\varepsilon}\right)\int_v \hat{D} \cdot \hat{D}\, d\tau,$$

where $\hat{D}$ is the electric displacement vector. In the second model the molecule is pictured as a point dipole contained within a spherical cavity of radius l in an infinite dielectric medium with dielectric constant ε and refractive index n. The total energy of the system is assumed to be the sum of the molecular energy and the solute–solvent interaction energy given by

$$U = -\hat{\mu} \cdot \hat{R},$$

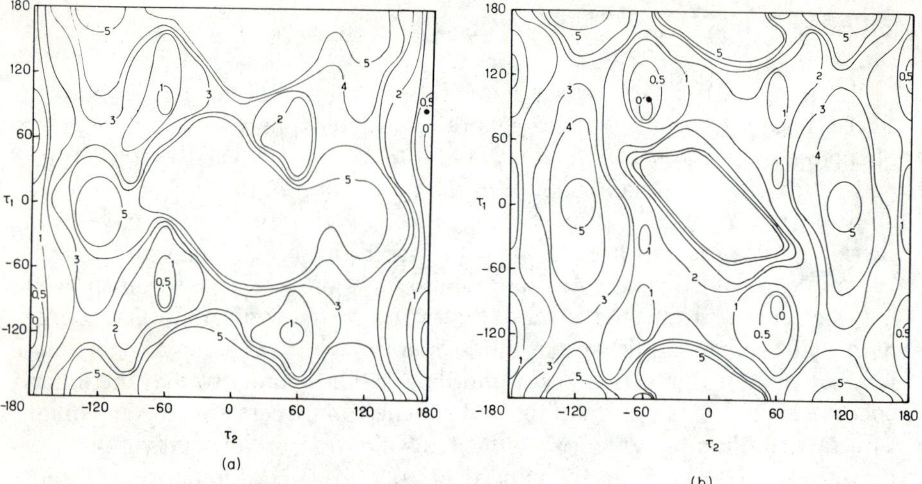

Fig. 35. Conformational energy maps of mescaline. Isoenergy curves in kilocalories per mole with respect to the global energy minimum taken as zero energy. (a) Input data from the crystal of mescaline hydrobromide (Ernst and Cagle, 1973); (b) input data from the crystal of mescaline hydrochloride (Tsoucaris *et al.*, 1973). (●) Experimental conformation.

where a reaction field,

$$\hat{R} = \left(\frac{2\hat{\mu}}{l^3}\right)\left[\frac{(\varepsilon - 1)}{(\varepsilon - 2)} - \frac{(n^2 - 1)}{(n^2 - 2)}\right],$$

is created within the cavity, $\hat{\mu}$ being the resultant molecular dipole.

The essential result of this study is to show that the effect of water on the conformation of noradrenaline is negligible. Unfortunately, since the author uses the CNDO/2 procedure which indicates that the trans conformation about τ_2 is appreciably more stable than the gauche ones for the free molecule (by 6 to 8 kcal/mole), his starting point is in contradiction with the indications of *ab initio* computations and represents probably an artifact of the method used.

Explicit PCILO computations have also been performed for mescaline (XXIV) (B. Pullman *et al.*, 1975b), which presents the curious phenomenon

(XXIV) MESCALINE

of having a gauche conformation in the crystal of its hydrobromide and a trans conformation in the crystal of its hydrochloride. The computations show (Fig. 35a,b) that the two different input geometries, taken from the two crystals, indicate in both cases close energy minima (separated by less than 1 kcal/mole) for the gauche and trans conformations, but, in each case, the global energy minimum corresponds to the crystalline conformation. An INDO computation with standard bond lengths and angles (Kang *et al.*, 1973b) predicts two energy minima at $\tau_1 = -90°$, $\tau_2 = 60°$ and $180°$.

VI. Acetylcholine

Acetylcholine (XXV) is the fundamental natural intercellular effector in nervous transmission systems, and its importance in this field has stimulated a great amount of experimental and, more recently, theoretical work on its structure both for its own interest and in comparison with related substrates of cholinergic systems. In fact, from the theoretical point of view, acetylcholine holds a record in that it has been examined by four empirical (Gill, 1965;

(XXV) ACETYLCHOLINE

$\tau_1 = \tau(C_6-O_1-C_5-C_4)$

$\tau_2 = \tau(O_1-C_5-C_4-N^+)$

Liquori *et al.*, 1968; Ajo *et al.*, 1972; Froimowitz and Gans, 1972) and eight quantum-mechanical computations: two carried out by the extended Hückel theory (Kier, 1967; Ajo *et al.*, 1973), one by the PCILO method (B. Pullman *et al.*, 1971), one by the INDO method (Beveridge and Radna, 1971), two by the CNDO/2 method (Saran and Govil, 1972; Ajo *et al.*, 1973), and two by the *ab initio* method (Genson and Christoffersen, 1973; Port and Pullman, 1973c; A. Pullman and Port, 1973b).

Acetylcholine is a flexible molecule containing eight single bonds. Nevertheless, there are only four important torsion angles (indicated as τ_0–τ_3 in structure XXV) and, in fact, only two essential such angles, τ_1 and τ_2. Thus, the methyl groups of the quaternary nitrogen may be taken in staggered positions, the trimethylethylammonium backbone being antiplanar with $\tau_3 = \tau(C_5-C_4-N^+-C_3) = 180°$. From other general studies on related situations (Maigret *et al.*, 1970), we may also fix the C_7 methyl group so that a C—H bond eclipses the $C_6{=}O_2$ double bond. Moreover, because of the partial double-bond character of the C_6-O_1 bond (similar to the partial double-bond character of the C—N bond of the peptide group

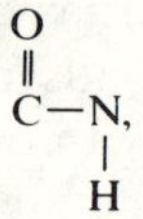

$$\overset{O}{\overset{\|}{C}}-\underset{H}{\underset{|}{N}},$$

the torsion angle $\tau_0 = \tau\,(C_7-C_6-O_1-C_5) = 180°$ (Pauling, 1968; Perricaudet and Pullman, 1973). We are, therefore, left with the two essential torsion angles: $\tau_1 = \tau\,(C_6-O_1-C_5-C_4)$ and $\tau_2 = \tau\,(O_1-C_5-C_4-N^+)$ which have, in fact, been the principal subject of the theoretical studies.

Let us start the description of the results with the presentation of the PCILO conformational energy maps, which were the first to give what now appears to be a satisfactory account of the conformational properties of this molecule. The problem was obscured for a short time by the question of the rule of the geometrical input data which was also solved by PCILO computations (B. Pullman *et al.*, 1971; B. Pullman and Courrière, 1972). Thus data are available in the literature corresponding to X-ray structure of two acetylcholine crystals: chloride (Herdklotz and Sass, 1970) and bromide

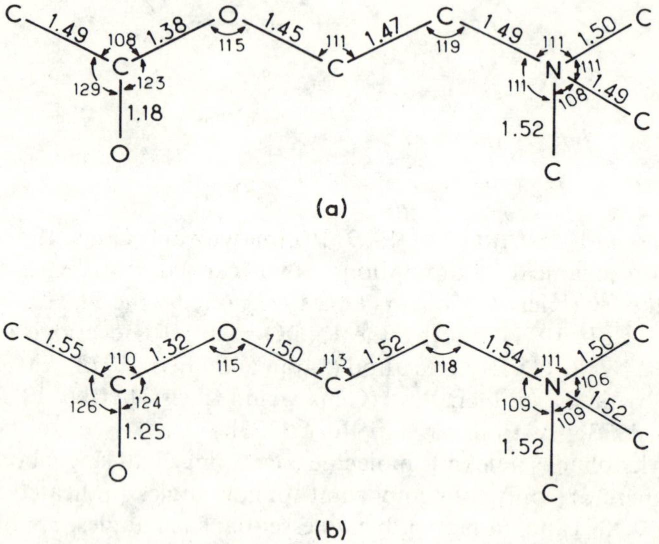

(a)

(b)

Fig. 36. Geometry of acetylcholine in crystals of the chloride (Herdklotz and Sass, 1970) and of the bromide (Canepa *et al.*, 1966).

(Canepa *et al.*, 1966). The geometry of the molecule differs appreciably in the two crystals (Fig. 36), and two PCILO conformational energy maps have been constructed using the two different input data. The results are presented in Figs. 37 and 38 (B. Pullman *et al.*, 1971; B. Pullman and Courrière, 1972b).

When considered within the same relatively large contour of, say, 6 kcal/mole above the global minimum isoenergy curve, the two maps resemble each other closely, both delimiting a comparable conformationally stable zone. On closer examination, however, distinct differences appear between the two. The essential one concerns the position of the global

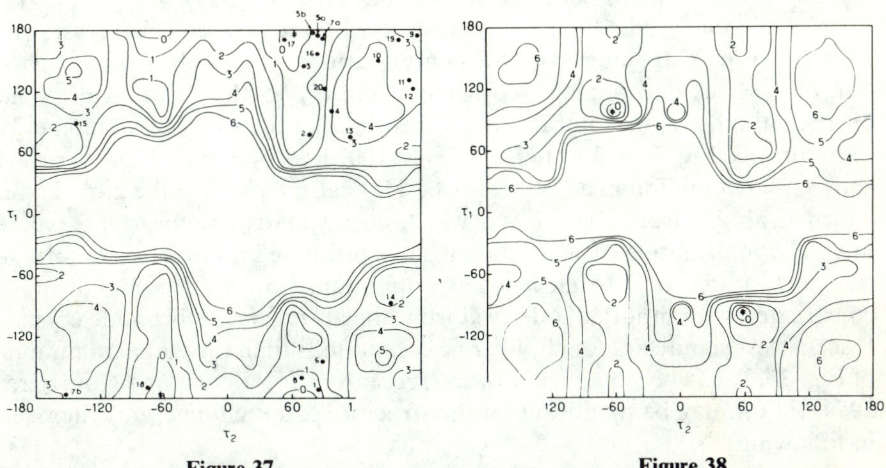

Fig. 37. The PCILO conformational energy map for acetylcholine, based on geometrical input data derived from crystal of acetylcholine chloride. Isoenergy curves expressed in kilocalories per mole with respect to the global minimum (‡), taken as zero energy. Shown are experimental conformations (●) in crystals of (1) acetylcholine chloride; (2) acetylcholine bromide; (3) L(+)-muscarine iodide; (4) L(+)-*cis*-2(*S*)-methol-4(*R*)-trimethylammonium methyl-1,3-dioxolane iodide; (5) 5-methylfurmethide iodide (a and b); (6) L(+)-*S*-acetyl-*β*-methylcholine iodide; (7) D(+)-*R*-acetyl-*α*-methylcholine iodide (a and b); (8) erythroacetyl-*α*-(*R*),*β*(*S*)-dimethylcholine iodide; (9) carbamoylcholine; (10) (+)-*trans*-2(*S*)-acetoxycyclopropyl-1(*S*)-trimethylammonium iodide; (11) acetylthiocholine bromide; (12) acetylselenocholine iodide; (13) (−)-*R*-3-acetoxyquinuclidine methiodide; (14) 2(*S*)-trimethylammonium-3(*S*)-acetoxy-*trans*-decahydronaphthalene iodide; (15) threoacetyl-*α*(*S*),*β*(*S*)-dimethylcholine iodide; (16) lactoylcholine iodide; (17) dimethylphenylpiperazine; (18) succinylcholine perchlorate; (19) muscarine iodide; (20) 2-methyl-4-trimethylammonium methyl-1,3-dioxolane iodide. The majority of the data are from a compilation by Baker *et al.* (1971, 1973) and by Pauling (1973).

Fig. 38. The PCILO conformational energy map for acetylcholine, based on geometrical input data derived from acetylcholine bromide crystal. Isoenergy curves expressed in kilocalories per mole with respect to the global minimum (‡) taken as zero energy.

minimum. This minimum is located in Fig. 37 at $\tau_1 = 180°$ and $\tau_2 = 60°$, which, together with the adopted values of $\tau_0 = \tau_3 = 180°$, correspond to a structure that may be described by the symbol TTGT, underscoring the gauche conformation of the O_1 and N^+ atoms and the trans arrangement of the remaining atoms of the backbone. A local minimum, about 1 kcal/mole above the global one, is found on Fig. 37 at $\tau_1 = 120°$ and $\tau_2 = -80°$. It may usefully be stressed that around these minima, and in particular around the global one, there are large plateaus of low energy, as illustrated by the relatively broad contour of the 1 kcal/mole isoenergy curve. The minima, and in particular the global one, are therefore associated with a considerable flexibility, especially as concerns the variation of τ_1.

The experimental conformation of acetylcholine in its chloride crystal, as shown in Fig. 37 together with the conformations of a number of acetylcholine derivatives and analogs, correspond closely to the theoretical global energy minimum. In contrast to the results of Fig. 37, the global energy minimum of Fig. 38 is located at $\tau_1 = -100°$ and $\tau_2 = 60°$. This map also shows a local minimum at $\tau_1 = \tau_2 = 60°$, 2 kcal/mole above the global one, which is absent from Fig. 37 and which corresponds closely to the experimental conformation of acetylcholine in its bromide crystal. This conformation corresponds only to a local energy minimum, however, even on the map constructed with input data derived from the acetylcholine bromide crystal. It seems less significant for cholinergic compounds than the global minimum of Fig. 37. It has been shown, however, recently (B. Pullman and Courrière, 1974) that it may be significant for the structure of anticholinergic drugs (see in following).

This result underscores the importance of the geometrical input data. The calculations indicate that fundamentally the structure of acetylcholine derived from the crystal of the chloride is slightly more stable (1–2 kcal/mole) than that derived from the crystal of the bromide. Moreover, the perturbing crystal forces seem to be more pronounced in the bromide. It is probably for this reason that the conformational energy map constructed with the chloride input data gives a better overall account of the conformations of acetylcholine and its derivatives than does the map based on the bromide crystal data. Under these circumstances, it may be stated that the essential result of the PCILO computations on acetylcholine in the free state is the clear prediction that the most stable conformation is a gauche one with respect to τ_2. The fully extended conformation ($\tau_1 = \tau_2 = 180°$) represents a local energy minimum 3 kcal/mole above the global one. The barrier between these two minima is about 4 kcal/mole which precludes an easy transition between the two.

This preference for a gauche conformation results to a large extent from the electrostatic interactions between the cationic head and the esteryl oxygen (see Section III for similar results in the series of antihistaminic drugs). The importance of these interactions for the production of a gauche conformation in the O—C—C—N^+ or similar systems was first emphasized by Sundaralingam (1968) and later verified by PCILO computations on acetylcholine. Since then it is frequently "rediscovered" by a number of authors (Ajo *et al.*, 1972; Eliel and Alcudia, 1974; Terui *et al.*, 1974).

We may now compare these findings with some theoretical results. Let us consider first the "empirical" computations, which, in fact, were the first performed for this molecule. Thus, the first empirical treatment of acetylcholine (Gill, 1965), which was based on steric requirements and interaction energies, indicated that the fully extended conformation should be much more stable than any folded one for this molecule. More elaborate empirical calculations carried out by Liquori *et al.* (1968) led to the prediction of four highly localized energy minima, none more than 0.7 kcal/mole above the global one. In order of increasing energy, Liquori *et al.* characterized these four preferred conformations as shown in Table II. According to these computations the most stable conformation corresponds also to the totally extended form, followed by one in which the O_1 atom is still trans to the N^+ atom. On the other hand, the third (γ) conformation on their scale corresponds to a gauche arrangement of the O—C—C—N^+ fragment. However, it is only 0.35 kcal/mole above the fundamental trans one, suggesting a relatively easy transition between the two forms.

TABLE II

CONFORMATIONAL ENERGY MINIMA FOR ACETYLCHOLINE FOLLOWING THE EMPIRICAL TREATMENT OF LIQUORI *et al.* (1968)

Symbol	τ_1	τ_2	ΔE (kcal/mole)
α	180°	180°	0.00
β	75°	180°	0.28
γ	177°	77°	0.35
δ	75°	77°	0.69

More recently, however, the same group of authors (Ajo *et al.*, 1972), obviously influenced by the results of the quantum-mechanical computations and their successful correlation with available X-ray data "refined" the empirical computations by adding electrostatic interactions to the van der Waals nonbonded interactions and torsional potentials, which were the only energy components taken into account in their previous empirical treatment. This refinement, although retaining the same four (α–δ) energy minima, reverses their relative order and indicates that the gauche form (minimum γ) may be expected to be more stable by 1–2 kcal/mole than the fully extended one. These refined empirical results show better agreement with the PCILO estimates.

On the other hand, in a still more recent empirical treatment (Froimowitz and Gans, 1972), in which the conformational energy is taken to be the sum of nonbonded and electrostatic pairwise interactions, results that give little credit to this type of treatment are obtained. Table III indicates the energy minima obtained and their relative values with respect to the global energy minimum taken as energy zero. This global energy minimum (A_1) at $\tau_1 = 70°$, $\tau_2 = -135°$ and the close local energy minimum (A_2) at $\tau_1 = -70°$, $\tau_2 = 235°$ do not correspond to any significant, observable conformation of acetylcholine. The gauche form, which represents the global energy minimum of the PCILO computations and the most common form observed in the crystal of acetylcholine derivatives, is only the fifth local energy minimum, 3.65 kcal/mole above the global one. Strange energy minima (D_2, E) occur at unusual values of τ_1 and τ_2. This situation clearly shows that because of their arbitrariness, diversity, and manifest technical deficiencies (e.g., large uncertainly about the value of the dielectric constant to be used in the computation of the electrostatic component of the energy), the empirical computations can be considered neither reliable nor unequivocal.

As to the remaining quantum-mechanical computations, their significance was somewhat obscured by the use of different type of geometrical input data, mentioned above. Taking these factors into account, it seems that at present the extended Hückel theory predicts a gauche minimum at $\tau_1 = 180°$, $\tau_2 = 80°$ as the global one, together with a local energy minimum

TABLE III

CONFORMATIONAL ENERGY MINIMA FOR ACETYLCHOLINE FOLLOWING THE EMPIRICAL COMPUTATIONS OF FROIMOWITZ AND GANS (1972)

Symbol	τ_1	τ_2	ΔE (kcal/mole)
A_1	70°	−135°	0.0
A_2	−70°	135°	0.29
B_1	70°	70°	2.74
B_2	−70°	−70°	2.75
C_1	−175°	75°	3.65
C_2	175°	−70°	3.70
D_1	180°	−175°	4.11
D_2	170°	−125°	4.27
E	180°	−5°	7.37

for the extended form ($\tau_1 = \tau_2 = 180°$), about 2 kcal/mole above the global one. However, in contradiction to the generally accepted stereochemical viewpoint, substantiated by a significant amount of theoretical and experimental data (Perricaudet and Pullman, 1973; Jones *et al.*, 1972), these extended Hückel theory calculations (Kier, 1967) indicate that the carbonyl group is free to rotate 60° to either side of the planar-cis arrangement with respect to the O(1)—C(5) bond; i.e., the torsion angle τ_0 shows a constant energy value between 120° and 240°.

The CNDO method (Saran and Govil, 1972) seems in this case to give results similar to the PCILO method. The INDO computations (Beveridge and Radna, 1971), which have been performed only with the geometry of the bromide crystal as input data, are not too different from the corresponding PCILO results.

The problem, which could thus have been considered as clarified by the above-described results, has been, however, somewhat reopened recently by an *ab initio* computation (Genson and Christoffersen, 1973) which surprisingly obtained the trans (extended) form as the most stable one for acetylcholine with, moreover, a relatively high energy (10–18 kcal/mole) with respect to the gauche forms. These results were obtained, however, by a particular *ab initio* procedure recently developed (Christoffersen, 1972) for the treatment of large molecules, where the molecular orbitals are built as linear combinations of predetermined Gaussian orbitals of simple molecular fragments. Another feature of this computation is the use of average bond lengths and idealized hybridization for bond angles in the input geometries. When computations are performed by a usual *ab initio* SCF method, using STO-3G basis set (program Gaussian 70) quite different results are obtained (A. Pullman and Port, 1973b; Port and Pullman, 1973c). These results are indicated in Fig. 39, where the isoenergy lines (in kilocalories per mole) are traced with respect to the global energy minimum taken as zero energy. This minimum corresponds to $\tau_1 = 150°$, $\tau_2 = 60°$ and represents a gauche arrangement about the C(4)—C(5) bond. It is a broad minimum and the only really significant one. The fully extended trans form ($\tau_1 = \tau_2 = 180°$) is in a plateau region about 4 kcal/mole above the global minimum. The agreement with the PCILO computation is substantial.

In an attempt to elucidate the reasons for the disagreement between the two *ab initio* treatments, STO-3G computations have been carried out (Port and Pullman, 1973c) for three essential conformations of acetylcholine, using the geometry of Genson and Christoffersen (1973). The results are given in Table IV. It is seen that even with these authors' geometry the near-gauche form is, within the usual *ab initio* procedure, the most stable one. Moreover, the energy differences among the trans, gauche, and near-gauche forms appear much smaller and more in line with other theoretical predictions.

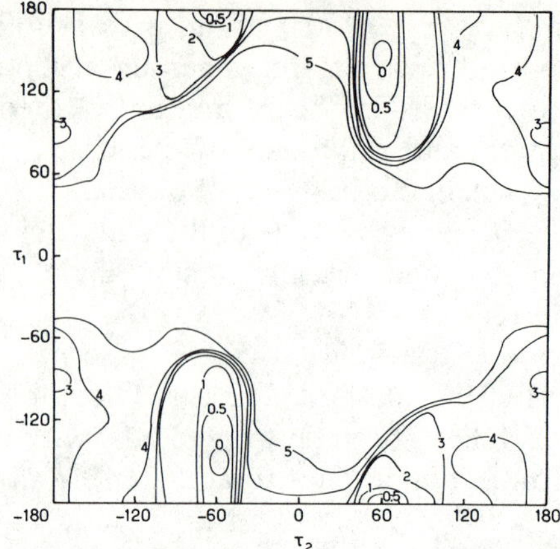

Fig. 39. *Ab initio* (STO-3G) conformational energy map of acetylcholine. Energy contours in kilocalories per mole above global minimum taken as zero energy.

A close examination of scale models corresponding to the various forms, as obtained by using Genson and Christoffersen's geometry, indicates that, in the gauche conformation, there is a very close approach of a methyl group of the onium head to the ester oxygen (the contact being, in this case, much closer than when the chloride geometry is used). It seems very likely that it is this close approach which is responsible for the appreciable rise in energy found by Genson and Christoffersen in the region of the gauche forms. It seems then probable that a gain in stability could be obtained for the gauche

TABLE IV

ENERGIES OF CONFORMERS OF ACETYLCHOLINE USING THE GEOMETRY OF GENSON AND CHRISTOFFERSEN (1973)

				ΔE (kcal/mole)	
Form	τ_1	τ_2	τ_3	SCF[a]	[b]
Trans	180°	180°	180°	0	0
Near-gauche	180°	80°	180°	−0.8	+10
Gauche	180°	60°	180°	+3.4	+19
Gauche	180°	60°	160°	−0.8	—

[a] SCF *ab initio* results.

[b] Results of Genson and Christoffersen (1973) (molecular fragments).

form by twisting the cationic head $[N^+(CH_3)_3]$ out of the standard conformation adopted for it in this and in all other calculations that correspond to $\tau_3 = 180°$. In fact, as shown in the last line of Table IV, a rotation of 20° of the cationic head ($\tau_3 = 160°$) stabilizes the gauche form ($\tau_1 = 180°$, $\tau_2 = 60°$) sufficiently to make it more stable than the trans form.

In summary, it may be stated that the preference of acetylcholine for a trans conformation with respect to τ_1 but gauche with respect to τ_2 is firmly established by computations for the free molecule.

VII. Conformation of Acetylcholine in Solution

In the preceding discussion, the results of the theoretical computations have been compared exclusively with the X-ray crystallographic data of the acetylcholine molecule. We may now usefully compare these data with the results of computations in solution, the more so as the conformation of acetylcholine and its derivatives in solution has been studied experimentally quite extensively by a number of investigators (Canepa, 1965; Culvenor and Ham, 1966, 1970; Cushley and Mautner, 1970; Casy *et al.*, 1971; Mautner *et al.*, 1972; Partington *et al.*, 1972; Chynoweth *et al.*, 1973; Terui *et al.*, 1974; Eliel and Alcudia, 1974; Lichtenberg *et al.*, 1974; Mautner, 1974), generally with the help of the NMR technique. As with the crystal, there are three different theoretical approaches for studying the solvent effect on molecular conformation.

In general, experimental studies indicate that the trans–gauche conformation with respect to τ_1 and τ_2, characteristic of the acetylcholine chloride crystal, is essentially preserved in water. The most recent and elaborate evaluation (Lichtenberg *et al.*, 1974) indicates the presence of 91% of this gauche form at room temperature. It also estimates at 65°–69° the value of τ_2 and at no more than 1 kcal/mole the value of ΔG^0 between the two forms at room temperature. A number of other authors indicate the proportion of the gauche form to be practically equal to 100% (Partington *et al.*, 1972; Terui *et al.*, 1974). The conformation of acetylcholine in solution seems to be independent of the counterion, Cl or Br^- (Terui *et al.*, 1974).

Noticeable differences seem, on the other hand, to characterize the results of the theoretical evaluation of the solvent effect on the conformational properties of acetylcholine. Three such studies, as just mentioned, are available.

1. An investigation by the continuum model due to Beveridge *et al.* (1974a,b). In order to understand fully the meaning of their results, it must be pointed out that these authors use a starting data for the free molecule the results of INDO computations, obtained with the bromide crystal structure

as input data. In this computation the global energy minimum for the free molecule at $\tau_1 = -90°$, $\tau_2 = 50°$ does not correspond to any known structure, the conformation of acetylcholine in its bromide crystal is represented by a local energy minimum at $\tau_1 = 50°$, $\tau_2 = 50°$, 3.7 kcal/mole above the global one, and its conformation in the chloride crystal by another local energy minimum at $\tau_1 = 180°$, $\tau_2 = 40°$, 6.3 kcal/mole above the global minimum, and the extended form ($\tau_1 = \tau_2 = 180°$) by a local minimum 10 kcal/mole above the global one. According to these authors, the effect of the solvent is to destabilize the various synclinal (60°) minima in both variables. The global minimum of the solvated molecule occurs at $\tau_1 = 160°$, $\tau_2 = 100°$, which the authors consider as representing a trans–gauche conformation, akin to the one found in the acetylcholine chloride crystal. A local energy minimum occurs for the fully extended hydrated conformation, 4 kcal/mole above the global minimum. The solvent has, thus, in this treatment, the double effect of favoring the increase in τ_1 and τ_2 and of leaving a relatively large energy difference between the gauche and trans forms.

2. An investigation by the hydration groups (shell) model due to Weintraub and Hopfinger (1974). These authors claim that the effect of water promotes the existence of the antiplanar, conformation with respect to both τ_1 and τ_2 so that, on the free energy scale, the extended form of acetylcholine $\tau_1 = \tau_2 = 180°$) should be by about 2–3 kcal/mole more stable than the gauche form and should thus predominate strongly in solution. It is easily seen that this result is due primarily to the strong hydration in these authors' methodology of the ester oxygen of acetylcholine and of the hydrogen atoms bound to two of the methyl carbons at the cationic head.

Because this result is an obvious contradiction to the experimental findings, the authors (Weintraub and Hopfinger, 1974) proposed, therefore, that an addition must be made to their molecule–solvent model when it is used to estimate conformational preferences in small molecules. The parameters used by these authors in their general theory (and applied, in the example quoted in Section V, to amphetamine) are, according to their new concept, derived for a static equilibrium state; that is, the water molecules are considered "bound" to the solvation groups. In reality, they say, there is a *dynamic* interaction between the water molecules and solvation groups. These dynamic interactions give rise to *modes of solvation*. If there is one highly favorable conformation in certain solvation modes and a second different conformation in other solvation modes, then, neglecting other considerations, the "average molecule" will preferentially adopt that conformation in which the product of the time in a given conformational state with the number of ways that state can be realized is largest. That is, if $\langle \tau_U \rangle$ and $\langle \tau_S \rangle$ are the average times the molecule spends in conformations U and S, respectively, and N_U and N_S are the number of solvation modes which promote conformations U and S, respectively, then the conformational weights $W(\mathrm{U})$ and $W(\mathrm{S})$ are

$$W(\mathrm{U}) = \left(\frac{\langle\tau_{\mathrm{U}}\rangle}{\langle\tau_{\mathrm{U}}\rangle + \langle\tau_{\mathrm{S}}\rangle}\right)\left(\frac{N_{\mathrm{U}}}{N_{\mathrm{U}} + N_{\mathrm{S}}}\right),$$

$$W(\mathrm{S}) = \left(\frac{\langle\tau_{\mathrm{S}}\rangle}{\langle\tau_{\mathrm{U}}\rangle + \langle\tau_{\mathrm{S}}\rangle}\right)\left(\frac{N_{\mathrm{S}}}{N_{\mathrm{U}} + N_{\mathrm{S}}}\right).$$

The molecule will preferentially adopt that conformational state corresponding to the larger of $W(\mathrm{U})$ and $W(\mathrm{S})$.

If such an addition is made to the authors' general theory, then the gauche form of solvated acetylcholine becomes probable, in agreement with experiment. The significance of this addition itself, to a scheme that dealt already with free energy and thus included entropy effects, seems, however, conceptually uncertain.

3. An investigation by the supermolecule model carried out in our laboratory. In this particular case, because of the absence of *strong* hydration sites, of the type present when N^+—H bonds are available on the solute molecule, it is not expected that the molecules of water of the first hydration shell will remain attached to the solute during conformational changes. It is rather supposed that they will float around the solute, forming weak, easily broken and re-formed bonds. For this reason, instead of computing conformational energy maps of the hydrated species, as was done for the molecules previously dealt with in this paper, the procedure was adopted to compute the form of the first hydration shell and its energy of interaction with the solute for both the trans and gauche conformations.

Thus, after fixing acetylcholine in these two conformations, the principal hydration sites were computed for both of them, and the energy of interaction in these two forms was evaluated by *ab initio* STO-3G computations (B. Pullman *et al.*, 1975a). The principal hydration sites thus obtained are shown in Figs. 40 and 41 for the trans ($\tau_1 = \tau_2 = 180°$) and gauche ($\tau_1 = 180°, \tau_2 = 60°$) conformers, respectively. It is seen that in both cases, 5 molecules of water seem to be bound, with medium strength, to the solute molecule. Three of them (on the right-hand side of the figure) are essentially bound to the cationic head, but much less strongly, of course, than in the previous cases of an N^+H_3 head. A fourth water molecule, at the upper part of the figure, is in a special situation. It is bound to the carbonyl oxygen but its relatively strong energy of interaction (−8.2 kcal/mole) with *trans*-acetylcholine is due to its simultaneous interaction with the cationic head. No such strong interaction is available to this water molecule with *gauche*-acetylcholine. On the other hand, a fifth water molecule feebly bound to the

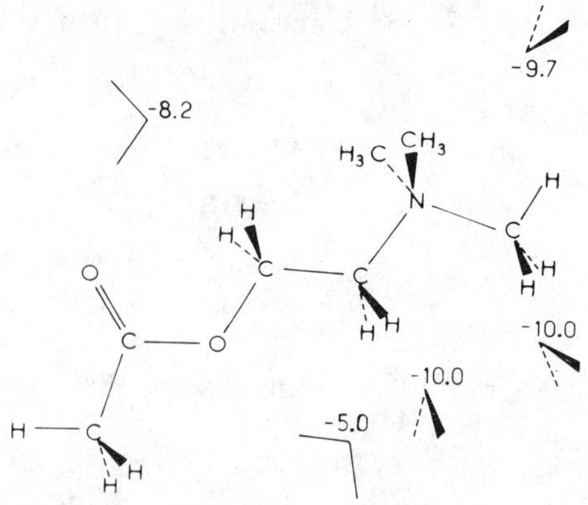

Fig. 40. Hydration scheme of *trans*-acetylcholine following the supermolecule model.

ester oxygen in the trans conformer has a somewhat higher energy of interaction with the gauche conformer due to its simultaneous attraction by the cationic head. Although it is known that the simultaneous fixation of a few water molecules on a solute modifies somewhat individual binding energies (Port and Pullman, 1973b), it seems probable that the modifications should be quite similar for the two forms of acetylcholine. With this assumption, it is found that the stabilization of the trans conformation of acetylcholine, due to its interaction with the first hydration shell of water molecules, should be about 2.2 kcal/mole greater than for the gauche one. When it is remembered that the *ab initio* computations on free acetylcholine indicate that the gauche form of this molecule is about 3 kcal/mole more stable than the trans one, it becomes obvious that the effect of the solvent is to bring the two forms much closer together, with still a preference for the gauche one, but which would only be less than 1 kcal/mole more stable than the trans one.

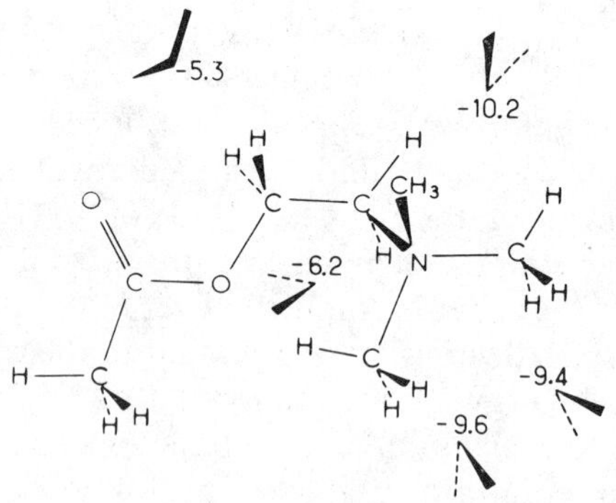

Fig. 41. Hydration scheme of *gauche*-acetylcholine following the supermolecule model.

Although it is indisputable that this particular supermolecule treatment as applied to acetylcholine is a very crude one and that its results cannot prejudge about its general significance, it is, nevertheless, interesting to remark that it seems to account satisfactorily, more so than the two other treatments, for the behavior of acetylcholine in solution.

Thus, although it is a common feature of the three procedures, and thus probably an indication of a real physical significance, that the interaction with water favors a displacement of the equilibrium toward the trans form, it appears probable that the displacement is overestimated in the Weintraub–Hopfinger procedure leading to a predominance on the free-energy scale of the trans form. As recognized during the discussion of their paper by one of the authors, this could well be due to their overevaluation of the free energy of hydration of the ester oxygen. The evaluation of the energies of interaction of water molecules with acetylcholine, illustrated in Figs. 40 and 41, shows also that allowing fixed contributions to the free energy of solvation by the different atoms and groups present in a solute is a rough approximation, which, by neglecting variations due to the effect of neighboring or sometimes even distant groups on the energies of interaction of water molecules with the solute, precludes the study of more sensitive cases. Actually, the authors establish agreement with experiment by introducing the concept of *solvation modes*, but, as their previous results were already obtained in terms of *free* energy, it is difficult to visualize the significance of this correction.

The results of Beveridge *et al.* (1974a, b) predict a gauche form in solution, which would be associated with $\tau_2 = 100°$ and, thus, quite distorted with respect to the value of 65°–69° indicated by Lichtenberg *et al.* (1974) as the plausible one on the basis of NMR spectroscopy. It is probable that this divergence springs from a defect of the methodology, namely, the use of the "dipole" approximation in the computation of the electrostatic component of the interaction energy. This component is by far the one whose contribution predominates in the overall value of the total solute–solvent interaction energies. As it is the square of the dipole moment that appears in the expression of the electrostatic component, the variation of the dipole moment has a decisive influence on the value of this component. For acetylcholine the value of this moment varies, according to Beveridge *et al.*, from 1.5 to 9.8 D, the latter value corresponding to the fully extended form. The dipole approximation in the computation of the electrostatic energy of solute–solvent interaction favors thus the trans form with respect to τ_1 and τ_2. Now, it is obvious that this is a rather unsatisfactory approximation because when a water molecule approaches acetylcholine closely, it does not "feel" the molecule as a dipole but rather as a series of local monopoles, the most effective being those which are close to the approaching water and whose charges do not vary much with changes in conformation. The overall variation of the interaction of a first hydration shell with conformation must, therefore, be much smaller than that computed with a dipole moment approximation. The variation of electrostatic energy of interaction in sterically allowed regions is 15 kcal/mole, as reported by Beveridge *et al.* From the supermolecule model, it is probably only of the order of 3 kcal/mole. Very likely, one may ascribe to this deficient approximation the results of the continuum model, which overestimate the value of τ_2 for the preferred solvated conformation as well as the differences in energy between different conformations in solution. The adoption of a spherical cavity for a rather elongated molecule such as acetylcholine may be another source of discrepancies.

On the whole, the supermolecule treatment of the solvation effect suggests a much smaller influence of the solvent on the conformation properties of acetylcholine than do the other procedures. This point of view seems more in line with the conclusions of the experimentalists, as explicitly expressed by Casy *et al.* (1971), who, after having studied by NMR spectroscopy the behavior of acetylcholine in a number of solvents, conclude that "The conformation of acetylcholine appears to be little influenced by the solvent. This result suggests that the conformation of AcCh is governed primarily by interactions within the molecule." This is also our viewpoint.

It may be remarked that the order of magnitude found by the supermolecule model for the *internal energy* difference between the gauche and trans forms of solvated acetylcholine ($\approx$ 1 kcal/mole) is of the order of magnitude of the difference in free energy between these conformers as found by Lichtenberg *et al.* (1974).

VIII. Acetylcholine Derivatives and Analogs

The conformational properties of a large number of acetylcholine derivatives have been studied both experimentally (in the crystal and in solution) and theoretically. Although nearly each such derivative presents its own specific and frequently most interesting problems, we cannot discuss them all here. We have selected a few compounds that are of particular significance for the problems discussed in this paper.

A. Acetylthiocholine (XXVI)

At first sight Fig. 37 accounts satisfactorily for the distribution of observed conformations in a large number of acetylcholine derivatives and analogs, most of them located in regions for which low conformational energies are obtained by computation. A closer examination shows,

$H_3C-C(=O)-S-CH_2-CH_2-N^+(CH_3)_3$

(XXVI) ACETYLTHIOCHOLINE

$H_2N-C(=O)-O-CH_2-CH_2-N^+(CH_2-CH_3)_2-CH_2-C_6H_5$

(XXVII)

$H_2N-C(=O)-O-CH_2-CH_2-N^+(CH_3)_2-CH_2-CH_3$

(XXVIII)

$C_6H_5-C(=O)-O-CH_2-CH_2-N(CH_3)_2$

(XXIX)

$C_6H_5-C(=O)-O-CH_2-CH_2-N^+H(CH_3)_2$

(XXX)

$H_2N-C_6H_4-C(=O)-O-CH_2-CH_2-N^+H(C_2H_5)_2$

(XXXI) PROCAINE

however, that from a conceptual point of view, the situation is more complicated. Although the majority of the derivatives adopt the gauche conformation in the central $-N^+-C-C-O-$ fragment, some of them, e.g., acetylthiocholine (compound 11 of Fig. 37), adopt a trans conformation in the crystal (Shefter and Mautner, 1969) and reside in a region of relatively high conformational energy (4 kcal/mole above the global minimum). A PCILO conformational energy map was constructed (B. Pullman and Courrière, 1972), therefore, for acetylthiocholine itself, with striking results (Fig. 42). The allowed conformational space (within the same limit of 3 kcal/mole above the global minimum) decreased considerably; moreover, the region of energy minimum connected with the gauche conformation disappeared completely, and a new global minimum appeared at $\tau_1 = 60°-80°$ and $\tau_2 = 180°$, corresponding to a trans conformation close to the observed one. This example shows the extreme usefulness of constructing *individual* conformational energy maps. It also confirms the electrostatic nature of the forces, in particular of the interaction between the ester oxygen and the cationic head, in determining the preference of acetylcholine for a gauche conformation.

B. Derivatives Modified at the Cationic Head

These particular modifications are interesting to consider not only from the conformational point of view but also because they touch upon an aspect of the structure of a large number of pharmacological compounds, including the majority of those discussed here, which is frequently misrepresented.

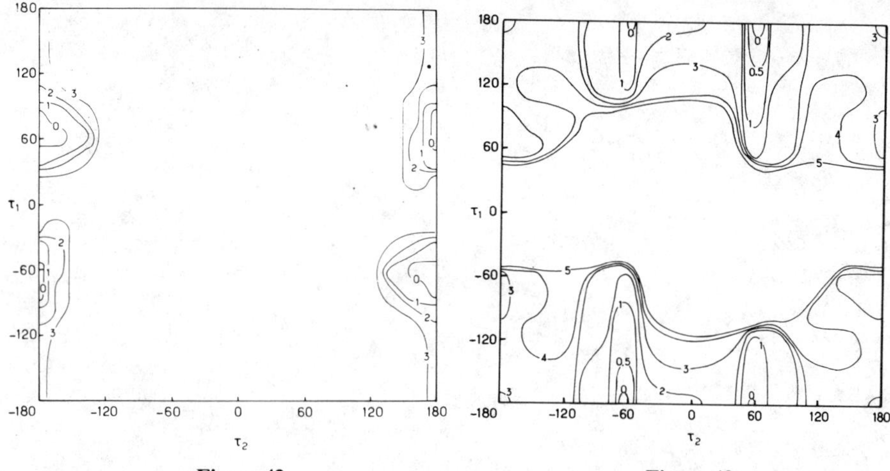

Figure 42 Figure 43

Fig. 42. The PCILO conformational energy map for acetylthiocholine. Isoenergy curves (kilocalories per mole) with respect to the global minimum taken as zero energy. (●) Crystallographic conformation.

Fig. 43. Conformational energy map for

$$CH_3-\overset{\overset{\displaystyle O}{\|}}{C}-O-CH_2-CH_2-N^+(C_2H_5)_5\,.$$

The most active compounds in the parasympathomimetic series contain the $-N^+(CH_3)_3$ group as cationic head. Successive replacement of the methyl groups by either hydrogen or ethyl leads to a steep decline in parasympathomimetic activity (Ariens *et al.*, 1964; Moran and Triggle, 1971). The influence of these structural modifications on the conformation of acetylcholine was investigated by the PCILO method (B. Pullman and Courrière, 1973).

Figure 43 represents the results of computations for the triethyl analog of acetylcholine. The conformational energy map appears very similar to that of acetylcholine. In particular the doubly degenerated, global energy minimum corresponds to a gauche arrangement of the N^+ and esteric O atoms ($\tau_2 = \pm 60°$, $\tau_1 = 180°$). The extended form (at $\tau_1 = \tau_2 = 180°$) represents a local energy minimum 3 kcal/mole above the global one.

This result must be considered as particularly significant, especially because it was presumed by some authors (Shefter, 1971) that the presence of larger alkyl groups on N^+ would diminish the tendency of the molecule to adopt a gauche conformation. Crystallographic evidence, available essentially in the series of carbamoylcholines, confirms that it need not be so: e.g., 2-*N*,*N*-diethyl-*N*-benzylammoniumethylcarbamate bromide (XXVII) exists in the crystal in the gauche conformation, as does also 2-*N*,*N*-dimethyl-*N*-ethyl-ammoniumethylcarbamate (XXVIII) (Babeau and Barrans, 1970).

It is therefore obvious that the decrease of parasympathomimetic activity in the N^+-triethyl derivative cannot be ascribed to a change in molecular conformation. On the other hand, it may reasonably be attributed, at least in part, to (1) the modification of the dimensions of the cationic head and (2) the modification in the electronic properties of the cationic head, both of which may profoundly perturb or even preclude the interaction of this head with the anionic receptor site. Whereas the role of the possible modification of the dimensions is straightforward, the nature of the possible modifications in the electronic properties raises a problem that has attracted much attention.

One of the most striking theoretical PCILO results (B. Pullman and Courrière, 1973) on the electronic structure of the cationic head of acetylcholine was to point out that the net positive charge, which in the usual chemical representation is localized on the quaternary N atom, is, in fact, distributed among the adjacent methyl and methylene groups, leaving the "N^+" atom nearly neutral (Fig. 44a). We are considering here, of course, the *total* net electronic charges, i.e., a summation of the net σ- and π-charges, where "net" charges denote an excess or deficit at each atom in relation to the number of electrons the atom would possess in an isolated state. Thus, in acetylcholine, 70% of the net positive charge is distributed among the three attached methyl groups, essentially among their hydrogens which carry thus each about 0.07 positive electronic charge as opposed to the usual positive charge of 0.04*e* found generally on H atoms linked to saturated carbons. These three methyl groups thus form a large ball of spread-out positive electricity to which the designation of a hydrophilic cationic center is appropriate. The remaining fraction of the positive charge seems to be concentrated on the two CH_2 groups of the backbone of acetylcholine, further enlarging the dimension of the cationic moiety of this molecule.

The general features of these results are confirmed by *ab initio* computations (Fig. 45) (A. Pullman and Port, 1973b). On the other hand, the INDO method (Beveridge and Radna, 1971) gives a somewhat different and probably less correct description: it attributes the highest positive charges ($\approx 0.12e$) to the C atoms of the CH_3 groups, a positive charge of about $0.11e$ to the quaternary nitrogen, and positive charges of only 0.03 to the hydrogens of the CH_3 groups.

We may now consider the electronic state of the cationic head in the $-N^+(C_2H_5)_3$ analog of acetylcholine (Fig. 44b). A drastic change is observed with respect to the situation in acetylcholine. The excess positive charge of the cationic head is now shared by a substantially increased number of hydrogen atoms with the result that the net positive charge

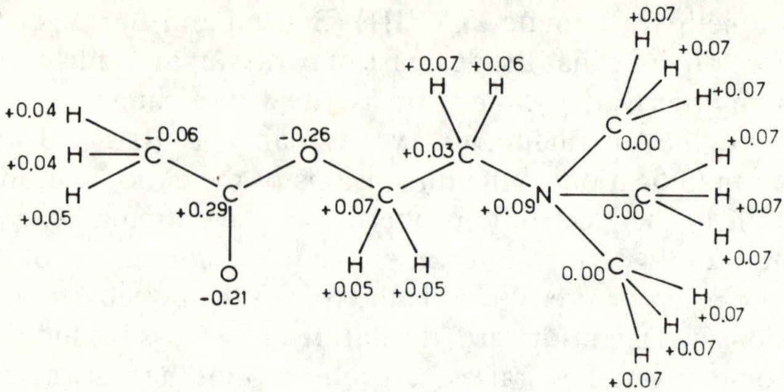

(a)

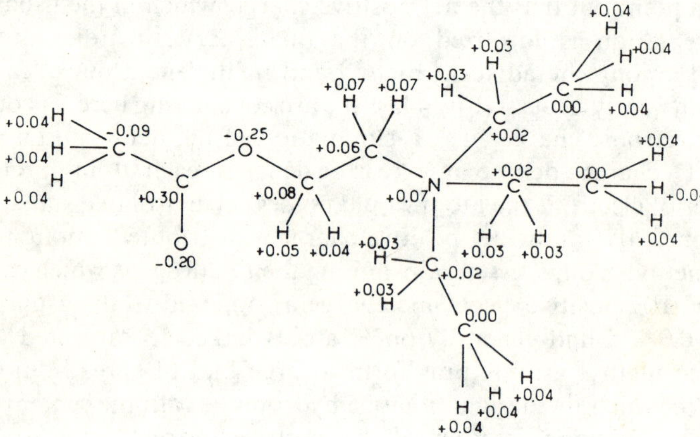

(b)

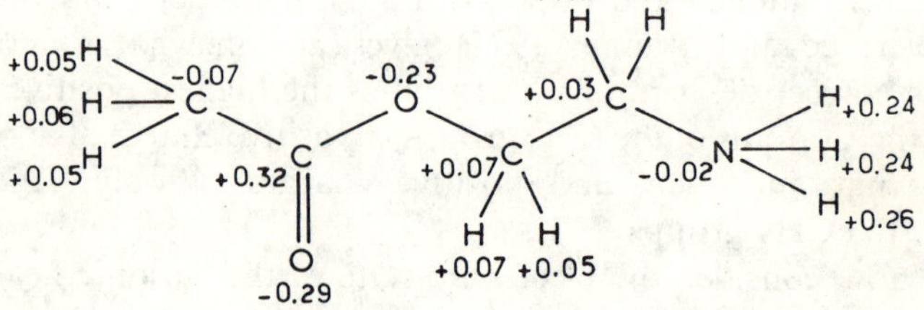

(c)

Fig. 44. Distribution of net electronic charges in (a) acetylcholine, (b) the triethyl analog, and (c) acetylethanolamine (PCILO method).

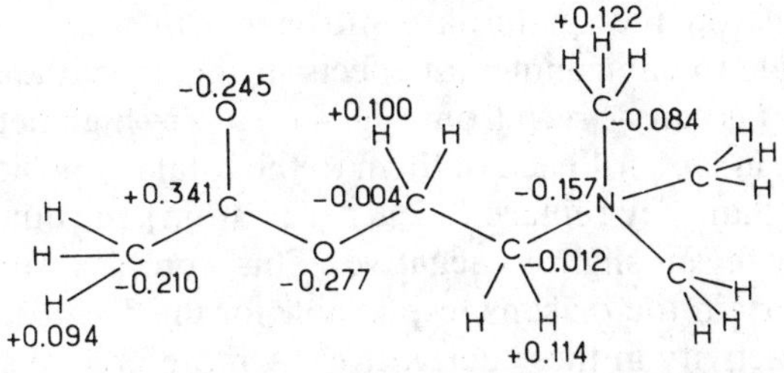

Fig. 45. Net electronic charges in acetylcholine by *ab initio* method (STO-3G).

carried by each of them (0.03*e* on the hydrogens of the CH_2 groups and 0.04*e* on the hydrogens of the CH_3 groups of the ethyl substituents) is now of the order of magnitude of the charge carried usually by hydrogens attached to saturated carbons. (It can be seen that the charge of the terminal H atoms of the onium head is identical to the charge carried by the hydrogens of the methyl group of the ester terminal.) These hydrogens loose thus their specific character, a transformation that implies a profound modification of the nature of the interaction of the onium group with its surroundings and with a potential receptor. To this transformation may probably be ascribed the transition from ionic to hydrophobic binding characteristic of the behavior of tetraalkylammonium ions upon the increase in size of the *N*-alkyl substituents; this change is particularly visible upon the replacement of methyl substituents by ethyl ones (Moran and Triggle, 1971).

We now consider the effect of decreasing the size of the cationic head, by replacing its methyl groups by hydrogen atoms, the conformational characteristics of the acetylcholine skeleton. Let us examine the effect of such a complete replacement (see Fig. 46). There is now a drastic change with respect to the conformational energy map of acetylcholine. The global minimum is transferred to the values of $\tau_1 = 60°$, $\tau_2 = -90°$ (and the symmetrical values $\tau_1 = -60°$, $\tau_2 = 90°$) which represent a gauche–gauche structure with respect to the torsion around both the C(5)—C(4) and O(1)—C(5) bonds. The gauche conformation characteristic of acetylcholine is now a local minimum, 5 kcal/mole above the global one. The fully extended form, which is at 3 kcal/mole above the global minimum in Fig. 43, is at 7 kcal/mole above the global minimum in Fig. 46. Moreover, the large zone between the gauche and the trans conformations ($\tau_1 \approx 180°$, $\tau_2 = 90° - 180°$) is in Fig. 46 at a relatively high-energy level, i.e., 7–8 kcal/mole above the minimum.

In this case we observe thus a triple effect of modifying simultaneously the dimension of the cationic head, the preferred conformation of the whole molecule, and the general morphology of important parts of the conformational energy map. It may also be expected that the direct presence of the hydrogen atoms on the quaternary nitrogen will make these compounds more susceptible to environmental effects in the crystal and in solution, as they carry now (as can be seen from Fig. 44c) a very high net positive charge, corresponding, in fact, for each of them to the total net positive charge of the methyl groups that they replace (0.24*e* per H atom), the quaternary nitrogen itself becoming even slightly negative. This complex situation makes it difficult to ascertain the reasons responsible for the decrease of the parasympathomimetic activity in these derivatives. A more precise knowledge of the nature of the different effects linked to the replacement of the CH_3 groups by H atoms should, however, help to determine the biological significance of each of them.

C. Local Anesthetics Derived by a Modification of the Ester Terminal

A particularly interesting group of structural analogs of acetylcholine is represented by 2-dialkylaminoethylbenzoates; compounds that are derived from acetylcholine by the replacement of the methyl group of the acetyl fragment by a phenyl ring (and a simultaneous suppression of one of the CH_3 groups of the cationic); these head molecules may, naturally, also be considered in the quaternary state ($R_1 = R_3 = CH_3$, $R_2 = H$, $R_4 = C_6H_5$). The introduction of the phenyl group at the esteric end of acetylcholine results in the decrease of the intrinsic parasympathomimetic activity (Ariëns *et al.*, 1964). The particular interest of these compounds resides, however, in their connection with the series of local anesthetics, the more so as it has

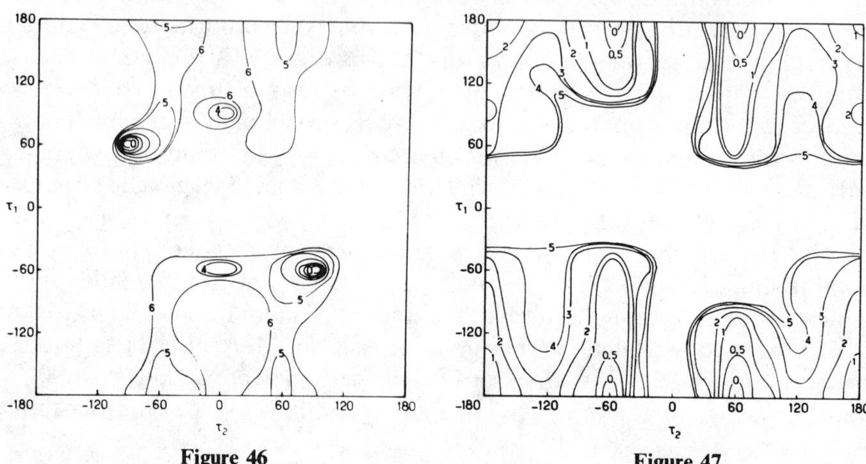

Fig. 46. Conformational energy map for acetylethanolamine. Isoenergy curves in kilocalories per mole with respect to the global minimum taken as zero energy.

Fig. 47. Conformational energy map for compound XXIX. Isoenergy curves in kilocalories per mole with respect to the global minimum taken as zero energy.

been proposed (Bartels, 1965; Bartels and Nachmansohn, 1965) that local anesthetics might block nerve conduction through attachment to axonal acetylcholine receptors. The study of the conformational relationship between these molecules and acetylcholine is, therefore, of obvious interest.

Figures 47 and 48 present the conformational energy maps for the neutral (XXIX) and ionized (XXX) forms, respectively, of the representative model of local anesthetics related to acetylcholine. [In building these maps and the following ones in this series, the phenyl group was fixed coplanar with the plane of the ester group following the indications of calculation by Coubeils and Pullman (1972) and in agreement with the results of X-ray studies on procaine (see in following).] Figure 47 indicates a preference of the neutral form for the gauche conformation ($\tau_1 = 180°, \tau_2 = \pm 60°$) identical to the preferred form of acetylcholine with, however, a local energy minimum for the fully extended form ($\tau_1 = \tau_2 = 180°$) only 1 kcal/mole above the global minimum and a reduced energy barrier (2–3 kcal/mole) between the two. The ionized form (Fig. 48) manifests the same global minimum and shows also a local energy minimum for the extended form, but it is now at 3 kcal/mole above the global one with a barrier of 4 kcal/mole between the two. From these results it may, thus, be inferred that the quaternization of the amino nitrogen increases the preference for the gauche conformation. It may be useful to remark here that, although the identification of the active form of this class of drugs is still controversial, recent evidence seems to favor the cationic form as the active one (for references, see Coubeils and Pullman, 1972).

Recent experimental data obtained with the use of NMR spectroscopy (Mautner *et al.*, 1972; Makriyannis *et al.*, 1972) confirm these theoretical conclusions: they indicate that, whereas compound XXIX exists 67% in the gauche and 33% in the trans forms in solution (a situation which, because of the possibility of two nearly equivalent gauche forms, indicates the near energetical equivalence of the three rotamers), compound XXX exists under the same conditions exclusively in the gauche form.

In this series of compounds, the PCILO conformational energy map was also computed for the effective local anesthetic procaine (XXXI) whose structure has been recently studied by X-ray crystallography by a number of authors (Beall *et al.*, 1970; Pletcher and Gustaffson, 1970; Dexter, 1972). Also studied by X-rays was the related 2-diethylaminoethyl-*p*-methoxybenzoate hydrochloride (Beall and Sass, 1970). By using as input data the crystallographic results for procaine hydrochloride, we obtained the conformational energy map of Fig. 49. The general features of the map (the contours of the stability zone within, say, the 3 kcal/mole isoenergy curve) are practically identical to that of acetylcholine. We observe, however, a degeneracy of the global energy minimum, which is associated both with the

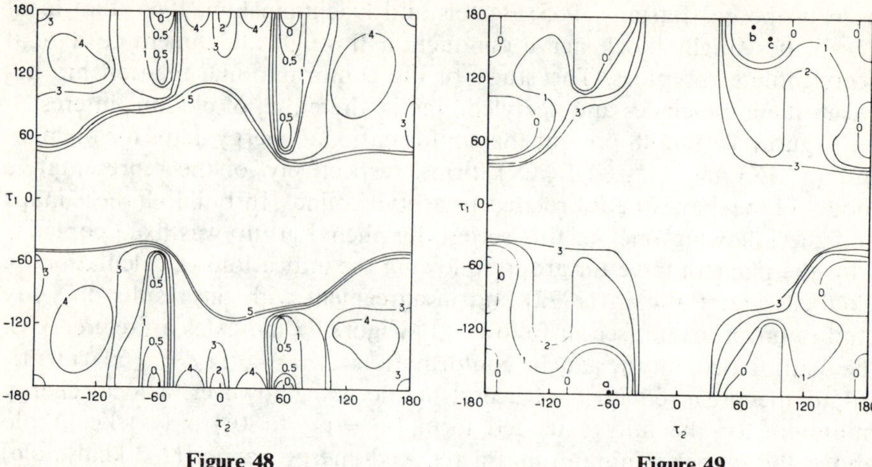

Fig. 48. Conformational energy map for compound XXX. Isoenergy curves in kilocalories per mole with respect to the global minimum taken as zero energy.

Fig. 49. Conformational energy map for procaine (XXXI). Isoenergy curves in kilocalories per mole with respect to the global energy minimum taken as zero energy. (●) Experimental conformations of procaine in crystals of (a) 1 : 1 procaine–bis-*p*-nitrophenyl phosphate complex (Pletcher and Gustaffson, 1970); (b) procaine hydrochloride (Beall *et al.*, 1970; Dexter, 1972); (c) conformation of 2-diethylaminoethyl-*p*-methoxybenzoate hydrochloride (Beall and Sass, 1970).

gauche and trans forms. The minimum for the gauche form covers a somewhat larger area and may, thus, perhaps be considered as more probable. Nevertheless, the experimental conformations are all, as indicated in Fig. 48, of the gauche type, and it is striking to observe that this major conformational feature of procaine is preserved in different crystal environments. It is also preserved in solution (Chu *et al.*, 1972).

There appears therefore to exist definite evidence, both theoretical and experimental, indicating that acetylcholine and protonated local anesthetics are conformationally similar. This, by itself, does not mean, however, that, these similar conformational features are involved in the activities of these two groups of molecules, the less so as no proofs exist that these conformations, although characteristic of the structure in crystal and in solution, are also the ones involved in the interaction with the biological receptor(s). In fact, it was proposed at one time by some investigators (Sundaralingam, 1968; Baker *et al.*, 1971) that the gauche conformation is essential for the ability of cholinergic compounds to initiate a nerve impulse and, similarly, by others (Bartels, 1965; Bartels and Nachmansohn, 1965) that the gauche conformation of local anesthetics may be an important feature necessary for the effector–receptor interaction and the blocking of the nerve impulse.

Biological studies, however, conducted in particular by Mautner *et al.* (1972) and by Partington *et al.* (1972) show no consistent correlation between the conformations of molecules triggering or blocking conduction of the nerve impulse or affecting electrically excitable membranes and their potency. Conformational factors alone, therefore, probably determine neither the activity of cholinergic agonists nor that of local anesthetics. Electronic structure may be quite important in this field. Actually, it is frequently argued (Ariëns *et al.*, 1964) that the distribution of the electronic charges, in particular in the carbonyl bond, is important in the action of local anesthetics and that only slight complementarity is needed between these drugs and their receptor.

D. Anticholinergic Compounds

A large number of the best-known anticholinergic agents have their cationic head incorporated into a relatively rigid ring so that τ_2 is blocked. Such is, for instance, the case of atropine (XXXII), scopolamine (hyoscine; XXXIII), quinuclidinyl benzylate (XXXIV) quinuclidinyl thienylglycolate (XXXV), glycopyrronium bromide (XXXVI), and hexapyrronium bromide (XXXVII). Thus, in these compounds, the essential conformational variable is τ_1, which makes the study of their conformational properties particularly easy.

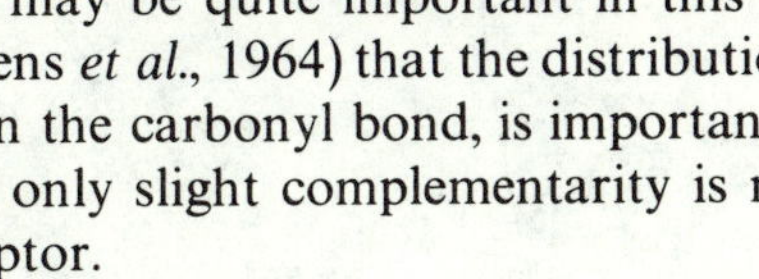

(XXXII) ATROPINE

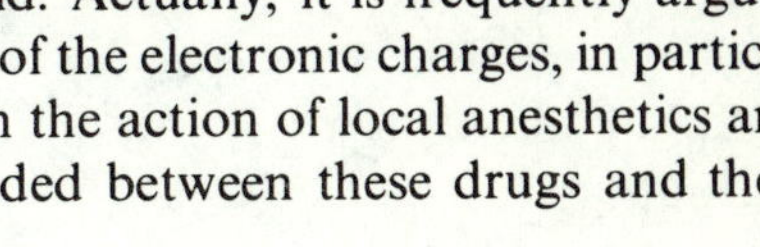

(XXXIII) HYOSCINE (scopolamine)

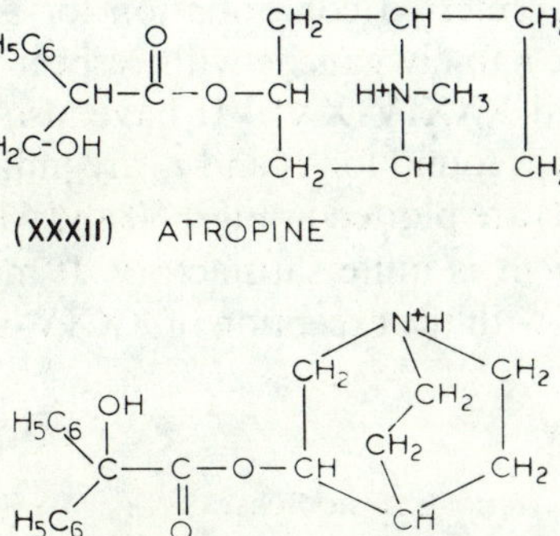

(XXXIV) QUINUCLIDINYL BENZYLATE

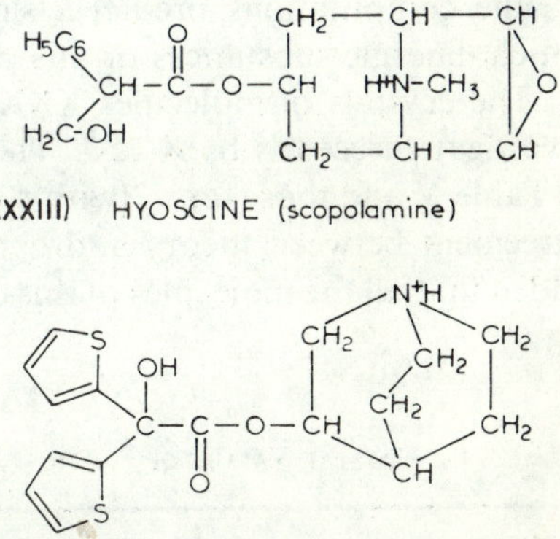

(XXXV) QUINUCLIDINYL DI-α,α'-THIENYLGLYCOLLATE

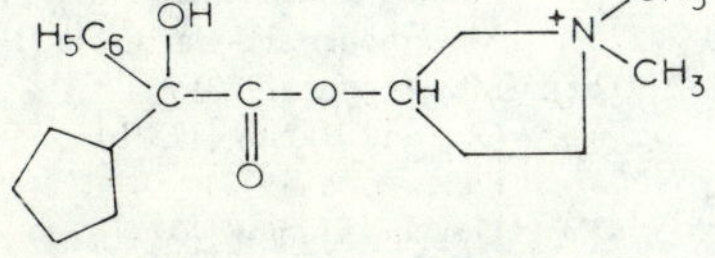

(XXXVI) GLYCOPYRRONIUM BROMIDE

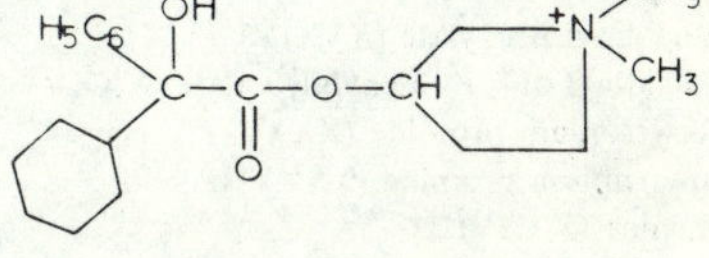

(XXXVII) HEXAPYRRONIUM BROMIDE

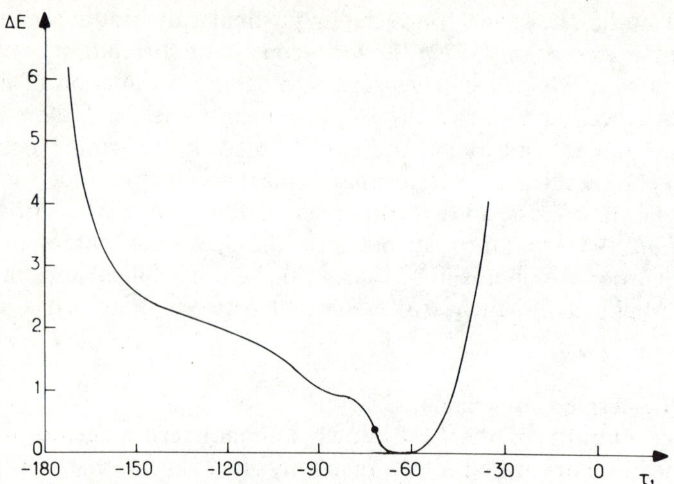

Fig. 50. The PCILO conformational energy curve for atropine (XXXII) as a function of τ_1 (kilocalories per mole) with τ_2 fixed in the crystallographic value (see Table V).

Figure 50 represents the PCILO conformational energy curve for atropine as a function of τ_1, and Fig. 51 shows this type of curve for compounds XXXIV–XXXVII (B. Pullman and Courrière, 1971). It is evident that in spite of the different nature of these "rigid" cationic heads and some variations of τ_2, the computations predict a similar preferred conformation for all the anticholinergic substances of this group, namely gauche with respect to τ_1.

The crystals of molecules XXXII and XXXIV–XXXVII have also been investigated recently by X-rays. The values found for τ_1 and τ_2 are indicated in Table V and those for τ_1 (with τ_2 fixed) are plotted in Figs. 50 and 51. The agreement between theory and experiment is quite satisfactory. It may be added that all the molecules of this class (with the exception of XXXV, which

TABLE V

Crystal Values of τ_1 and τ_2 in Anticholinergic Substances

Compound	τ_1	τ_2	Reference
Atropine (XXXII)	−74.1°	−76.9°	Kusäther and Haase (1972)
Hyoscine (XXXIII)	Gauche		Pauling and Petcher (1969)
Quinuclidinyl benzylate (XXXIV)	65.4°	125°	Meyerhöffer and Carlström (1969)
Quinuclidinyl di-α,α′-thienylglycollate (XXXV)	85.2°	111.8°	Meyerhöffer (1970)
Glycopyrronium bromide (XXXVI)	79.5°	96.6°	Guy and Hamor (1973b)
Hexapyrronium bromide (XXXVII)	82°	93°	Baker *et al.* (1973)
Adiphenine (XXXVIII)	166.5°	83.3°	Guy and Hamor (1973a)
Parpanit (XXXIX)	±170°	±85.7°	Griffith and Robertson (1972)
Benactyzine (XL)	−177.6°	45.4°	Petcher (1974)

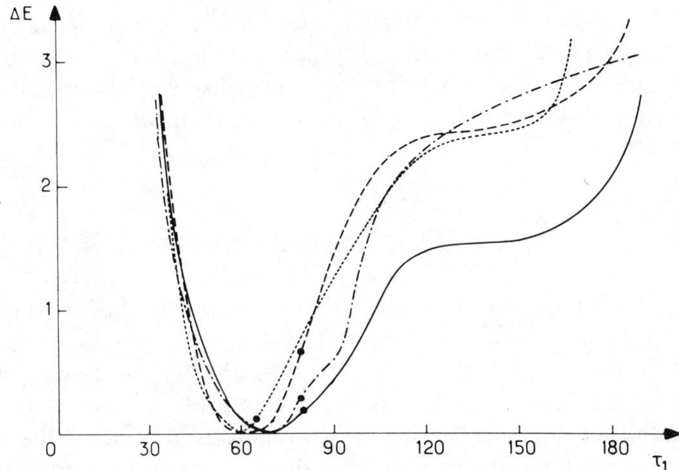

Fig. 51. Conformational energy curves for compounds XXXIV–XXXVII as a function of τ_1 (kilocalories per mole) with τ_2 fixed in the crystallographic value (see Table V). (···) Compound XXXIV; (——) compound XXXV; (- - -) compound XXXVI; (—·—) compound XXXVII.

is neutral) are accompanied in their crystal by the counterion Br^- and that their conformation is similar to that of acetylcholine in the crystal of its bromide (see Section VI), i.e., gauche–gauche.

There is another group of anticholinergic drugs that are devoid of such rigidity and differ essentially from acetylcholine by the presence of large substituents at the cationic head and in place of the methyl group at the acetyl extremity. Representative of this type of molecules are adiphenine (XXXVIII), Parpanit (XXXIX), and benactyzine (XL). For these molecules,

$(H_5C_6)_2CH{-}C({=}O){-}O{-}CH_2{-}CH_2{-}{}^+NH(C_2H_5)_2$

(XXXVIII) ADIPHENINE

$(C_5H_8)(H_5C_6)C{-}C({=}O){-}O{-}CH_2{-}CH_2{-}{}^+NH(C_2H_5)_2$

(XXXIX) PARPANIT

$HO{-}C(C_6H_5)_2{-}C({=}O){-}O{-}CH_2{-}CH_2{-}N^+H(C_2H_5)_2$

(XL) BENACTYZINE

the PCILO computations predict a conformation similar to that of acetylcholine in the chloride crystal (trans–gauche). As indicated in Table V, this is the conformation found for these molecules in their crystals. It may be observed that all the molecules of this class contain the counterion Cl^- in their crystals.

IX. Conclusions

After this rapid survey of the conformational studies in the different classes of biogenic amines and related systems, we may now try to reach conclusions as to the significance of the results for the general problems raised in the Introduction to this paper. These problems are concerned with the evaluation of the different theoretical methods utilized for the computation of conformations, the unraveling of the conformational possibilities and preferences of this type of molecules, and the significance of the results obtained for the determination of the possible role of conformations in the activity of drugs.

A major problem of immediate interest to quantum chemists is, of course, the appraisal of the validity in conformational analysis of the different computational procedures. The problem has, in fact, a broader issue because of the tendency of a large number of scientists who are not experts in quantum chemistry to use available programs for carrying out such computations themselves, a situation particularly frequent in relation to pharmacology.

The large amount of computations carried out by different methods for the compounds discussed in this paper has produced definite indications in this respect. The availability of nonempirical, SCF *ab initio* calculations is of particular significance as a reference standard. From that point of view, this survey seems to us to show the nearly complete agreement between the PCILO and the SCF *ab initio* results for the free molecules, as exemplified by the cases of histamine, serotonine, and acetylcholine. Sometimes the *ab initio* results go somewhat further than the PCILO ones by resolving degeneracies, as is the case, e.g., for the series of phenethylamines. Although the *ab initio* results are generally obtained for practical reasons with a small basis set (STO-3G in general), it was shown at least in one case (phenethylamine) that more elaborate basis sets do not modify the essential results found with the smaller set. This convergence is a welcome affirmation of the validity of the PCILO computations.

By contrast, the very profound and frequent disagreements found between the results of the PCILO and *ab initio* computations, on the one hand, and the EHT calculations, on the other, practically disqualify this last procedure as a valid tool for conformational studies. A prediction, such as that of a similar conformation for free mono- and dications of histamine is hardly acceptable on even purely qualitative grounds. In general the EHT method seems to underestimate intramolecular attractive interactions and to predict an extended conformation as the preferred one in many cases in which all other methods point to a folded conformation as the most plausible. An extreme example of such erratic behavior of the EHT method may be found in its predictions (Kier and Truitt, 1970) that the intrinsically preferred conformation of free flexible zwitterions of the type γ-aminobutyric acid (GABA, $COO^-{-}CH_2{-}CH_2{-}N^+H_3$, an important transmitter in central nervous system) should be an extended one, again in contradiction to *ab initio* and PCILO results (B. Pullman and Berthod, 1974, 1975) and to simple electrostatic considerations (Warner and Steward, 1974). In fact, these examples in the field of pharmacology only corroborate others that have already been found in conformational studies in the field of protein (B. Pullman and Pullman, 1974) and nucleic acids constituents (B. Pullman and Saran, 1975) or of membrane components (B. Pullman and Berthod, 1974). This does not mean, of course, that the conformations which are predicted as the most stable ones by EHT are of no practical significance. We have seen in particular in our discussion on histamine that they may even happen to describe appropriately the conformations of that molecule in water. The point is that in the case of *isolated* molecules, it has been established that these computations frequently give an incorrect answer about conformational preferences. By doing so, they obscure the effect of environmental forces. Moreover, by repeating constantly their errors, for example, in a series of related molecules, they may lead to apparent correlations based in fact on artifacts. As it seems difficult to predict *a priori* the cases in which EHT computations will or will not yield an appropriate answer, the utilization of this procedure is greatly discouraged.

The CNDO and INDO methods have been used to a limited extent only in molecular pharmacology. In the examples quoted in this paper, their performance is generally not unsatisfactory. One should, however, bear in mind the well-established deficiency of the CNDO procedure in dealing with conformation of conjugated systems (Gropen and Seip, 1971); as opposed to the success of the PCILO method in this field (Perahia and Pullman, 1973), and also its much higher cost as compared to that of PCILO.

Finally, concerning the empirical, partitioned, potential function computations, the examples quoted of their applications to acetylcholine indicate disagreements among reports of different authors or among consecutive papers of the same authors. This procedure has not yet reached a stage of development so that it may be considered a clear-cut, well-established, unambiguous approach, yielding consistent, reliable information.

In terms of the results themselves, the computations, at least those that we consider as the most representative and reliable ones, point to the existence of patterns indicating a tendency toward a set of intrinsically preferred conformations for the free amines and analogs discussed here. These patterns become only visible, however, when these molecules are subdivided into appropriate subgroups corresponding to their different states of ionization (neutral, monocation, dication, zwitterion) and to the different possible structures of their amino terminal group [whether $—NH_2$ or $—N(alkyl)_2$, $—N^+H_3$ or $—N^+H(alkyl)_2$ or $—N^+(alkyl)_3$]. Thus, for example, free monocations with an $—N^+H_3$ cationic head (histamine, serotonin, the corresponding phenethylamines, acetylethanolamines) present a preference for a gauche conformation with respect to the torsion angle τ_2. Monocations with an $—N^+(CH_3)_3$ cationic head have a tendency toward an extended conformation (trimethylhistamine, acetylthiocholine) with, however, the important exception of the $O—C—C—N^+(CH_3)_3$ system, characteristic of acetylcholine-like structure; the latter shows a marked intrinsic preference for a gauche conformation. In the monocations with the intermediate $—N^+H(CH_3)_2$ cationic head, the situation depends to some extent on the value of τ_3. Dications, on the other hand, show an increased preference for an extended structure. Although we have not considered here neutral forms, available computations indicate a small intrinsic preference, according to the compound involved, for a gauche (histamine, serotonin) or trans (bufotenine) conformation. The preference is smaller in this case than for the ionic species.

The question may now be raised as to the relationship between the intrinsically preferred conformations of the free molecules obtained by computations and from the experimentally observed structures. In the absence of data on the gaseous phase and the great scarcity of data on conformations in nonpolar solvents (see, however, the case of pheniramine in Section III), we may turn first to X-ray crystal data.

A comparison of the theoretical predictions with experiment clearly shows that in the case of the biogenic amines considered here the situation is relatively complex. For example, whereas cationic serotonin, acetylcholine, procaine, and some other compounds conserve in their crystals the intrinsically preferred gauche forms (with, however, serotonin existing also in one of its crystals in an extended form), other substances, in particular the phenethylamines, are found in their crystals in extended forms in spite of their intrinsic preference for the folded one. Generally speaking, if a molecule exhibits an intrinsic preference for an extended structure, such a structure will most probably be preserved in the crystals. On the other hand, if the intrinsically preferred structure is a folded or gauche one, there is no certainty that it will be preserved in the crystal. The crystal-packing forces seem to have a general tendency to elongate the conformations of the biogenic amines and their analogs. In amines with an (hetero)aromatic ring, the possible influence of the crystal-packing forces on the coplanarity of the ring with the side chain (torsion angle τ_1) may also be added to this complex situation. The detailed mechanism of action of these forces has not yet been elucidated, but work is in progress in our laboratory on this subject.

By contrast, the problem of conformations in aqueous solution may be dealt with in a more precise way. In the first place, all the examples described in this paper show that, contrary to the usual state in the crystal, a number of conformations generally coexist in solution. Experimental information refers generally to the torsion angle τ_2 only and indicates in most cases coexistence of trans and gauche forms, in proportions variables from one molecule to another. In some cases, such as that of zwitterionic GABA, (B. Pullman and Berthod, 1975), a large number of different trans, gauche, and intermediate forms are predicted and seem actually (Ham, 1974) to coexist in water solution.

Computations have been carried out for a number of biogenic amines (mono- and dication of histamine, serotonin, amphetamine, norepinephrine, acetylcholine) in water, and a more direct comparison is possible between theory and experiment in this environment than it was in the crystal. From the discussions in the course of this paper and in spite of the limited number of cases studied, it seems obvious that at present the results of the supermolecule approach seem to describe better the situation in water than do the continuum or the hydration group models and that, on a qualitative level, the effect of the solvent seems, indeed, quite satisfactorily accounted for by the supermolecular model. Of course, nobody would dare claim that this model, in particular in the restricted form in which it was used till now, is an appropriate way of describing the total effect of the solvent. Nevertheless, by correctly indicating the major features of the modifications produced by the immersion of the compound into water, it opens the way for deeper exploration by an extension of the procedure. Individualization of "bound" water molecules seems significant, and the hypothesis that in some hydrates (serotonin, histamine, phenethylamines), these strongly bound solvent molecules follow conformational movements of bonds seems verified. It remains to refine the procedure by an appropriate introduction of the more remote water shells and of the "bulk" of the solvent. Work in this direction is also in progress in our laboratory.

There remains now to discuss the possible significance of these results for molecular pharmacology. To some extent the discussion transcends the problem of quantum-mechanical computations and concerns all physicochemical studies, theoretical and experimental, whether on pharmacological molecules themselves or on model compounds. In all cases, work is being carried out on molecules in an environment that bears only a remote relation to the conditions in the biophase and still less, perhaps, to the conditions at the receptor. Nevertheless, far as we still may be from the ultimate goal, these studies do contribute to bringing us closer to it. In this type of investigation quantum-mechanical computations have a definite role to play, as well as also having certain advantages.

One of the principal recognized goals of studies on conformation of pharmacological molecules was the hope that it may lead to the "mapping" of the receptor. Now in a very large number of papers published in this field, it was implicitly assumed that there is only one conformation that corresponds to a given receptor and, moreover, that this is the "preferred" conformation of the drug. As stated by one of the leading scientists in this field (Kier, 1973b): "It is our basic hypothesis that at the most remote distance of drug receptor engagement it is the preferred conformation of the drug or a conformation very close in energy to the preferred conformation which is 'recognized' by the receptor." Concerning this preferred conformation, it was implicitly admitted by many that it could be the conformation observed in crystals, as this was for a number of years the only available information. More recently, attention has focused on the intrinsically preferred conformation as predicted by empirical or quantum-mechanical computations for the free molecule. When the two types of data converge, there does not seem to be much doubt in the opinion of some authors that they have found the "active" conformation. Unfortunately, in work using this approach, the intrinsically preferred conformations were mostly evaluated by the extended Hückel theory (Kier, 1971).

Recent developments in computations of conformations in solution and the concomitant increase in application of NMR spectroscopy to pharmacological compounds led to a modification of this simple viewpoint. In the first place, it becomes increasingly obvious that, for a large number of pharmacological molecules, there are present in solution a number of conformations that frequently differ only little in energy. The most stable among them may be different from the intrinsically most stable conformation for the free molecule or from the one present in the crystal of the compound. Although, there is no *a priori* certitude that conformations which are preferentially stabilized in water should be the ones that are favored in the biophase and still less the ones that are interacting with the receptor, there is certainly no reason to consider them as less probable from that point of view than the conformations favored under vacuum or in the crystal. The possibility also exists that the microenvironment at the receptor site may favor a particular tautomeric form (in case tautomerism is possible) or produce a type of ionization different from that prevalent in water. A recent suggestion (Ganellin, 1974) that tautomeric changes in the imidazole ring (between the N^{τ}—H and N^{π}—H tautomers) may be related to the dual activity of histamine at the H_1 and H_2 receptors (Black *et al.*, 1972) is an illustration of such a probability.

Another recent demonstration, namely that neutral histamine exists in the crystal in the N^{τ}—H form (Bonnet and Ibers, 1973), although it exist in the N^{π}—H form in the equally neutral 6-histamine purine dihydrate (Thewalt and Bugg, 1972b), indicates the duality of possibilities that may occur if the neutral form of the molecule was in some ways involved in the microenvironment of the receptor at one or another stage of the interactions leading to activity.

Altogether these recent results underline the extreme complexity of the situation and the necessity of proceeding with extreme caution when proposing conformational models for pharmacophores. In particular, it is obvious that in order to have a chance of being persuasive the studies have to involve a large series of related molecules for which one is able to make a parallel evaluation of changes in activity with changes in conformations and, if feasible, other properties possibly involved in drug–receptor activity. The utilization of rigid or, better, semirigid analogs may be useful in this respect but must again be treated with caution as the introduction of too much rigidity even in an appropriate conformation may in some cases deny the molecular skeleton the amount of flexibility that it may need for a proper interaction with (the also more or less flexible) receptor. This is the new reality that must be faced if one wants to avoid a proliferation of oversimplified and frequently unfounded proposals.

Now an extensive theoretical investigation of conformational properties of pharmacological molecules has advantages which may contribute noticeably to progress in this field. An obvious and particularly important advantage of theoretical computations on conformations (in particular nowadays when they may be carried out both for free molecules and for molecules in solution and will be carried out probably in the near future for molecules in the solid state; moreover, refined methods are available) is to produce *complete conformational energy maps* instead of giving data, as most experimental techniques do, for only selected conformations, occurring under certain conditions. The maps produce an overall image of the conformational possibilities and preferences of the molecules, they give (or may easily give) information on the barriers separating the different (more-or-less) stable forms and the energy variations along the paths between these forms. Thus, they may provide information regarding the dynamics of possible conformational changes, about which very little practical information is available. In some instances, they may give easy access to the preferred values of certain torsion angles for which no experimental results exist or for which they are difficult to obtain. Thus, in the biogenic amines discussed in this paper, current NMR studies in solution provide information only on the torsion angle τ_2. No information seems to be available in the literature on the value of the torsion angle τ_1 in solution (see, however, Giessner-Prettre and Pullman, 1975). Theoretical computations apply, of course, as easily to τ_1 as to τ_2 and the $\tau_1 - \tau_2$ conformational energy maps may, thus, compensate the absence of experimental data.

The relatively large scope of conformational possibilities of molecules, as given by the quantum-mechanical conformational energy maps, also enables the critical appraisal of proposals concerning the role of conformational analogies in drug activity. We have mentioned in the Introduction to this paper the contradictory proposals that were made about the role of possible structural and conformational analogies among LSD, mescaline, and psilocin for their similar hallucinogenic activity. One of the hypotheses underlined the structural resemblance between LSD and the extended forms of mescaline and psilocin in that it may be possible to make a more-or-less strict spatial superposition of the aromatic rings and protonated N atoms of these compounds (Fig. 4). The proposals were originally made on the basis of simple two-dimensional drawings.

It became obvious more recently that crystallographically observed conformations, which were looked for as possible confirmation of such analogies, are, in fact, substantially different in these three molecules, as can be judged from the data reproduced in the appropriate sections of this review. Recently, Green and co-workers (Kang *et al.*, 1973a) proposed a more subtle study of a possible "congruence" between the three drugs. Considering that such a congruence would correspond to the superposition of the phenyl ring of mescaline with the A ring of LSD or that of the indole ring of psilocin with the indole ring of LSD [whose X-ray structure has recently been established (Baker *et al.*, 1972)] with the concomitant superposition of the protonated nitrogens of the compounds, these authors propose that the best "congruence" would occur for $\tau_1 = -33°$, $\tau_2 = 160°$ in mescaline and for $\tau_1 = -134°$, $\tau_2 = -30°$ in psilocin. These conformations are the most stable ones neither on the INDO nor on the PCILO maps for the free molecules. They do not correspond to any outstanding energy minima. Actually, they represent conformations only about 3–5 kcal/mole above the global ones predicted for these molecules. The corresponding deformations may probably be easily brought about by interaction with a complex receptor. On the other hand, the folded structures postulated for mescaline and psilocin by Snyder and Richelson (1968) in order to mimic the configurations of rings A + B or C of LSD (Fig. 4) would imply $\tau_1 \approx \tau_2 \approx 0°$ in mescaline and $\tau_1 \approx 180°$, $\tau_2 \approx 0°$ in psilocin. Such conformations correspond to relatively very high energies on the conformational energy maps of these compounds. Although it is not possible, of course, to eliminate definitely their participation at the active site of the drug, it does not seem probable that they are implicated.

Other indications that low-energy conformations, although not necessarily associated with energy minima on the conformational energy maps of the free molecules, may be involved in interactions with receptors come from the convergence of systematic theoretical and experimental studies in some series of the compounds discussed here. Thus, a large series of X-ray crystal studies on cholinergic substances and various studies of model compounds suggest that different conformations could be involved in interactions with the muscarinic or nicotinic receptors and in binding to esterase. The estimations differ from one author to another. For the muscarinic receptor, $\tau_1 = 180°$, $\tau_2 = 90°$, Chothia (1970); $\tau_2 = 140°–180°$, Chiou *et al.* (1969). For the nicotinic receptor, $\tau_1 = 180°$, $\tau_2 = 75°$, Chothia and Pauling (1970); $\tau_1 = 150°$, $\tau_2 = 180°$, Martin-Smith *et al.* (1967). For the esterase receptor, $\tau_1 = 150°$, $\tau_2 = 180°$, Chothia and Pauling (1967); $\tau_1 = \tau_2 = 180°$, Hasbun *et al.* (1973). Whatever value these different suggestions may have, they all have in common that all the "active" conformations are located in low-energy regions of the conformational map for acetylcholine. We have ourselves noticed conformational analogies, at least for the free and crystallographic compounds between the low-energy conformations of acetylcholine and those of anticholinergic agents. A certain resemblance seems also to exist between the conformations in solution for histamine and for antihistaminic drugs.

Although these and other similar results point to the possible importance of low-energy conformations related to the computed ones in pharmacological activity, the contribution of this particular factor must not be overestimated. Outstanding examples show that this cannot be the only factor involved in structure–activity correlations. We have already indicated in Section VIII,B of this paper some examples of the insufficiency of considerations based on conformation alone for explaining, for example, the decrease in parasympathomimetic activity upon the replacement of the CH_3 groups by C_2H_5 groups at the cationic and of acetylcholine. Similarly, as shown in Section VIII,C, the analogy between the conformational properties of cholinergic agents and local anesthetics does not warrant a common mechanism of action. Also, the similarity of the conformational properties of norepinephrine and isoproterenol demonstrated in Section V, shows that this factor cannot account for the specificity of these compounds for α- and β-adrenergic receptors, respectively. In the interesting series of LSD, mescaline, and psilocin, it should not be forgotten that the tedious search for conformational analogies may overlook preponderant structural factors in regions of the molecules not considered in such a search. This could be a possible role of the diethylamide group of LSD: its replacement by a dimethylamide group lowers appreciably the activity of LSD, and lysergic acid itself has practically no psychotomimetic activity at all (Isbell *et al.*, 1959).

Factors other than conformations must, therefore, be taken into consideration in structure–activity correlations and they may concern events taking place both before and after the conformational complementarity becomes established. The first concerns aspects such as solubility and metabolism. Little quantum-mechanical work has been done in this respect with the exception perhaps of some early studies by our group on the electronic aspects of some fundamental metabolic reactions of drugs (B. Pullman, 1964, and the references indicated therein; also see Green *et al.*, 1974). The second concerns aspects such as the nature and the strength of the drug–receptor interactions and their reactional consequences. It is what is happening after the allowed contact between the drug and the receptor has been established that may decide the issue. Only isolated attempts have been carried out so far in the quantum-mechanical studies along these directions. Thus, although a large amount of quite refined computation has been performed for evaluating energies of intermolecular interactions between important biomolecules (for reviews and references, see B. Pullman, 1968; A. Pullman and Pullman, 1968; B. Pullman and Pullman, 1969), only a few explicit applications have been made in the field of pharmacology (e.g., Hoyland and Kier, 1972; Höltje and Kier, 1974; Kier and Ardrich, 1974). As the methodology exists more progress may be expected along these lines in the near future.

The same situation prevails with respect to the "reactivity" of drugs with the receptor. Only isolated attempts are available in this direction also, such as the discussion from that point of view of the possible differences between the α- and β-adrenergic catecholamines (George *et al.*, 1971) or an attempt to relate the dual receptivity of acetylcholine in muscarinic and nicotinic activity to distinct reactive features of the molecule rather than to distinct conformations (Chothia, 1970; Beers and Reich, 1970). In an approach along these lines, Weinstein *et al.* (1973a,b; Loew *et al.*, 1974) propose to employ the electrostatic molecular potential method (Bonaccorsi *et al.*, 1970; A. Pullman, 1974) for the evaluation of an "interaction pharmacophore." A danger, however, in the use of such a methodology, excellent in itself for the study of certain types of reactivity of biomolecules (Bonaccorsi *et al.*, 1972; A. Pullman, 1974), is to see analogies where there are none.

Theoretical considerations on drug–receptor interactions and on conformation–activity relationships are very complex problems. They have been for some time a beautiful field for creative imagination. The accumulation of more precise data, theoretical and experimental, transformed this into a field of hard labor. It is certainly more prosaic but also more secure.

ADDENDUM

Studies on the conformational basis of molecular pharmacology are continuing actively, and it seems useful to list here some of the more interesting publications that appeared after this paper was completed and which are relevant to the subjects discussed.

Histamine. Molecular orbital computations on the role of counterion in conformational studies (Abraham, 1976; Abraham and Birch, 1975).

Antihistamines. Further NMR studies in solution (Ham, 1976).

Phenethylamines and phenethanolamines. A series of X-ray studies by Andersen and colleagues (1975). *Ab initio* computations (Martin *et al.*, 1975; Hall *et al.*, 1975).

Acethylcholine. Reinvestigation of the crystal structure of acetylcholine bromide (Svinning and Sorum, 1975). Crystal structure of acetylcholine derivatives—a series of X-ray studies (Jensen, 1975).

ACKNOWLEDGMENT

The author acknowledges with thanks the permission of John Wiley and Sons and the *International Journal of Quantum Chemistry* to reproduce Figs. 1, 2, 9, 15 and 16 from *Quant. Biol. Symp.* (1974), **1**, 73; of Springer-Verlag and *Theoretica Chimica Acta* to reproduce Figs. 1 and 3 from *Theoret. Chim. Acta* (1973), **32**, 77 and Figs. 1, 2, 5, 8, 9, and 12 from *Theoret. Chim. Acta* (1973), **31**, 19; of *Molecular Pharmacology* to reproduce Figs. 2–8 from *Mol. Pharmacol.* (1974), **8**, 612 and Fig. 3 from *Mol. Pharmacol.* (1972), **8**, 371; and of the American Chemical Society to reproduce Figs. 1, 5, 6, and 8 from *J. Med. Chem.* (1974), **17**, 439.

REFERENCES

Abraham, R. J. (1976). *In* "Environmental Effects on Molecular Structure and Properties" (B. Pullman, ed.), p. 41. Reidel Publ., Dordrecht, Netherlands.

Abraham, R. J., and Birch, D. (1975). *Mol. Pharmacol.* **11**, 663.

Ajo, D., Bossa, M., Damiani, A., Fidenzi, R., Gigli, S., Lanzi, L., and Lapiccirella, A. (1972). *J. Theor. Biol.* **34**, 15.

Ajo, D., Bossa, M., Damiani, A., Fidenzi, R., Gigli, S., and Ramunni, G. (1973). *In* "Conformation of Biological Molecules and Polymers" (E. D. Bergmann and B. Pullman, eds.), p. 571. Academic Press, New York.

Alagona, G., Pullman, A., Scrocco, E., and Tomasi, J. (1973). *Int. J. Pept. Protein Res.* **5**, 251.

Andersen, A. M. *et al.* (1975). *Acta Chem. Scand., Ser. B* **29**, 75, 239, and 871.

Arïens, E. J., Simonis, A. M., and Van Rossum, J. M. (1964). *Mol. Pharmacol.* **1**, 119.

Babeau, A., and Barrans, Y. (1970). *C. R. Hebd. Seances Acad. Sci. Ser. C* **270**, 609.

Baker, R. W., Chothia, C. H., Pauling, P., and Petcher, T. J. (1971a). *Nature (London)* **230**, 439.

Baker, R. W., Chothia, C. H., Pauling, P., and Petcher, T. J. (1971b). *Nature (London), New Biol.* **234**, 439.

Baker, R. W., Chothia, C. H., Pauling, P., and Weber, H. P. (1972). *Science* **178**, 614.

Baker, R. W., Chothia, C. H., and Pauling, P. (1973a). *Mol. Pharmacol.* **9**, 23.

Baker, R. W., Datta, N., and Pauling, P. (1973b). *J. Chem. Soc., Perkin Trans. 2* p. 1963.

Bartels, E. (1965). *Biochim. Biophys. Acta* **109**, 194.

Bartels, E., and Nachmansohn, D. (1965). *Biochem. Z.* **342**, 359.

Beall, R., and Sass, R. L. (1970). *Biochem. Biophys. Res. Commun.* **40**, 833.

Beall, R., Herdklotz, J., and Sass, R. L. (1970). *Biochem. Biophys. Res. Commun.* **39**, 329.

Beers, W. H., and Reich, E. (1970). *Nature (London)* **228**, 917.

Bennett, I., Davidson, A. G. H., Harding, M. M., and Morelle, I. (1970). *Acta Crystallogr., Sect. B* **26**, 1722.

Bergin, R., and Carlström, D. (1971). *Acta Crystallogr., Sect. B* **27**, 2146.

Beveridge, D. L., and Radna, R. J. (1971). *J. Am. Chem. Soc.* **93**, 3759.
Beveridge, D. L., Kelly, M. M., and Radna, R. J. (1974a). *J. Am. Chem. Soc.* **96**, 3769.
Beveridge, D. L., Radna, R. J., Schnuelle, W., and Kelly, M. M. (1974b). *In* "Molecular and Quantum Pharmacology" (E. D. Bergmann and B. Pullman, eds.), p. 153. Reidel Publ., Dordrecht, Netherlands.
Black, J. W., Duncan, W. A. M., Duraut, C. J., Ganellin, C. R., and Parsons, E. M. (1972). *Nature (London), Teil B* **23**, 385.
Bonaccorsi, R., Petrongolo, C., Scrocco, E., and Tomasi, J. (1970). *In* "Quantum Aspects of Heterocyclic Compounds in Chemistry and Biochemistry" (E. D. Bergmann and B. Pullman, eds.), p. 181. Academic Press, New York.
Bonaccorsi, R., Pullman, A., Scrocco, E., and Tomasi, J. (1972). *Theor. Chim. Acta* **24**, 51.
Bonnet, J. J., and Ibers, J. A. (1973). *J. Am. Chem. Soc.* **95**, 4829.
Bonnet, J. J., and Jeannin, Y. (1972). *Acta Crystallogr., Sect. B* **28**, 1079.
Bustard, T. M., and Egan, R. S. (1971). *Tetrahedron* **27**, 4457.
Caillet, J., and Claverie, P. (1974). *Biopolymers* **13**, 601.
Candlin, R., and Harding, M. J. (1970). *J. Chem. Soc. A* **384**. p.
Canepa, F. G. (1965). *Nature (London)* **207**, 1152.
Canepa, F. G., Pauling, P., and Sörum, H. (1966). *Nature (London)* **210**, 907.
Carlström, D. (1973). *Acta Crystallogr., Sect. B* **29**, 161.
Carlström, D., and Bergin, R. (1967). *Acta Crystallogr.* **23**, 313.
Carlström, D., Bergin, R., and Falkenberg, G. (1973). *Q. Rev. Biophys.* **6**, 257.
Carroll, D. G., Armstrong, A. T., and McGlynn, P. S. (1966). *J. Chem. Phys.* **44**, 1865.
Casy, A. F., and Ison, R. R. (1970). *J. Pharm. Pharmacol.* **22**, 270.
Casy, A. F., Hassan, M. M. A., and Wu, E. C. (1971). *J. Pharm. Sci.* **60**, 67.
Chiou, C. Y., Long, J. P., Cannon, J. G., and Armstrong, P. D. (1969). *J. Pharmacol. Exp. Ther.* **166**, 243.
Chothia, C. (1970). *Nature (London)* **225**, 36.
Chothia, C., and Pauling, P. (1967). *Nature (London)* **223**, 919.
Chothia, C., and Pauling, P. (1969). *Proc. Natl. Acad. Sci. U.S.A.* **63**, 1063.
Chothia, C., and Pauling, P. (1970). *Proc. Natl. Acad. Sci. U.S.A.* **65**, 447.
Christoffersen, R. E. (1972). *Adv. Quantum Chem.* **6**, 333.
Chu, S. H., Hilman, G. R., and Mautner, H. G. (1972). *J. Med. Chem.* **15**, 760.
Chynoweth, K. R., Ternai, B., Simeral, L. S., and Maciel, G. E. (1973). *Mol. Pharmacol.* **9**, 144.
Clark, G. R., and Palenik, G. J. (1972). *J. Am. Chem. Soc.* **94**, 4005.
Clementi, E., Clementi, H., and Davis, D. R. (1967). *J. Chem. Phys.* **46**, 4725.
Coiro, V. M., Giaconnello, P., and Giglio, E. (1971). *Acta Crystallogr., Sect. B* **27**, 2112.
Coiro, V. M., Giglio, E., and Quagliata, C. (1972). *Acta Crystallogr., Sect. B* **28**, 3601.
Coiro, V. M., Giglio, E., Lucano, A., and Puliti, R. (1973). *Acta Crystallogr., Sect. B* **29**, 1404.
Coubeils, J. L., and Pullman, B. (1972). *Mol. Pharmacol.* **8**, 278.
Coubeils, J. L., Courrière, P., and Pullman, B. (1971). *C. R. Hebd. Seances Acad. Sci.* **272**, 1813.
Courrière, P., Coubeils, J. L., and Pullman, B. (1971). *C. R. Hebd. Seances Acad. Sci.* **272**, 1697.
Culvenor, C. C. J., and Ham, N. S. (1966). *Chem. Commun.* p. 537.
Culvenor, C. C. J., and Ham, N. S. (1970). *Chem. Commun.* p. 242.
Culvenor, C. C. J., and Ham, N. S. (1974). *Aust. J. Chem.* **27**, 2191.
Cushley, R. J., and Mautner, H. G. (1970). *Tetrahedron* **26**, 2151.
Decon, D. F., Jr. (1964). Dissertation No. 64-9987. University of Michigan, Ann Arbor.
Dexter, D. D. (1972). *Acta Crystallogr., Sect. B* **28**, 77.
Diner, S., Malrieu, J. P., Jordan, F., and Gilbert, M. (1969). *Theor. Chim. Acta* **15**, 100.
Ditchfield, R., Hehre, W. J., and Pople, J. A. (1971). *J. Chem. Phys.* **54**, 724.
Donohue, J., and Caron, A. (1964). *Acta Crystallogr.* **17**, 1178.

Eliel, E. E., and Alcudia, F. (1974). *J. Am. Chem. Soc.* **96**, 1939.
Ernst, S. R., and Cagle, F. W., Jr. (1973). *Acta Crystallogr., Sect. B* **29**, 1543.
Falkenberg, G. (1972a). *Acta Crystallogr., Sect. B* **28**, 3075.
Falkenberg, G. (1972b). *Acta Crystallogr., Sect. B* **28**, 3219.
Ferro, D. R., and Hermans, J. (1972). *Biopolymers* **11**, 105.
Froimowitz, M., and Gans, P. Y. (1972). *J. Am. Chem. Soc.* **94**, 8020.
Ganellin, C. R. (1973a). *J. Pharm. Pharmacol.* **25**, 787.
Ganellin, C. R. (1973b). *J. Med. Chem.* **16**, 620.
Ganellin, C. R. (1974). *In* "Molecular and Quantum Pharmacology" (E. D. Bergmann and B. Pullman, eds.), p. 43. Reidel Publ., Dordrecht, Netherlands.
Ganellin, C. R., Pepper, E. S., Port, G. N. J., and Richards, W. (1973a). *J. Med. Chem.* **16**, 610.
Ganellin, C. R., Port, G. N. J., and Richards, W. G. (1973b). *J. Med. Chem.* **16**, 616.
Ganellin, C. R., Port, G. N. J., and Richards, W. G. (1973c). *In* "Conformation of Biological Molecules and Polymers" (E. D. Bergmann and B. P. Pullman, eds.), vol. 5, p. 579. Academic Press, New York.
Gavuzzo, E., Mazza, F., and Giglio, E. (1974). *Acta Crystallogr., Sect. B* **30**, 1351.
Genson, D. W., and Christoffersen, R. F. (1973). *J. Am. Chem. Soc.* **95**, 362.
George, J. M., Kier, L. B., and Hoyland, J. R. (1971). *Mol. Pharmacol.* **7**, 328.
Germer, H. A., Jr. (1974). *J. Pharm. Pharmacol.* **26**, 467.
Giessner-Prettre, C., and Pullman, B. (1975). *J. Magn. Reson.* **18**, 564.
Gill, E. W. (1965). *Prog. Med. Chem.* **4**, 39.
Graven, B. M., and Hite, G. (1973). *Acta Crystallogr., Sect. B* **29**, 1132.
Green, J. P., Johnson, C. L., and Kang, J. (1974). *Annu. Rev. Pharmacol.* **14**, 319.
Griffith, E. A. H., and Robertson, B. E. (1972). *Acta Crystallogr., Sect. B* **28**, 3377.
Gropen, O., and Seip, H. M. (1971). *Chem. Phys. Lett.* **1**, 445.
Guy, J. J., and Hamor, T. A. (1973a). *J. Chem. Soc., Perkin Trans. 2* p. 442.
Guy, J. J., and Hamor, T. A. (1973b). *J. Chem. Soc., Perkin Trans.* p. 1875.
Hagler, A. T., and Lifson, S. (1974). *Acta Crystallogr., Sect. B* **30**, 1336.
Hall, G., Miller, C. J., and Schnuelle, G. W. (1975). *J. Theor. Biol.* **53**, 475.
Ham, N. S. (1971). *J. Pharm. Sci.* **60**, 1764.
Ham, N. S. (1974). *In* "Molecular and Quantum Pharmacology" (E. D. Bergmann and B. Pullman, eds.), p. 261. Reidel Publ., Dordrecht, Netherlands.
Ham, N. S. (1976). *J. Pharm. Sci.* **65**, 612.
Ham, N. S., Casy, A. F., and Ison, R. R. (1973). *J. Med. Chem.* **16**, 470.
Hamodrakas, S., Geddes, A. J., and Sheldrick, B. (1974). *Acta Crystallogr., Sect. B* **30**, 881.
Hanna, P. E., and Ahmed, A. E. (1973). *J. Med. Chem.* **16**, 963.
Hasbun, J. A., Barker, K. K., and Merles, M. P. (1973). *J. Med. Chem.* **16**, 847.
Hehre, W. G., Steward, R. F., and Pople, J. A. (1969). *J. Chem. Phys.* **51**, 2657.
Herdklotz, J. K., and Sass, R. L. (1970). *Biochem. Biophys. Res. Commun.* **40**, 583.
Hoffmann, R. (1963). *J. Chem. Phys.* **39**, 1397.
Hoffmann, R. (1964). *J. Chem. Phys.* **40**, 2745.
Höltje, H. D., and Kier, L. B. (1974). *J. Med. Chem.* **17**, 814.
Hopfinger, A. J. (1973). "Conformational Properties of Macromolecules." Academic Press, New York.
Hoyland, J. R., and Kier, L. B. (1972). *J. Med. Chem.* **15**, 84.
Hylton, J., Christoffersen, R. E., and Hall, G. (1974). *Chem. Phys. Lett.* **24**, 501.
Isbell, H., Miner, E. J., and Logan, C. R. (1959). *Psychopharmacologia* **1**, 20.
Ison, R. R. (1974). *In* "Molecular and Quantum Pharmacology" (E. D. Bergmann and B. Pullman, eds.), p. 54. Reidel Publ., Dordrecht, Netherlands.
Ison, R. R., Partington, P., and Roberts, G. C. K. (1972). *J. Pharm. Pharmacol.* **24**, 84.

Ison, R. R., Franks, F. M., and Son, K. S. (1973a). *J. Pharm. Pharmacol.* **25**, 887.

Ison, R. R., Partington, P., and Roberts, G. C. K. (1973b). *Mol. Pharmacol.* **9**, 756.

James, M. N. G., and Williams, G. J. B. (1971). *J. Med. Chem.* **14**, 670.

James, M. N. G., and Williams, G. J. B. (1974a). *Can. J. Chem.* **52**, 1872.

James, M. N. G., and Williams, G. J. B. (1974b). *Can. J. Chem.* **52**, 1880.

Jensen, B. (1975). *Acta Chem. Scand., Sect. B* **29**, 115, 531, and 891.

Johnson, C. L., Kang, S., and Green, J. P. (1973). *In* "Conformation of Biological Molecules and Polymers" (E. D. Bergmann and B. Pullman, eds.), p. 517. Academic Press, New York.

Jones, G. P., Roberts, R. T., Anderton, K. J., and Ahmed, A. M. I. (1972). *J. Chem. Soc., Faraday Trans. 2* **68**, 400.

Kang, S., and Cho, M. H. (1971). *Theor. Chim. Acta* **22**, 176.

Kang, S., and Green, J. P. (1970). *Proc. Natl. Acad. Sci. U.S.A.* **67**, 62.

Kang, S., Johnson, C. L., and Green, J. P. (1973a). *Mol. Pharmacol.* **9**, 640.

Kang, S., Johnson, C. L., and Green, J. P. (1973b). *J. Mol. Struct.* **15**, 453.

Karle, I. L., Dragonette, K. S., and Brenner, S. A. (1965). *Acta Crystallogr.* **19**, 713.

Katz, R., Heller, S. R., and Jacobson, A. E. (1973). *Mol. Pharmacol.* **9**, 486.

Katz, R., Heller, S. R., Jacobson, A. E., Rotman, A., and Creveling, C. R. (1974). *In* "Molecular and Quantum Pharmacology" (E. D. Bergmann and B. Pullman, eds.), p. 67. Reidel Publ., Dordrecht, Netherlands.

Kier, L. B. (1967). *Mol. Pharmacol.* **3**, 487.

Kier, L. B. (1968a). *J. Med. Chem.* **11**, 441.

Kier, L. B. (1968b). *J. Pharm. Sci.* **57**, 1188.

Kier, L. B. (1968c). *J. Pharmacol. Exp. Ther.* **164**, 75.

Kier, L. B. (1969). *J. Pharm. Pharmacol.* **21**, 93.

Kier, L. B. (1971). "Molecular Orbital Theory in Drug Research." Academic Press, New York.

Kier, L. B. (1973a). *J. Theor. Biol.* **40**, 211.

Kier, L. B. (1973b). *Pure Appl. Chem.* **35**, 509.

Kier, L. B., and Aldrich, H. S. (1974). *J. Theor. Biol.* **46**, 529.

Kier, L. B., and Truitt, E. B. (1970). *J. Pharmacol. Exp. Ther.* **26**, 988.

Kitaigorodsky, A. I. (1973). "Molecular Crystals and Molecules." Academic Press, New York.

Kneale, G., Geddes, A. J. and Sheldrick, B. (1974a). *Acta Crystallogr., Sect. B* **30**, 878.

Kneale, G., Geddes, A. J., and Sheldrick, B. (1974b). *Cryst. Struct. Commun.* **3**, 351.

Krimm, S. and Venkatachalam, C. M. (1971). *Proc. Natl. Acad. Sci. U.S.A.* **68**, 2468.

Kussäther, von E., and Haase, J. (1972). *Acta Crystallogr., Sect. B* **28**, 2896.

Lehmann, M. S., Koetzle, T. F., and Hamilton, W. C. (1972). *Int. J. Pept. Protein Res.* **4**, 229.

Lichtenberg, D., Kroon, P. A., and Chan, J. I. (1974). *J. Am. Chem. Soc.* **96**, 5934.

Liquori, A. M., Damiani, A., and DeCoen, J. L. (1968). *J. Mol. Biol.* **33**, 445.

Loew, G. H., Berkowitz, D., Weinstein, H., and Srebrenik, S. (1974). *In* "Molecular and Quantum Pharmacology" (E. D. Bergmann and B. Pullman, eds.), p. 355. Reidel Publ., Dordrecht, Netherlands.

McGuire, R. F., Vanderkooi, G., Momany, F. A., Ingwall, R. T., Crippen, G. M., Lotan, N., Tuttle, R. W., Kashuba, K. L., and Scheraga, H. A. (1971). *Macromolecules* **4**, 112.

Madden, J. J., McGandy, E. L., and Seeman, N. C. (1972a). *Acta Crystallogr., Sect. B* **28**, 2377.

Madden, J. J., McGandy, E. L., Seeman, N. C., Harding, M. M., and Hoy, A. (1972b). *Acta Crystallogr., Sect. B* **28**, 2382.

Maigret, B., Pullman, B., and Dreyfus, M. (1970). *J. Theor. Biol.* **26**, 321.

Makriyannis, A., Sulivan, R. F., and Mautner, H. G. (1972). *Proc. Natl. Acad. Sci. U.S.A.* **69**, 3416.

Margolis, S., Kang, S., and Green, J. P. (1971). *Int. J. Clin. Pharmacol.* **5**, 279.

Martin, M., Carbo, R., Petrongolo, C., and Tomasi, T. (1975). *J. Am. Chem. Soc.* **97**, 1338.
Martin-Smith, M., Smail, G. A., and Stenlake, J. B. (1967). *J. Pharm. Pharmacol.* **19**, 561.
Mautner, H. G. (1974). *In* "Molecular and Quantum Pharmacology" (E. D. Bergmann and B. Pullman, eds.), p. 119. Reidel Publ., Dordrecht, Netherlands.
Mautner, H. G., Dexter, D. D., and Low, B. W. (1972). *Nature (London), New Biol.* **238**, 87.
Meyerhöffer, A. (1970). *Acta Crystallogr., Sect. B* **26**, 341.
Meyerhöffer, A., and Carlström, D. (1969). *Acta Crystallogr., Sect. B* **25**, 1119.
Momany, F. A., Carruthers, L. M., McGuire, R. F., and Scheraga, H. A. (1974a). *J. Phys. Chem.* **78**, 1595.
Momany, F. A., Carruthers, L. M., and Scheraga, H. A. (1974b). *J. Phys. Chem.* **78**, 1621.
Moran, J. F., and Triggle, D. J. (1971). *In* "Cholinergic Ligand Interactions" (D. J. Triggle, J. F. Moran, and E. A. Barnard, eds.), p. 119. Academic Press, New York.
Motherwell, W. D. S., and Isaacs, N. W. (1972). *J. Mol. Biol.* **71**, 231.
Motherwell, W. D. S., Riva di Sanseverino, L., and Kennard, O. (1973). *J. Mol. Biol.* **80**, 405.
Paiva, T. B., Tominaga, M., and Paiva, A. C. M. (1970). *J. Med. Chem.* **13**, 689.
Partington, P., Feeney, J., and Burgen, A. S. V. (1972). *Mol. Pharmacol.* **8**, 269.
Pauling, P. (1968). *In* "Structural Chemistry and Molecular Biology" (A. Rich and N. Davidson, eds.), p. 555. Freeman, San Francisco, California.
Pauling, P. (1973). *In* "Conformation of Biological Molecules and Polymers" (E. D. Bergmann and B. Pullman, eds.), p. 505. Academic Press, New York.
Pauling, P., and Petcher, T. J. (1969). *Chem. Commun.* p. 1001.
Pedersen, L., Hoskins, R. E., and Cable, H. (1973). *J. Pharm. Pharmacol.* **23**, 218.
Perahia, D., and Pullman, A. (1973). *Chem. Phys. Lett.* **19**, 73.
Perricaudet, M., and Pullman, A. (1973). *Int. J. Pept. Protein Res.* **5**, 99.
Petcher, T. J. (1974). *J. Chem. Soc., Perkin Trans. 2* p. 1151.
Petcher, T. J., and Weber, H. P. (1974). *J. Chem. Soc., Perkin Trans. 2* p. 946.
Petrongolo, C., Tomasi, J., Macchia, B., and Macchia, F. (1974). *J. Med. Chem.* **17**, 501.
Pletcher, M. S. J., and Gustaffson, B. (1970). *Acta Crystallogr., Sect. B* **26**, 114.
Pople, J. A., and Segal, G. A. (1965). *J. Chem. Phys.* **43**, S136.
Pople, J. A., and Segal, G. A. (1966). *J. Chem. Phys.* **44**, 3289.
Pople, J. A., Beveridge, D. L., and Dobosh, P. A. (1967). *J. Chem. Phys.* **47**, 2026.
Port, G. N. J., and Pullman, A. (1973a). *FEBS Lett.* **31**, 70.
Port, G. N. J., and Pullman, A. (1973b). *Theor. Chim. Acta* **31**, 231.
Port, G. N. J., and Pullman, A. (1973c). *J. Am. Chem. Soc.* **95**, 4059.
Port, G. N. J., and Pullman, A. (1974a). *Int. J. Quantum. Chem., Symp. Quant. Biol.* **1**, 33.
Port, G. N. J., and Pullman, B. (1974b). *Theor. Chim. Acta* **33**, 275.
Prout, K., Critchley, S. R., and Ganellin, C. R. (1974). *Acta Crystallogr., Sect. B* **30**, 2884.
Pullman, A. (1974). *In* "Chemical and Biochemical Reactivity" (E. D. Bergmann and B. Pullman, eds.), p. 1. Reidel Publ., Dordrecht, Netherlands.
Pullman, A., and Armbruster, A. M. (1974). *Int. J. Quantum Chem., Symp.* **8**, 169.
Pullman, A., and Port, G. N. J. (1973a). *C. R. Hebd. Seances Acad. Sci.* **277**, 2269.
Pullman, A., and Port, G. N. J. (1973b). *Theor. Chim. Acta* **32**, 77.
Pullman, A., and Pullman, B. (1968). *Adv. Quantum Chem.* **4**, 267.
Pullman, A., and Pullman, B. (1975). *Q. Rev. Biophys.* **7**, 505.
Pullman, A., Alagona, G., and Tomasi, J. (1974). *Theor. Chim. Acta* **33**, 87.
Pullman, B., ed. (1964). "Electronic Aspects of Biochemistry," p. 559. Academic Press, New York.
Pullman, B., ed. (1968). "Molecular Associations in Biology," p. 1. Academic Press, New York.
Pullman, B., and Berthod, H. (1974a). *C. R. Hebd. Seances Acad. Sci., Ser. D* **278**, 1433.
Pullman, B., and Berthod, H. (1974b). *FEBS Lett.* **44**, 266.

Pullman, B., and Berthod, H. (1975). *Theor. Chim. Acta* (in press).
Pullman, B., and Courrière, P. (1972a). *Mol. Pharmacol.* **8**, 371.
Pullman, B., and Courrière, P. (1972b). *Mol. Pharmacol.* **8**, 612.
Pullman, B., and Courrière, P. (1973). *Theor. Chim. Acta* **31**, 19.
Pullman, B., and Courrière, P. (1974). *C. R. Hebd. Seances Acad. Sci., Ser. D* **278**, 1785.
Pullman, B., and Port, G. N. J. (1974). *Mol. Pharmacol.* **10**, 360.
Pullman, B., and Pullman, A. (1969). *Prog. Nucleic Acid Res. Mol. Biol.* **9**, 327.
Pullman, B., and Pullman, A. (1974). *Adv. Protein Chem.* **28**, 347.
Pullman, B., and Saran, A. (1976), *Prog. Nucleic Acid Res. Mol. Biol.* **18**, 215.
Pullman, B., Courrière, P., and Coubeils, J. L. (1971). *Mol. Pharmacol.* **7**, 397.
Pullman, B., Coubeils, J. L., Courrière, P., and Gervois, J. P. (1972). *J. Med. Chem.* **15**, 17.
Pullman, B., Berthod, H., and Courrière, P. (1974a). *Int. J. Quantum Chem.* (*Quant. Biol. Symp.*), **1**, 93.
Pullman, B., Courrière, P., and Berthod, H. (1974b). *J. Med. Chem.* **17**, 439.
Pullman, B., Berthod, H., and Gresh, N. (1975a). *C. R. Hebd. Seances Acad. Sci.* **280**, 174.
Pullman, B., Berthod, H., and Pullman, A. (1975b). *An. Quim.* **70**, 1204.
Pullman, B., Courrière, P., and Berthod, H. (1975c). *Int. J. Quant. Chem.* (*Quant. Biol. Symp.*), **1**, 93.
Rein, R., Fukuda, N., Win, H., Clarke, G. E., and Harris, F. E. (1966). *J. Chem. Phys.* **45**, 4743.
Roberts, G. C. K. (1974). *In* "Molecular and Quantum Pharmacology" (E. D. Bergmann and B. Pullman, eds.), p. 77. Reidel Publ., Dordrecht, Netherlands.
Roothaan, C. C. J. (1951). *Rev. Mod. Phys.* **23**, 69.
Saran, A., and Govil, G. (1972). *J. Theor. Biol.* **57**, 181.
Shefter, E. (1971). *In* "Cholinergic Ligand Interactions" (D. J. Triggle, J. F. Moran, and E. A. Barnard, eds.), p. 83. Academic Press, New York.
Shefter, E., and Mautner, H. G. (1969). *Proc. Natl. Acad. Sci. U.S.A.* **63**, 1253.
Snyder, S. H., and Richelson, E. (1968). *Proc. Natl. Acad. Sci. U.S.A.* **60**, 206.
Sundaralingam, M. (1968). *Nature* (*London*) **217**, 35.
Svinning, T., and Sorum, H. (1975). *Acta Crystallogr., Sect. B* **31**, 1581.
Terui, Y., Ueyama, M., Satoh, S., and Tori, K. (1974). *Tetrahedron* **30**, 1465.
Testa, B. (1974). *In* "Molecular and Quantum Pharmacology" (E. D. Bergmann and B. Pullman, eds.), p. 241. Reidel Publ., Dordrecht, Netherlands.
Thewalt, U., and Bugg, C. E. (1972a). *Acta Crystallogr., Sect. B* **28**, 82.
Thewalt, U., and Bugg, C. E. (1972b). *Acta Crystallogr., Sect. B* **28**, 1767.
Tsoucaris, D., de Rango, C., Tsoucaris, G., Zelwer, C., Parthasarathy, R., and Cole, F. E. (1973). *Cryst. Struct. Commun.* **2**, 193.
Veidis, M. V., Palenik, G. J., Schaffren, R., and Trotter, J. (1969). *J. Chem. Soc. A* p. 2659.
Venkatachalam, C. M., and Krimm, J. (1973). *Jerusalem Symp.* **5**, 141.
Wakahara, A., Fujiwara, T., and Tomita, K. (1973). *Bull. Chem. Soc. Jpn.* **46**, 2481.
Warner, D., and Steward, E. G. (1974). *J. Mol. Struct.* (in press).
Weber, H. P., and Petcher, T. J. (1974). *J. Chem. Soc., Perkin Trans. 2* p. 942.
Weinstein, H., Apfelder, B. Z., Cohen, S., Maayani, S., and Sokolowsky, M. (1973a). *In* "Conformation of Biological Molecules and Polymers" (E. D. Bergmann and B. Pullman, eds.), p. 531. Academic Press, New York.
Weinstein, H., Maayani, S., Srebrenik, S., Cohen, S., and Sokolowsky, M. (1973b). *Mol. Pharmacol.* **9**, 820.
Weintraub, H. J. R., and Hopfinger, A. J. (1973). *J. Theor. Biol.* **41**, 53.
Weintraub, H. J. R., and Hopfinger, A. J. (1974). *In* "Molecular and Quantum Pharmacology" (E. D. Bergmann and B. Pullman, eds.), p. 131. Reidel Publ., Dordrecht, Netherlands.
Yamane, T., Ashida, T., and Kakudo, M. (1973). *Acta Crystallogr., Sect. B* **29**, 2884.

Subject Index

A

Acetylcholine
  conformation of, 289–314
    derivatives and analogs, 302–314
    in solution, 297–302
  structure of, 303
μ-Acetylene dicyclopentadienyldinickel, charge densities in, 18
Acetylethanolamine, conformation of, 306, 308
Adiphenine
  conformational studies on, 312
  structure of, 313
Amphetamine
  conformational properties of, 282, 285, 286
  structure of, 283
Anticholinergic compounds, conformational studies on, 311–314
Antihistamines, conformational studies on, 268–275
Atomic orbitals
  Dirac-Fourier transformation of, 44–58
  scaled, Hund's rule application to, 147–153
Atropine
  conformation of, 311, 312
  structure of, 311
Azide ion, charge density in, 18, 19

B

Balmer-α intensity, expression for, 131
Benactyzine
  conformational studies on, 312
  structure of, 313
Benzenechromium tricarbonyl, charge densities in, 17, 27–28
Biogenic amines, general representation of, 257
Bloch-Horowitz operator, 224, 226
  reduction of diagrams of, 229–231
Brillouin theorem, diagrammatic form of, 214
Brillouin-Wigner perturbation theory, 188, 190–191
Brueckner-Goldstone formalism, for open-shell systems, 220
Bufotenine
  conformational studies on, 280
  structure of, 279

C

Carbon, isoelectronic sequence of, 156
Carbon bonds, deformation densities of, 9
Cartesian Gaussian-type orbitals, Dirac-Fourier transformation of, 55–58
Charge density distribution, X-ray diffraction measurement of, 1–35
Chebyshev-Gauss quadrature scheme, 64
Chromium hexacarbonyl, charge densities of, 8
Clebsch-Gordon coefficients, 111
Coester-Kümmel-Čižek formalism, in linked-cluster perturbation theory, 209–213, 221
Compton profiles
  for atomic and molecular systems, 37–76
    evaluation of, 62–70
Conformation, quantum mechanics of, in molecular pharmacology, 251–328
Coulomb repulsion elements, 241
Covalent bonds, charge densities of, 9
Cyanide ion, charge density in, 18
Cyclobutane, charge densities in, 22

D

Deformation density, definition of, 2–4, 7
Density functions, definition of, 2–4
Diamond, charge densities of, 11

Dimethyl tryptamine structure of, 279
Diphenhydramine
  conformational studies on, 271, 273
  structure of, 268
Dirac-Fourier transformation
  of atomic orbitals, 44–58
    Cartesian and ellipsoidal Gaussian type, 55–58
    Gaussian-type orbitals, 52–53
    hydrogen-like orbitals, 48–52
    spherical harmonic orbitals, 44–58
  of position space wavefunctions, 40–44
Dopamine
  conformational studies on, 281, 282
  structure of, 283

E

Elastic scattering, in electron-atom collisions, 92–94
Electron-atom collisions, 77–141
  electron-photon coincidences in, 114–130
  polarization of impact radiation in, 78–92
  scattering and excitation amplitudes in, 92–101
  spin effects in, 92–113
π-Electron Hamiltonions, 244
Electron-photon coincidences, angular correlations and, 114–130
Electron scattering, from laser-excited atoms, 132–139
Ellipsoidal Gaussian-type orbitals, Dirac-Fourier transformation of, 55–58
Ephedrine
  conformational studies on, 281, 282
  structure of, 283
Epinephrine
  conformational properties of, 282
  structure of, 283
Extended Hückel theory, application to pharmacological conformation, 251, 261–262

F

Factorization theorem, 197
Fano-Macek orientation parameter, variation of, 125
Folded diagrams, examples of, 227
Fourier transform, 40

G

Gaunt coefficients, 59
Gaussian-type atomic orbitals, Dirac-Fourier transformation of, 54–55
Gegenbauer polynomials, 50, 52
Glycopyrronium bromide
  conformation of, 311, 312
  structure of, 311
Glycylglycine
  charge densities of, 10, 16, 27, 29
  orbital exponents of, 29
Goldstone's adiabatic derivation, 215–220
Goldstone diagrams, 192
Green's function, 190

H

Hallucinogens, conformational studies on, 281
Hankel transform, 45, 51–53
Hartree-Fock approximation, 188
Hartree-Fock potential, 213
Heisenberg spin Hamiltonian, 237
Helium, isoelectronic sequence of, 157
Hellmann-Feynman theorem, 9, 159
Hanne-Kessler apparatus, 112
Helium, polarization studies on, 90, 91
Hermite polynomial, 53, 57
Hermiticity effective operators and, 231–233
Hexapyrronium bromide
  conformation of, 311, 312
  structure of, 311
Histadyl
  conformational studies on, 271–273
  structure of, 268
Histamine conformational studies on, 259–268
Hydrogen
  atomic, electron-impact excitation of, 131–132
  polarization studies on, 84
Hydrogen-like atomic orbitals, Dirac-Fourier transformation of, 48–53
Hydrogen peroxide, charge densities of, 13

Hund's rule
theoretical interpretation of, 143–185
correlation correction in, 156–160
pair distribution function in, 162–175
role of inner shells in, 160–162
summary, 176–180
scaled atomic orbitals and, 147–153
validity within independent particle model, 144–147

I

Impact radiation, polarization of, in electron-atom collisions, 78–92
Indolealkylamines, conformational studies on, 275–281
Inelastic scattering, in electron-atom collisions, 94–97
Inner shells, in interpretation of Hund's rule, 160–161
Iterative extended Hückel theory application to pharmacological conformation, 251
Isoproterenol
conformational studies on, 281, 282
structure of, 283
Isotropic Compton profile, spherically averaged momentum density and, 71–73

J

Jacobi polynomials, 50
JILA-Stirling experiment, 100

K

Kronecker tensor, 107
Kummer's confluent hypergeometric series, 53

L

Laguerre function atomic orbitals, Dirac-Fourier transformation of, 48–52
Laguerre polynomials, 51
Laser, atoms excited by electron scattering from, 132–139
LCAO method, for Compton profile evaluation, 73
Linked-cluster perturbation theory
adiabatic derivation of Goldstone in, 215–220
Brillouin-Wigner theory, 190–191
cancellation of unlinked diagrams in, 196–197
for closed- and open-shell systems, 187–249
Coester-Kümmel-Čižek formalism in, 209–213
core and valence contributors to, 223–224
degenerate Rayleigh-Schrödinger perturbation theory in, 225–226
effective spin Hamiltonians for, 237–246
exclusion-violating diagrams in, 199–200
expection values, 204–205
factorization theorem in, 197–199
folded-diagram expansion in, 226–229
ground-state wavefunction in, 205–207
hermiticity and effective operators in, 231–233
nonorthoganality castastrophes in, 237–239
physical interpretation and applications of, 201–220
relation to other many-body theories, 233–236
single-particle potentials in, 213–215
spin Hamiltonian derivation in, 237–246
unlinked cluster problem in, 195–196
Lithium, polarization studies on, 84
Local anesthetics, acetycholine analog-type, conformation of, 308–314
Lone-pair regions, charge densities of, 15
LSD, structure of, 258

M

"Madison convention," 102
Mercury, excitation of, exchange scattering in, 110–113
Mescaline
conformational properties of, 288
structure of, 258, 288
Molecular dipole moments, from charge density measurements, 29
Molecular pharmacological conformation, 251–328

acetylcholine, 289–314
anticholinergic compounds, 311–314
antihistamines, 268–275
histamine, 259–268
indolealkylamines, 275–281
phenethylamines and
phenethanolamines, 281–289
serotonin, 275–281
Momentum distributions
evaluation of, 37–76
from basis overlap integrals, 69–70
Cartesian Gaussian-type functions, 60–62
from momentum densities, 63–69
spherical harmonic basis functions, 58–60

N

Nitrate ion charge density of, 22
p-Nitropyridine N-oxide
charge densities of, 15
orbital exponents of, 29
Norepinephrine
conformational studies on, 281, 282, 286
structure of, 283
N.Y.U. recoil experiment, 98

O

Optimized atomic orbitals
Hund's rule application to, 153–156
for minimal basis set, 153–154
SCF computations, 154–156
Orbital exponents, from charge density measurements, 29–32
Oxalic acid, orbital exponents of, 29

P

Pair distribution function, in application of Hund's rule, 162–175, 181–184
Parpanit, conformational studies on, 312
PCILO method for conformational energy calculation, 260–265
Phenethanolamines
conformational studies on, 281–289
structures of, 283
Phenethylamines
conformational studies on, 281–289
structures of, 283
Pheniramine
conformational studies on, 271, 274
structure of, 268
Procaine
conformation of, 309
structure of, 303
Pseudoatoms, angular deformation of, 32
Psilocin
conformational studies on, 280, 281
structure of, 258, 279
Psilocybin, structure of, 279
Pyrrobutamine, structure of, 269

Q

Quantum mechanics, of molecular pharmacological conformation, 251–328
Quinuclidinyl derivatives
conformation of, 311, 312
structure of, 311

R

Rayleigh-Schrodinger perturbation theory, degenerate form of, 225–226
Recoil experiment (N.Y.U.), 98
Russel-Saunders coupling, 111

S

Scaled atomic orbitals, Hund's rule application to, 147–153
SCF computations, in applications of Hund's rule, 154–156
Scopolamine
conformation of, 311, 312
structure of, 311
Serotonin
conformational studies on, 276
structure of, 276
Schrödinger's equation, 44, 221
Shull-Löwdin radial function, 52
Silicon, charge densities of, 11
Slater's multiplet theory, 158

Sodium-23, polarization studies on, 84, 85
Sodium atoms, laser-excited, electron scattering from, 130–139
Sodium azide, charge densities of, 8
Solute-solvent interaction energy, of pharmacological molecules, 255
Solvation groups (shell) model, for pharmacological molecules, 255
Spherical harmonic orbitals, Dirac-Fourier transformation of, 46–48
Spin effects, in electron-atom collisions, 92–113
Sulfur, orthorhombic, charge densities of, 8, 12, 29, 30

T

Tetracyanocyclobutane
  charge densities in, 22
  deformation density of, 31
Tetracyanoethylene, charge densities of, 14, 24
Tetracyanoquinodimethanide, charge density studies on, 28
Tetraphenylbutatriene, charge densities of, 15
Tetrathiofulvalene, charge density studies on, 28
Thermal smearing, in charge density measurement, 22–25
Thiocyanate ion, charge density in, 18
Torsion angle, definition of, 257
Transition metals, compounds containing, charge densities of, 17
Triprolidine, structure of, 269
Two-electron operators, expectation values of, in application of Hund's rule, 174–175

U

Uronium nitrate, charge density of, 22

V

"Vacuum convention," 194
Valence shell deformation methods, in charge density studies, 26–27

W

Waller-Hartree theory of X-ray scattering, 4–5
Wannier functions, 240, 241
Wavefunction diagrams, vacuum convention for, 205

X

X-ray diffraction
  charge density distribution by, 1–35
    experimental data, 8–19
    experimental errors, 7–8
    functions based on, 25–32
    in small molecules, 18
    theory, 20–22
X-ray scattering, theory of, 4
X-N map, 7
X-X densities, 7

CONTENTS OF PREVIOUS VOLUMES

Volume 1

The Schrödinger Two-Electron Atomic Problem
Egil A. Hylleraas

Energy Band Calculations by the Augmented Plane Wave Method
J. C. Slater

Spin-Free Quantum Chemistry
F. A. Matsen

On the Basis of the Main Methods of Calculating Molecular Electronic Wave Functions
R. Daudel

Theory of Solvent Effects on Molecular Electronic Spectra
Sadhan Basu

The Pi-Electron Approximation
Peter G. Lykos

Recent Developments in the Generalized Hückel Method
Y. I. Haya

Accuracy of Calculated Atomic and Molecular Properties
G. G. Hall

Recent Developments in Perturbation Theory
Joseph O. Hirschfelder, W. Byers Brown, and Saul T. Epstein

AUTHOR INDEX—SUBJECT INDEX

Volume 2

Quantum Calculations, Which Are Accumulative in Accuracy, Unrestricted in Expansion Functions and Economical in Computation
S. F. Boys and P. Rajagopal

Zero Differential Overlap in π-Electron Theories
Inga Fischer-Hjalmars

Theory of Atomic Hyperfine Structure
S. M. Blinder

The Theory of Pair-Correlated Wave Functions
R. McWeeny and E. Steiner

Quantum Chemistry and Crystal Physics. Stability of Crystals of Rare Gas Atoms and Alkali Halides in Terms of Three-Atom and Three-Ion Exchange Interactions
Laurens Jansen

Charge Fluctuation Interactions in Molecular Biophysics
Herbert Jehle

Quantum Genetics and the Aperiodic Solid. Some Aspects on the Biological Problems of Heredity, Mutations, Aging, and Tumors in View of the Quantum Theory of the DNA Molecule
Per-Olov Löwdin

AUTHOR INDEX—SUBJECT INDEX

Volume 3

Approximate Hartree–Fock Calculations on Small Molecules
R. K. Nesbet

Single-Center Molecular Wave Functions
David M. Bishop

Molecular Orbital Theory
Frank D. Harris

Nonadditivity of Intermolecular Forces
H. Margenau and J. Stapmer

Quantum Theory of Chemical Reactivity
R. Daudel

Electronic Theories of Hydrogen Bonding
S. Bratož

Molecular Orbital Calculations of π Electron Systems
Kimio Ohno

Quantum Theory of Time-Dependent Phenomena Treated by the Evolution Operator Technique
Per-Olov Löwdin

AUTHOR INDEX—SUBJECT INDEX

Volume 4

Functional Analysis for Quantum Theorists
Bela A. Lengyel

The Symmetric Group Made Easy
A. J. Coleman

Field Theoretic Approach to Atomic Helium
Donald H. Kobe

Atomic Intensities: A Comparison of Theoretical and Experimental *f*-Values for Zn I, Cd I, and Hg I
T. M. Bieniewski, T. K. Krueger, and S. J. Czyzak

Probability of Singlet-Triplet Transitions
Lionel Goodman and Bernard J. Laurenzi

Molecular Orbital Theories of Inorganic Complexes
J. P. Dahl and C. J. Ballhausen

Paramagnetic Properties and Electronic Structure of Iron in Heme Proteins
Masao Kotani

Aspects of the Electronic Structure of the Purine and Pyrimidine Bases of the Nucleic Acids and of Their Interactions
Alberte Pullman and Bernard Pullman

AUTHOR INDEX—SUBJECT INDEX

Volume 5

Some Aspects of the Quantum Theory of Photochemical Reactivity of Organic Molecules
R. Daudel

The Origin of Binding and Antibinding in the Hydrogen Molecule-Ion
M. J. Feinberg, Klaus Ruedenberg, and Ernest L. Mehler

Adiabatic Approximation and Its Accuracy
W odzimierz Kolos

The Theory of Nonadiabatic Transitions: Recent Development with Exponential Models
E. E. Nikitin

On the Nonorthogonality Problem
Per-Olov Löwdin

Symmetry Properties of Reduced Density Matrices and Natural p-States
W. A. Bingel and W. Kutzelnigg

Symmetry Adaptation to Sequences of Finite Groups
D. J. Klein, C. H. Carlisle, and F. A. Matsen

The Method of Optimized Valence Configurations: A Reasonable Application of the Multiconfiguration Self-Consistent-Field Technique to the Quantitative Description of Chemical Bonding
Arnold C. Wahl and G. Das

AUTHOR INDEX—SUBJECT INDEX

Volume 6

Statistical Exchange-Correlation in the Self-Consistent Field
John C. Slater

Aspects of the Localizability of Electrons in Atoms and Molecules: Loge Theory and Related Methods
Claude Aslangul, Raymond Constanciel, Raymond Daudel, and Philemon Kottis

Magnetic Circular Dichroism and Diamagnetic Molecules
Dennis J. Caldwell and Henry Eyring

Collective Electron Oscillation in Pi-Electron Systems
Sadhan Basu and Purnendranath Sen

Molecular Orbital Theory of Chemical Reactions
Hiroshi Fujimoto and Kenichi Fukui

Unified Treatment of van der Waals Forces between Two Molecules of Arbitrary Sizes and Electron Delocalizations
Bruno Linder and David A. Rabernold

Natural Orbitals
Ernest R. Davidson

Matrix Elements and Density Matrices for Many-Electron Spin Eigenstates Built from Orthonormal Orbitals
Klaus Ruedenberg and Ronald D. Poshusta

Upper and Lower Bounds to Quantum-Mechanical Properties
F. Weinhold

Ab Initio Calculations on Large Molecules
Ralph E. Christoffersen

AUTHOR INDEX—SUBJECT INDEX

Volume 7

Rotation and Translation of Regular and Irregular Solid Spherical Harmonics
E. O. Steinborn and K. Ruedenberg

Molecular Integrals between Real and between Complex Atomic Orbitals
E. O. Steinborn

The Symmetric Groups and Calculation of Energies of n-Electron Systems in Pure Spin States
G. A. Gallup

Scattered-Wave Theory of the Chemical Bond
Keith H. Johnson

Projection Operators in Hartree–Fock Theory
Sigeru Huzinaga, Denis McWilliams, and Antonio A. Cantu

An Analytic Independent Particle Model for Atoms
I. Initial Studies
A. E. S. Green

An Analytic Independent Particle Model for Atoms
II. Modified Hartree–Fock Calculations for Atoms
J. N. Bass, A. E. S. Green, and J. H. Wood

An Analytic Independent Particle Model for Atoms
III. Ionization of Rare Gas Atoms by Electrons in the Born Approximation
R. A. Berg and A. E. S. Green

Solvent-Shift Effects on Electronic Spectra and Excited-State Dipole Moments and Polarizabilities
A. T. Amos and B. L. Burrows

Thermochemistry in the Hartree–Fock Approximation
A. C. Hurley

On Physical Properties and Interactions of Polyatomic Molecules: With Application to Molecular Recognition in Biology
Robert Rein

Quantum Theory of DNA Summary of Results and Study Program
Janos J. Ladik

AUTHOR INDEX—SUBJECT INDEX

Volume 8

Symmetry Groups of Nonrigid Molecules
J. Serre

Genealogical Electronic Spin Eigenfunctions and Antisymmetric Many-Electron Wavefunctions Generated Directly from Young Diagrams
William I. Salmon

Current Problems and Perspectives in the MO-LCAO Theory of Molecules
Giuseppe Del Re

Photoelectron Spectra Showing Relaxation Effects in the Continuum and Electrostatic and Chemical Influences of the Surrounding Atoms
Christian Klixbüll Jørgensen

Some Recent Developments in the Theory of Coordination Compounds of Metals
G. Berthier

Effective Hamiltonian Methods for Molecular Collisions
David A. Micha

SUBJECT INDEX

Volume 9

Utilization of Transferability in Molecular Orbital Theory
Brian O'Leary, Brian J. Duke, and James E. Eilers

A Series of Electronic Spectral Calculations Using Nonempirical CI Techniques
Sigrid D. Peyerimhoff and Robert J. Beunker

Time-Independent Diagrammatic Approach to Perturbation Theory of Fermion Systems
J. Paldus and J. Čižek

Coupled-Channel Studies of Rotational and Vibrational Energy Transfer by Collision
William A. Lester, Jr.

Theory of Low Energy Electron Scattering by Complex Atoms
R. K. Nesbet

SUBJECT INDEX

A 7
B 8
C 9
D 0
E 1
F 2
G 3
H 4
I 5
J 6